The Agronomy and Economy of Important Tree Crops of the Developing World

AF849230

The Agronomy and Economy of Important Tree Crops of the Developing World

K.P. Prabhakaran Nair
Distinguished Visiting Scientist,
Indian Council of Agricultural
Research New Delhi, India

ELSEVIER

AMSTERDAM • BOSTON • HEIDELBERG • LONDON • NEW YORK • OXFORD
PARIS • SAN DIEGO • SAN FRANCISCO • SINGAPORE • SYDNEY • TOKYO

Elsevier
32 Jamestown Road London NW1 7BY
30 Corporate Drive, Suite 400, Burlington, MA 01803, USA

First edition 2010

Copyright © 2010 Elsevier Inc. All rights reserved

No part of this publication may be reproduced or transmitted in any form or by any means, electronic or mechanical, including photocopying, recording, or any information storage and retrieval system, without permission in writing from the publisher. Details on how to seek permission, further information about the Publisher's permissions policies and our arrangement with organizations such as the Copyright Clearance Center and the Copyright Licensing Agency, can be found at our website: www.elsevier.com/permissions

This book and the individual contributions contained in it are protected under copyright by the Publisher (other than as may be noted herein).

Notices
Knowledge and best practice in this field are constantly changing. As new research and experience broaden our understanding, changes in research methods, professional practices, or medical treatment may become necessary.

Practitioners and researchers must always rely on their own experience and knowledge in evaluating and using any information, methods, compounds, or experiments described herein. In using such information or methods they should be mindful of their own safety and the safety of others, including parties for whom they have a professional responsibility.

To the fullest extent of the law, neither the Publisher nor the authors, contributors, or editors, assume any liability for any injury and/or damage to persons or property as a matter of products liability, negligence or otherwise, or from any use or operation of any methods, products, instructions, or ideas contained in the material herein.

British Library Cataloguing in Publication Data
A catalogue record for this book is available from the British Library

Library of Congress Cataloging-in-Publication Data
A catalog record for this book is available from the Library of Congress

ISBN: 978-0-323-16505-1

For information on all Elsevier publications
visit our website at elsevierdirect.com

This book has been manufactured using Print On Demand technology. Each copy is produced to order and is limited to black ink. The online version of this book will show colour figures where appropriate.

Working together to grow
libraries in developing countries

www.elsevier.com | www.bookaid.org | www.sabre.org

ELSEVIER BOOK AID International Sabre Foundation

Transferred to Digital Printing in 2016

I dedicate this book, which was compiled under very trying circumstances, to the memory of my late parents, my father, Kuniyeri Pookkalam Kannan Nair and my mother Kodoth Padinhareveetil Narayani Amma, both of whom left me an orphan at a very young age, but, whose boundless love and blessings made me what I am today

Contents

Introduction xiii

1 **Arecanut (*Areca catechu* L.)** 1
Origin, History, and Geographical Distribution 1
The Botany and Taxonomy of the Areca Plant 2
The Cytogenetics of the Areca Palm 6
Genetic Resource Program 8
Areca Crop Improvement 10
The Agronomy of Arecanut 11
Arecanut Physiology 14
Mixed Cropping Systems in Areca 15
Arecanut Pathology 16
Arecanut Entomology 17
Arecanut Nematology 18
Harvesting and Processing of Arecanut 18
A Look into Arecanut's Future 19

2 **Cashew Nut (*Anacardium occidentale* L.)** 21
Area and Global Production 22
The Indian Scenario 24
World Trade in Cashew 26
The Tale of Cashew Trade in India Over the Years 27
The History, Evolutionary Origin, and Distribution of Cashew 28
Economic Botany of Cashew 31
Taxonomy of the Cashew Plant 31
Cytogenetics of the Cashew Plant 33
Collection, Conservation, and Cataloging of Genetic Resources of the Cashew Plant 33
The Genetic Improvement of the Cashew Plant 34
Establishing and Managing a Cashew Orchard 42
The Relevance of "The Nutrient Buffer Power Concept" in Cashew Nutrition 48
Some Salient Aspects of Raising Soft Wood Grafting 48
Planting Technology 49
Controlling Pests and Diseases in Cashew Plantations 52
Effect of Inclemental Weather Conditions 59

Cashew End Products 59
Organization of Cashew Research in India and Overseas 63
A Look into Cashew's Future 64
Biodynamic Cashew 66

3 The Coconut Palm (*Cocos nucifera* L.) 67
Origin and Evolution 67
The Evolution of Coconut along the Drifting Coastlines 67
Development of Wind-Resistance 68
The "Swimming" Coconut Fruit 69
True Palm Traits 69
Human Influence on Coconut Evolution 70
Botany of Coconut 71
Morphology 73
The Inflorescence 76
The Fruit 76
The Seed and Seedling 77
Cytogenetics of Coconut 78
Genetic Improvement of Coconut 78
Constraints in Coconut Breeding 81
Selection and Its Progress 82
Hybrids and Their Future 83
Commercial Production of Hybrid Seeds 84
Agronomy of Coconut 85
Coconut-Based Mixed Cropping Systems and Their Management 89
Seed and Seedling Management 89
Germination Rate 90
Poly Bag Seedlings 90
Seedling Selection 91
Field Management 91
Productive Palms 91
Adaptation to Biotic Factors 92
The Range of Coconut Pests 92
Adaptation in Coconut Palm 94
A Devastating Virus 94
Adapting the Coconut to Market Needs 95
Yield Potential of the Coconut Palm 95
Quality Traits 95
The Fatty Acid Mix 96
Coconut as a Food Item 96
Research and Development in Coconut Production 97
Global Coordination 97
National Research Centers 98
Research in India and Sri Lanka 98
Research in the Second Half of Twentieth Century 98

	A Look into the Future of Coconut	**99**
	Protection of the Production Base	**100**
	Advances in Processing Technology	**100**
	Contact Information for Research Centers and Institutes	**101**
4	**Cinchona (*Cinchona* sp.)**	**111**
	Origin and Use of Cinchona	**111**
	Further Voyage of Cinchona to the East	**113**
	Introduction of Cinchona to India and Sri Lanka	**114**
	The Other Major Cinchona Producing Countries	**115**
	The Characteristics and Chemistry of Cinchona Plantation Species	**116**
	Taxonomy: Diversity Analysis of Source Species	**116**
	Chemistry of Cinchona	**118**
	Extraction, Production, and Trade	**119**
	Research and Development Efforts in Cinchona	**123**
	Raising a Cinchona Plantation	**124**
	Role of Biotechnology in Cinchona Improvement	**127**
	A Look into Cinchona's Future	**128**
5	**Cocoa (*Theobroma cacao* L.)**	**131**
	Origin and Distribution	**131**
	Area and Production	**133**
	The Botany of Cocoa	**134**
	Crop Improvement and Management	**137**
	Climatic Requirements for Cocoa	**143**
	Soil Type	**144**
	Propagation	**144**
	Field Maintenance	**147**
	Flowering and Fruit Set	**152**
	Plant Protection	**155**
	Harvesting	**164**
	Primary Processing	**165**
	Recovery of Cured Beans	**172**
	Cleaning and Bagging	**172**
	Storage	**172**
	Secondary Processing	**173**
	Nutritional Value of Chocolates	**177**
	By-Products of Chocolate	**177**
	Research and Development Organizations	**178**
	Manufacturers of Cocoa Products	**179**
	A Look into Cocoa's Future	**179**
6	**Coffee**	**181**
	Genetic Resources	**182**
	Breeding Programs	**187**

Field Management **190**
Diseases of Coffee **197**
Pests of Coffee **202**
Consumer Choice and Coffee Adaptation **205**
Research and Developmental Setup **206**
A Look into the Future of Coffee **206**

7 Oil Palm (*Elaeis guineensis* Jacquin) **209**
Historical Origin and Global Distribution **211**
Taxonomy and Botany **212**
Growing Conditions **213**
Collection of Oil Palm Germplasm **215**
Characterization of Germplasm **218**
Genetic Improvement Through Breeding and Selection **221**
Prospects for Biotechnological Interventions **224**
Pests and Diseases **231**
Oil Palm Plantations and the Environment **233**
A Look into the Future of Oil Palm **235**

8 Rubber (*Hevea brasiliensis*) **237**
The Importance of Natural Rubber **237**
Origin and Distribution of Rubber **238**
The Rubber Industry—A Historical Perspective **238**
Area, Production, and Productivity **240**
The Botany of *Hevea* Plant **243**
The Morphology of the *Hevea* plant **244**
Cytogenetics of the *Hevea* plant **245**
Commercial Adoption of Vegetative Propagation **245**
Bark Anatomy **246**
The Physiology of Latex Flow **247**
Biochemistry of Latex and Rubber Production **250**
Genetic Improvement of Rubber **250**
Genetic Studies **254**
Genetic Base and Germplasm Resources **254**
Biotechnological Interventions **256**
Crop Management Under Different Cropping Systems **259**
Ideal Soils for Rubber and Manuring Schedule **261**
Diseases and Pests of Rubber **261**
Technological Developments **262**
Modified Forms of Rubber **263**
Discovery of Vulcanization **264**
Product Development **265**
The "Ecofriendliness" of Rubber **267**
Useful Information on Research and Development **268**
A Look into the Future of Rubber **272**

9	**Tea (*Camellia sinensis* L.)**	**275**
	The Origin and History of Tea	**275**
	The Different Practices of Tea Consumption	**276**
	The Botany of Tea	**276**
	Taxonomy and Genetics	**277**
	Growing Conditions	**278**
	Nutrient Requirements	**278**
	Nursery Practices	**279**
	Crop Management	**280**
	Propagation and Genetic Improvement of Tea Stocks	**282**
	The Evolution of Rational Pest and Disease Management	**284**
	The Variations in Tea Manufacture	**286**
	Antioxidants in the Main Types of Tea	**287**
	The Economics of Tea: Production and Global Trade	**287**
	Research and Development Institutions Related to Tea	**291**
	A Look into Tea's Future	**295**
	SWOT: The Final Word	**300**
10	**Wattle**	**301**
	The Different Kinds of Wattle	**301**
	Growing conditions	**302**
	Natural Regeneration and Rotation	**307**
	Rotation	**307**
	Utility of Bark and Wood and Maintaining Soil Fertility	**307**
	Yield	**308**
	The Processing of Bark	**310**
	Pests and Diseases	**311**
Bibliography		**313**

Introduction

Industrial tree crops contribute substantially to the economy of many developing countries on the Asian, African, and Latin American continents. With the World Trade Organization (WTO) substantially focusing on agriculture, the commercial aspects of growing these crops assume considerable economic significance. Within the developing world, there are countries whose sole economic sustenance depends on these crops. Even within the geographical boundary of a country, there are states whose economy is exclusively linked to certain crops. For instance, within India, arecanut, coconut, and rubber contribute substantially to the economy of the State of Kerala in Southern India. Within the Asian continent, oil palm contributes substantially to the economy of Malaysia and Indonesia.

Palm oil, a cheap source of cooking oil, is fast replacing fossil fuel as "green fuel" from which diesel is extracted. Currently, while a ton of crude oil costs around US$600 (though the price surge seems unstoppable as this book is being written), palm oil is quoted at more than US$800 a ton. The global commercial impact of these developments can well be imagined. With global warming becoming a very serious issue of human survival, as has recently been amply demonstrated in the Copenhagen Summit on climate change, with fossil fuel consumption considered the main culprit, there is an ever-growing need for green fuel. Palm oil fits the bill.

Within the African continent, tea, coffee, and cocoa contribute substantially to the economy of countries like Kenya, and The Republic of Cameroon. On the Latin American continent, rubber is a very valuable foreign export. India grows some of these crops that contribute substantially to the country's economy. The ministry of commerce under the Government of India deals with several aspects, and there are commodity boards like the Coconut Board, the Rubber Board, the Tea and Coffee Boards, and others that coordinate research, development, and commerce in these crops. The Rubber Board in India has played a very effective role in the research, marketing, and development of natural rubber. Historically, tea, coffee, and rubber were raised as "plantation crops" on the Asian and African continents by the colonial powers. One can see very large estates, running to hundreds of acres, of these crops in the countries on these continents. For instance, in India, huge estates of tea exist in Northeastern India and of rubber in the State of Kerala to the south. And most of these estates were controlled by the colonial powers, primarily the British. With the political changes that took place in these continents starting late 1950s and early 1960s and the emergence of independence from the colonial powers that followed, the pattern of ownership changed to native hands. Simultaneously a large number of small holders came into existence. This has also happened with arecanut, coconut, and rubber in Kerala.

Arecanut is a masticatory nut that is paired with "betel leaf" (an annual twiner) and tobacco along with lime (CaO), which gives a red color to the saliva, for the practice of "chewing," a popular habit in many parts of India and Pakistan. In fact, an important market for arecanut from India is Pakistan. In addition, the dried nut is processed into a scented end-product known locally as *Supari*, which is very popular both in India and Pakistan.

Coconut is known as *Kalpavriksha*, a term derived from the ancient Indian language Sanskrit meaning "Heaven's Tree." Coconut provides materials for culinary purposes from its endocarp (the grated pulp), which is essential in South Indian food preparations, especially those in Kerala. The edible oil—most of the cooking in Kerala is done in coconut oil, which has a high percentage of unsaturated fats, now considered "unhealthy" by the medical fraternity, though the opinion is divided—has a large market. The oil extracted from the shell is an industrial lubricant. The tender coconut water is a highly nutritive and invaluable health drink. It can even be used to culture cells.

Tea and coffee are beverage crops. Cocoa is not only a beverage crop, but is the main source for chocolate manufacture. Cashew is turning out to be a very important industrial tree crop of India. A highly nutritive nut, free of cholesterol, it has a global market and finds its use in bakery, sweet (particularly Oriental type) preparation, and the cashew nut shell oil (CSL) finds its use in many industrial purposes. Of late, its false fruit is increasingly used in the production of ethanol, another green fuel. In rural India, the false fruit goes into the manufacture of illicit alcohol. Wattle produces the bark from which tannin is extracted for tannery industry.

In summary, the book is an invaluable compendium of scientific and developmental information and that would serve a large audience of academics, researchers, developmental policymakers, and the millions of students of agriculture from the developing world.

1 Arecanut (*Areca catechu* L.)

The arecanut palm is the source of a widely used masticatory nut, popularly known as arecanut, betelnut, or *supari*. While supari is a processed and scented nut powder for mastication, highly popular in northern India and Pakistan, the term betelnut is derived from the fact that arecanut is used along with betel leaf (a twiner) for chewing purposes. Since ancient times, the habit of "chewing" is a symbol of friendship and general well being. Even in Hindu temples during festivals, betelnut and arecanut are offered to the deities as materials of worship. No Hindu auspicious occasion, such as marriage or a betrothal ceremony, is complete without the offering of arecanut and betel leaves to the guests. Arecanut palm is a popular crop in India in the states of Kerala, Karnataka, and Tamil Nadu in southern India and in Assam, Meghalaya, and West Bengal in northeastern India. The areca palm is a monocot that belongs to the family Palmae. The commonly cultivated species is *Areca catechu* in most of the countries where it is used for chewing. In Sri Lanka, the fruits of *Areca concinna* are occasionally chewed. In the world, the majority of production is in India, followed by China.

Origin, History, and Geographical Distribution

There is no definitive record on the origin of arecanut. No fossil remains of the genus *Areca* exist, but the fossil records of closely related genera indicate its presence during the tertiary period. The maximum diversity of species that number 24 and other indicators suggest that the original habitat is in contiguous regions of Borneo, Celebes, and Malaya (Bavappa, 1963; Raghavan, 1957). There are innumerable references to arecanut palm, arecanut, and its various uses in ancient *Sanskrit*. The antiquity of such references has been shown, the most important being *Anjana Charitra* (Sisy Mayana, 1300 BC), where the reference had been made to groups of arecanut palms full of inflorescence and branches presenting an exquisite appearance (Bhat and Rao, 1962). Although it is not precisely known when arecanut found its way to the Indian subcontinent, innumerable pieces of evidence exist of its antiquity (Mohan Rao, 1982). Arecanut is mentioned in various Sanskrit scriptures (650–1300 BC), and its medicinal properties were known to the famous Indian scholar Vagbhatta (500 AD). A well-known cave in central India from around 200 BC to 900 AD features an exquisitely painted arecanut palm providing a backdrop to the *Padmapani* Buddha. According to Furtado (1960) one of the earliest references to arecanut dates back to

The Agronomy and Economy of Important Tree Crops of the Developing World. DOI: 10.1016/B978-0-12-384677-8.00001-1
Copyright © 2010 by Elsevier Inc. All rights of reproduction in any form reserved.

1510 AD. The abundant uses of arecanut in chewing and auspicious religious functions of the Hindus of India were indicated even during the times of the *Aryans*, the early conquerors of India, who were supposed to have migrated from Europe.

In terms of area, India has 57 percent of the world's total and 53 percent (Table 1.1) of its production. China occupies the second place, which is followed by Bangladesh and Myanmar. In Asia, the countries of Indonesia, Malaysia, Thailand, Philippines, and Vietnam also grow arecanut. Productivity is highest (3752 kg/ha) in China followed by Malaysia (1667 kg/ha), Thailand (1611 kg/ha), and India (1189 kg/ha). Details on production and productivity are given in Table 1.1. In India, the crop is grown largely in the Western Ghats and the northeastern regions. Details on area and production, in terms of regions in India, are given in Table 1.2. As much as 90 percent of the area and 95 percent of the production are from three states, namely Kerala, Karnataka, and Assam. Within India, productivity is highest (3947 kg/ha) in the state of Maharashtra in central India.

The Botany and Taxonomy of the Areca Plant

Though unsatisfactory, as they were not based on real affinities, the earliest attempts to restrict the genus *Areca* was that of Martius (1832–1850). Subsequently, various species grouped under *Areca* were separated into different genera and limited the genus to close relatives of the type of the genus *Areca catechu* (Blume, 1836). Furtado (1933) described the limits of the genus *Areca* and its sections. A list of *Areca* species and their geographical distribution is given in Table 1.3. Detailed morphology, floral biology, and embryology have been described (Murthy and Pillai, 1982). The arecanut palm is a graceful looking, erect, and unbranched palm, sometimes reaching a height of 20 m. Shorter varieties have been bred lately. The stem has scars of fallen leaves in regular annulated forms. The girth of the stem depends

Table 1.1 Country-Wise Area and Production of Arecanut

Country	Area ('000 ha)					Production (mt)				
	1961	1971	1981	1991	1998	1961	1971	1981	1991	1998
Bangladesh	82.60	40.10	36.43	35.81	36.00	62.99	23.36	25.05	24.12	28.00
China	0.62	1.33	2.83	26.96	46.00	3.71	10.07	24.35	111.09	172.57
India	135.00	167.30	185.20	217.00	270.00	120.00	141.00	195.90	258.50	310.00
Indonesia	65.00	75.00	90.00	95.76	75.38	13.00	15.00	18.00	22.81	32.60
Malaysia	6.00	2.50	1.30	2.20	2.40	6.50	3.00	2.50	4.00	4.00
Maldives	0.003	0.003	0.006	0.030	0.030	0.001	0.001	0.005	0.016	0.016
Myanmar	11.13	24.68	26.47	28.93	29.50	8.00	19.20	25.80	92.27	31.50
Thailand	–	–	–	8.50	9.00	0.00	0.00	0.00	13.25	14.50
Global	300.55	310.92	342.25	415.20	468.31	214.21	211.64	291.62	446.15	593.29

Source: Food and Agriculture Organizations (FAO) of the United Nations; mt, million tons.

Table 1.2 State-Wise Area and Production of Arecanut in India

State	Area ('000 ha)					Production ('000 t)				
	1966–67	**1971–72**	**1981–82**	**1991–92**	**1997–98**	**1966–67**	**1971–72**	**1981–82**	**1991–92**	**1997–98**
Andhra	0.00	0.20	0.20	0.20	0.20	0.00	0.20	0.20	0.20	0.50
Andaman and Nicobar	0.00	0.00	0.00	0.00	3.60	0.00	0.00	0.00	0.00	5.20
Assam	26.20	25.90	47.20	66.00	74.10	25.20	29.20	48.10	50.50	64.00
Goa Daman and Diu	0.00	1.40	1.70	1.30	1.50	1.20	1.70	1.30	1.50	1.80
Karnataka	34.80	43.20	55.20	65.40	88.40	52.70	56.30	80.20	96.00	128.50
Kerala	71.20	86.80	61.20	63.51	76.10	44.30	53.40	66.00	65.14	94.00
Maharashtra	0.00	2.30	2.10	1.90	1.90	0.00	2.80	2.40	2.60	7.50
Meghalaya	0.00	6.30	6.50	8.90	9.50	0.00	4.20	4.90	8.70	12.10
Mizoram	0.00	0.10	0.40	0.00	0.00	0.00	0.00	0.00	0.10	0.10
Puduchery	0.00	0.00	0.00	0.10	0.10	0.00	0.00	0.00	0.10	0.20
Tamilnadu	0.00	5.00	4.30	2.80	2.70	0.00	5.00	2.80	3.40	4.30
Tripura	0.00	0.00	0.70	1.20	1.80	0.00	0.00	1.00	2.30	3.50
West Bengal	0.00	3.10	3.10	5.90	8.10	0.00	0.80	0.80	7.90	12.40

Source: Directorate of Economics and Statistics, Government of India.

Table 1.3 Geographical Distribution of the *Areca* Species

Country	Species
India	*Areca catechu, Areca triandra*
Andaman Islands (India)	*Areca catechu, Areca laxa*
Sumatra	*Areca catechu, Areca triandra, Areca latiloba*
Sri Lanka	*Areca catechu, Areca concinna*
Malaysia	*Areca catechu, Areca triandra, Areca latiloba, Areca montana, Areca ridleyana*
Borneo	*Areca catechu, Areca borneensis, Areca kinabaluensis, Areca arundinacea, Areca bongayensis, Areca amojahi Areca mullettii, Areca minuta, Areca furcata*
Java	*Areca catechu, Areca latiloba*
Celebes	*Areca celebica, Areca oxicarpa, Areca paniculata, Areca henrici*
Australia	*Areca catechu, Areca alicae*
Solomon Islands	*Areca niga-solu, Areca rechingeriana, Areca torulo, Areca guppyana*
New Guinea	*Areca congesta, Areca jobiensis, Areca ladermaniana Areca macrocalyx, Areca nannospadix Areca warburgiana*
Philippines	*Areca catechu, Areca hutchinsoniana, Areca vidaliana Areca costulata, Areca macrocarpa, Areca parens Areca caliso, Areca whitfordii, Areca camariensis, Areca ipot*
Moluccas	*Areca glandiformis*
Bismarck Islands	*Areca novo hibernica*
Laos	*Areca laosensis*
Cochin China	*Areca triandra*
Lingga Islands	*Areca hewittii*

on genetic variation, soil condition, and plant vigor. The palm has an adventitious root system. The crown of an adult palm normally has 7–12 leaves. The apical bud produces leaves in succession once every 2 months and the leaves live up to about 2 years. The leaves are pinnatisect and consist of a sheath, a rachis, and leaflets. The leaf sheath completely encircles the stem. It is about 54 cm in length and 15 cm broad. Average length of leaf is 1.65 m bearing about 70 leaflets. The leaflets are 30–70 cm in length and 5.8–7.0 cm in breadth, depending on the position of the leaf. When grown under favorable conditions, flowering occurs once 5 years, appearing at the tenth node in the "*South Kanara*" cultivar (Murthy and Bavappa, 1960b). In another semi-tall cultivar "*Mangala*," flowering occurs earlier in about 5 years

(Bavappa, 1977). The inflorescence is a spadix produced at the leaf axils. The number of spadices produced depends upon the number of leaves. The mean number of spadices produced is 3 to 4, depending on the age of the plant (Murthy and Bavappa, 1960a). The fruit of arecanut is a monocular one-seeded berry that is orange-red to scarlet in color when ripe, encircled by a thick fibrous outer layer, akin to a husk. Each branch bears 100–250 fruits.

Arecanut has a fairly wide root distribution, and the distribution pattern has been described by Shama Bhat and Leela (1969). Within a radius of about 50 cm, close to 70 percent of the roots are distributed. Almost 80 percent of the roots do not go beyond a radius of 85 cm. All the cultural operations are confined within this radius. Depthwise, 0–50 cm contained about 70 percent of the roots and further from 51 to 100 cm about 18–24 percent. When the palms are planted close, roots penetrate deeper than when planted at wider spacing.

Arecanut is a crosspollinated, monoecious palm, and both male and female flowers are present on the same spadix (Bavappa and Ramachander, 1967). Male flowers last 4–7 weeks, and the female flowers, which are cream-colored at flowering, turn green within a week. Flowers open between 2 a.m. and 10 a.m., and female flowers last up to 10 days. The stigma remains receptive up to a week (Murthy and Bavappa, 1960a; Shama Bhat et al., 1962b). Wind carries the pollen. Fruit development takes place in three stages (Shama Bhat et al., 1962a). Increase in size is rapid in the first phase, followed by increase in volume and dry matter accumulation in kernel in the second phase. During this period, the embryo becomes macroscopic and develops rapidly. In the third stage, the fruit swells, and gradually the green color fades giving rise to a bright yellow color. In all, it takes about 9 months to a full year for fruit to develop for harvest.

The Cytogenetics of the Areca Palm

Venkatasubban (1945) was the first to report the number of chromosomes of *Areca catechu L.* at $2n = 32$. This was later confirmed (Sharma and Sarkar, 1956; Raghavan and Baruah, 1958; Abraham et al., 1961; Bavappa and Raman, 1965). The chromosome number of *Areca triandra* Roxb. as $2n = 32$ reported by Darlington and Janaki Ammal (1945) was subsequently confirmed (Bavappa and Raman, 1965; Sharma and Sarkar, 1956). Nair and Ratnambal (1978) determined the meiotic chromosome number of *Areca macrocalyx* Becc. at $2n = 16$. Meiotic abnormalities such as nondisjunction, lagging chromosomes, univalents, and pentads were reported by Sharma and Sarkar (1956) in *Areca catechu*. Bavappa and Raman (1965) observed in the meiosis of four eco types of *Areca catechu* abnormalities such as univalents at diakinesis and metaphase I, nonsynchronization of orientation, clumping, delayed disjunction, chromosomes bridges and laggards at anaphase I and II, chromosome mosaics, and supernumerary spores. The meiotic division was found to be quite normal in *Areca triandra*, except for the occasional presence of 14 and 18 chromosomes during metaphase II (Sharma and Sarkar, 1956). Regular meiotic division in the types of *Areca triandra* was also reported by Bavappa and Raman (1965). Intracultivar variation in meiotic behavior of the *Areca* species was reported by Bavappa (1974) and Bavappa and Nair (1978). While normal bivalent formation was observed in a few plants, hexavalent, octovalent, and even decavalent formation was observed in others. Bridges and laggards, which are nothing but abnormalities, and disorientation of chromosomes at anaphase I and II were also reported in this species.

Bavappa and Nair (1978) reported intrapalm variation in chromosome number in the pollen mother cells of *Areca catechu* and *Areca triandra* and their hybrids. Cytomixis to an extent of 39 percent appeared to have contributed to this abnormality. Despite the presence of a high degree of multivalents in *Areca catechu*, pollen fertility was very high. These authors suggested the possibility of the frequency of multivalent formation and disjunction being under genotypic control and being subjected to selection. Partial desynapsis of chromosomes at diakinesis was reported by Bavappa (1974). Bavappa and Nair (1978) reported the same in *Areca triandra* and *Areca catechu* × *Areca triandra* hybrids. Desynapsis observed at diakinesis was followed by an increase in pairing at metaphase I as reflected by the frequency of bivalents in *Areca triandra* and *Areca catechu* × *Areca triandra* hybrids. This was attributed to distributive pairing, a mechanism that has been possibly adopted for ensuring their regular segregation (Bavappa and Nair, 1978). The extent of desynapsis was higher in the F1 hybrids of *Areca catechu* and *Areca triandra* as compared to *Areca triandra*, suggesting that the gene controlling this character may be dominant. The large number of univalents observed in the hybrid as compared to *Areca triandra* parent has been attributed to reduced homology of the parental chromosomes (Bavappa and Nair, 1978).

Venkatasubban (1945) observed two pairs of short satellite chromosomes in the somatic chromosome complement of *Areca catechu*, while Sharma and Sarkar (1956) observed three pairs of long chromosomes, six pairs of medium-sized chromosomes, and seven pairs of short chromosomes in the same species. These authors categorized the chromosomes into seven groups based on their morphology and relative length. Two pairs of long chromosomes next to the longest were found to have secondary constrictions. The chromosomes of *Areca triandra* were found to be longer than that of *Areca catechu* (Sharma and Sarkar, 1956), while Bavappa and Raman (1965) found the chromosomes of *Areca catechu* and *Areca triandra* to be different in size, total chromatin length, position of primary and secondary constrictions, and the number and position of satellites. These authors concluded that *Areca catechu* is more advanced in evolution compared to *Areca triandra* based on the observations of Sharma and Sarkar (1956), which showed that gradual reduction in chromatin matter had taken place in the evolution from primitive to advanced forms of different genera and tribe of Palmae. The pachytene chromosomes of *Areca catechu* were found to be morphologically similar to the somatic chromosomes, though pachytene chromosomes were about 10 times longer than the somatic chromosomes (Bavappa and Raman, 1965). Raghavan (1957) reported the chromosome morphology of some cultivars of *Areca catechu* from Assam in northeastern India. There were only minor variations in structure and length of individual chromosomes, total length of the complement, and the position of constrictions among the types. On the basis of morphology, the author recognized nine groups in the somatic chromosomes of the cultivars.

Bavappa (1974) and Bavappa et al. (1975) investigated the karyotypes of eight cultivars of *Areca catechu* and four ecotypes of *Areca triandra,* which showed substantial differences in their gross morphological characteristics. The karyotypes of *Areca triandra* ecotypes showed higher frequency of median and submedian chromosomes compared to *Areca catechu.* A classification of the karyotypes of the two species according to their degree of asymmetry, which recognizes three grades of size differences and four grades of asymmetry in centromere position, was made by Stebbins (1958). This showed that karyotypes 1B, 2A, 2B, and 3B are represented in *Areca catechu* cultivars, while in that of *Areca triandra* only 1A, 2A, and 2 were represented in its ecotypes. Two different types of asymmetry in karyotypes were observed within the same cultivar of *Areca catechu,* whereas no such variations were seen to occur in the ecotypes of *Areca triandra.* This clearly shows that Areca *triandra* has a more symmetrical karyotype compared to *Areca catechu.* This also led to the conclusion that delineating the cultivars of *Areca catechu* on the basis of a standard karyotype is rather difficult compared to that of *Areca triandra.* Obviously, the lesser chromatin matter and an asymmetrical karyotype in *Areca catechu* compared to *Areca triandra* point to the fact that the former is more evolved than the latter.

Genetic Resource Program

The Regional Research Station located in Vittal, Karnataka State, affiliated to the Central Plantation Crops Research Institute situated in Kasaragod, Kerala State, under the overall administrative control of the Indian Council of Agricultural Research, New Delhi, is mandated to carry out research in arecanut. One of its primary tasks is the collection of germplasms. Several cultivars have been recognized in Karnataka State (Rau, 1915) and Philippines (Beccari, 1919), based on kernel and fruit characteristics. Based on stomatal characteristics, nut size, leaf shedding pattern, female flowers, and so on, the areca species can be separated (Bavappa, 1966; Bavappa and Pillai, 1976). A collection of five species, namely *Areca catechu*, *Areca triandra*, *Areca macrocalyx*, *Areca normanbyii*, and *Areca concinna* and two genera, *Actinorhytis* and *Pinanga dicksonii*, are available (Anuradha, 1999). The germplasm collection now holds 113 accessions (Ananda, 1999b). Among these, 23 exotic accessions were introduced from Fiji, Mauritius, China, Sri Lanka, Indonesia, Vietnam, Singapore, Solomon Islands, and Australia. From the different arecanut growing states within India, 90 collections have been sourced. About 39 of these have been described based on the various descriptors. The different cultivars show wide variations in stem height, inter node length, leaf size and shape, and fruit characters. Nuts obtained from Malnad, parts of Shimoga, and Chickmagalur districts in the State of Karnataka are small in size, whereas those from North Kanara and Ratnagiri from the same state are bigger (Murthy and Bavappa, 1962). These traits lead to wide variations in yield, earliness in bearing, fruit bunches, quality, and dwarfness.

The exotic and indigenous collections have been under evaluation since 1957 for morphology, nut characters, and yield attributes (Ananda 1999b; Bavappa and Nair, 1982). Yield evaluation resulted in the release of four high-yielding cultivars, of which three are selections from exotic collections. The characteristics of these varieties have been described (Ananda, 1999a; Ananda and Thampan, 1999). Among the exotic collection, cultivar VTL-3 introduced from China was released and named *Mangala* (Bavappa, 1977). This cultivar has bearing earliness, greater number of female flowers, high yield, and shorter stem height compared to other accessions. Other varieties released for cultivation are *Sumangala* and *Sreemangala*, which are introductions from Indonesia and Singapore, respectively. One of the indigenous collections from the State of West Bengal was found to be a high yielder and was released with the name *Mohitnagar*. Table 1.4 summarizes the characteristics of these varieties. Other promising cultivars are SAS-I, *Tirthahalli,* and *Calicut-17.*

Table 1.4 Yield and Nut Characteristics of Arecanut Varieties

Variety	Growth Habit	Shape and Size of Nut	Dry Nut Yield (kg/plant)	Remarks
South Kanara	Tall	Round, Bold	2.00	Coastal Karnataka and Kerala State
Mangala	Semi-tall	Round, Small	3.00	Coastal Karnataka and Kerala State
Sumangala	Tall	Oval, Medium	3.20	Karnataka and Kerala States
Sreemangala	Tall	Round, Bold	3.18	Karnataka and Kerala States
Mohitnagar	Tall	Round, Medium	3.67	West Bengal, Kerala and Karnataka States

Source: Ananda (1999a).

The morphological, anatomical, and yield characteristics of 13 cultivars of *Areca catechu* and four ecotypes of *Areca triandra* have been recorded by Bavappa (1974). These data were with reference to the years 1963, 1966, and 1972. The analysis of variance of the results obtained in 1963 showed that the differences between the cultivars are highly significant for all the six morphological characters. Combined analysis of the data for the 2 years for the 24 common characters recorded during 1967 and 1972 also revealed significant interactions between cultivars for all the characters. Significant interaction between years and cultivars was seen with regard to plant height, girth, internodal distance, bunch number, inflorescence, length and breadth of leaf sheath, length and volume of nut, length, breadth, weight, and volume of kernel. From D2 studies, Bavappa (1974) concluded that detection of genetic divergence in the early years of the productive phase is of considerable advantage in formulating breeding programs in a perennial crop like arecanut. The investigations showed that mean nut volume and kernel breadth are the traits of primary importance contributing to the overall genetic divergence in areca. For divergence between *Areca catechu* and *Areca triandra*, mean fruit length, nut, and kernel characters are the most important for differentiation within *Areca catechu* cultivars and between *Areca catechu* and *Areca triandra* types. Results obtained from canonical analysis were also in broad agreement with the clustering pattern found from D2 analysis. However, canonical analysis is only of limited utility in view of the fact that the first two canonical roots accounted for only 85 percent of the variation or less (Bavappa, 1974).

The grouping obtained by D2 analysis showed that the three cultivars each from Saigon and Solomon Islands and the two ecotypes of *Areca triandra* from Indonesia were invariably in one cluster each. As against this, close similarity between the cultivars from different countries has also been observed. The cultivar from Singapore was grouped with the three cultivars from Saigon in one cluster. A similar affinity between the two geographically distant cultivars was shown by "Ceylon-1" and "Indonesia-6," both always coming within the same cluster. The local cultivar was found to be invariably associated with the Singapore cultivar in forming the cluster. One of the cultivars of *Areca catechu* from Sri Lanka, "Ceylon-2" was always forming a separate cluster, indicating its distant nature of divergence. The clustering pattern of cultivars and ecotypes revealed that geographic diversity need not always be related to genetic diversity (Bavappa, 1974).

Areca Crop Improvement

The crop can be improved through hybridization. By nature, the areca palm is tall and this is a definite constraint to spraying the plant to control pest attack, in addition to harvest of the nuts. Hence, dwarfing the plant is a primary objective through hybridization. The hybridization program was initiated with the primary objective of exploiting the variability in the *Areca* germplasm (Bavappa and Nair, 1982). Primary concern was to evolve high-yielding varieties with the regular bearing traits, with nuts of superior quality, and to evolve semi-tall ideotypes. Interspecific hybrids of

Areca catechu × *Areca triandra* had only one stem, as in *Areca catechu*, which indicates dominance of this character (Bavappa, 1974). Hybrids mostly equaled parents in internodal length and further exhibited hybrid vigor for a number of characters such as male and female flower number, spadix length, and stem girth. As dwarfing is a primary objective, attempts were made to identify such types, and one such, "*Hirehalli Dwarf*" *(H. Dwarf)*, was identified (Naidu, 1963). This is a low yielder with poor quality nuts, which are only suitable for "chewing"—the widespread practice in India and Pakistan. High-yielding varieties were crossed with "*Hirehalli Dwarf*" to exploit the dwarfing nature (Ananda, 2000). Maximum dwarfs and intermediates were recovered in crosses of "*Sumangala*" × "*H. Dwarf*," "*Mohitnagar*" × "*H. Dwarf*," and "*Mangala*" × "*H. Dwarf*" from among the 12 combinations.

The Agronomy of Arecanut

Soil and Climatic Requirement

A wide spread of soils are used to grow the arecanut palm. The largest area under the crop, however, is in the gravelly laterite soils of red clay in northern Kerala and coastal belt of Karnataka state (Nambiar, 1949). The deep-black, fertile, clay-like loams of the plains of Karnataka State are very suitable for the crop. Sandy, alluvial, brackish, or calcareous and sticky clay soils are unsuitable.

Subhumid tropical climate suits the crop best, and it thrives very well in the regions 28° north and 28° south of the equator. Latitude determines the altitude at which the crop grows (Shama Bhat and Abdul Khader, 1982). In the northeastern regions of India, the crop is mostly grown in the plains. At higher elevations, lower temperature limits the crop growth. An altitude up to 1000 m above mean sea level (MSL) is alright, beyond which the quality of the nut is adversely affected (Nambiar, 1949). The crop thrives in a range of temperature between 14°C and 36°C. It can also withstand very low minimum temperatures (4°C) as found in winter in India's West Bengal state. The crop requires very high rainfall ranging from 300 to 450 cm per annum. However, in the plains of Karnataka and Tamil Nadu states, where rainfall can be as low as 75 cm supplemental irrigations especially in the summer months, irrigations is required to sustain the areca plantation. In coastal India, the climatic conditions are different. Southern Kerala gets better distributed rainfall than northern Kerala and coastal Karnataka, which receives most of the rainfall during the months between June and September. This is followed by drought. Hence, in these regions, arecanut is grown principally as an irrigated crop.

Management Aspects

The areca palm is a seed-propagated plant. In selecting the planting material, both the age of the mother palm and seed size are important. While selecting seeds, it must be ensured that they are obtained from trees that are already stabilized in their yielding pattern, at least more than 10 years in maturity. Heavy, mature seeds are selected, which ensures good germination. Selected seeds are sown 5 cm apart with their stalk ends pointing upwards. Regular irrigation, on alternate days, is crucial to maintain healthy seedlings. Three-month-old saplings can be transferred to a secondary nursery and planted 30 cm apart. Seedlings can also be raised in PVC (polyvinyl chloride) bags of 25 cm × 15 cm size, 150 G in a mixture of top soil, cattle manure, and sand in the ratio of 7:3:2, respectively. Twelve- to 18-month-old seedlings are selected for field planting.

Spacing depends primarily on depth and soil fertility. The effect of spacing on cumulative yield is given in Table 1.5. Results show that 2.7 m × 2.7 m is the best spacing (Shama Bhat and Abdul Khader, 1982). Sixty centimeter cubic pits, filled with soil to a depth of 50 cm, are used for planting seedlings.

Irrigation

The soil water depletion profile is the major deciding factor for the frequency of irrigation. Therefore, rather than using a fixed irrigation schedule, considerable flexibility should be given to accommodate varying crop-evapotranspirational (ET) losses. Irrigation is done mostly by splashing and basin filling. Several researchers have investigated irrigational requirements for the areca crop in various regions. However, all of these were done by taking into account frequency and the quantum of applied water, considering the ET demand and ambient temperature. Systematic studies revealed that at IW/CPE (irrigation water/cumulative pan evaporation) ratio of 1, that is, 300 mm depth of water, resulted in the best yields (Abdul Khader and Havanagi, 1991; Yadukumar et al., 1985). Based on a modified Penman model, a net irrigation requirement of 899 mm during post-monsoon season and an irrigation interval of 6–7 days was found to be sufficient (Sandeep Nayak, 1996). Using a modified Penman model, Mahesha et al. (1990) estimated that the ET rates of arecanut increased from 4.60 mm/day in December to 6.25 mm/day in April, which fell to 5.78 mm/day in May due to pre-monsoon showers.

In this connection, results of the drip irrigation system need to be researched. Drip irrigation has the added advantage of fertigation. An investigation carried over 10 years

Table 1.5 Effect of Spacing on Cumulative Yield of Arecanut (kg/ha)

Spacing (m × m)	Vittal (Coastal Karnataka)	Hirehalli (Plains of Karnataka)	Kahikuchi (Assam)	Peechi (Kerala)
1.8 × 1.8	6290	3130	2.00	1749
1.8 × 2.7	8167	3705	3.00	1766
1.8 × 3.6	7829	4132	3.20	1710
2.7 × 2.7	10722	3867	3.18	1867
3.6 × 3.6	6169	2417	3.67	1448

Source: Sannamarappa (1990).

(1996–2006) to evaluate the efficacy of four fertigation levels (25, 50, 75 and 100 percent of the recommended fertilizer dose, 100:18:117 g N:P:K/palm/year), three frequencies of fertigation (10, 20 and 30 days), and two control treatments (control 1, i.e., drip irrigation without fertilizer application and control 2, i.e., drip irrigation with 100 percent NPK soil application) on productivity and resource use efficiency of arecanut indicated that adoption of fertigation not only increases productivity, but ensures higher efficiency of use of the two most critical inputs, that is, water and nutrients (Ravi Bhat et al., 2007). Data on comparative merit of drip irrigation over conventional surface irrigation in influencing areca yield are given in Table 1.6 (Chinnappa and Hippargi, 2005), which clearly indicates an advantage of more than 25 percent for the former over the latter. The crucial point to be examined is the economics of drip irrigation over surface irrigation. The data reported here refers to the response of 90 arecanut farmers from 15 villages in the southern transitional zone of Karnataka State. It was observed that installation of drip irrigation with ISI (Indian Standards) materials required Rs 50,394 (approximately US$1260) per hectare. Adoption of drip irrigation in arecanut gardens has resulted in additional output of 5.03 qtl/ha (1 qtl = 100 kg) worth Rs 67,972 (approximately US$1700). The maintenance cost of drip irrigated gardens was found to be lower at Rs 29,109 (approximately US$730) as compared to surface irrigated gardens at Rs 37,856 (approximately US$950). There is substantial saving in the cost of cultivation at Rs 8837/ha (US$220) in drip irrigated gardens, to increased use of labor and inputs cost in surface irrigated gardens. These results reported by Chinnappa and Hippargi (2005) clearly establish the economic viability of drip irrigation. Further, there is economy of water use to the tune of 42.18 acre inches. With this saving in water use, it is possible to irrigate an additional area of 2.5 ha. The saving equivalent of more than US$200/ha is substantial on the Indian economic scale compared to that of the United States.

Balasimha et al. (1996) reported that the photosynthetic parameters and arecanut yield increased with increases in drip irrigation levels from 10 l/day to 30 l/day in a mixed cropping system in which cocoa is interplanted with arecanut. The results are given in Table 1.7. The saved irrigation water can be used for other crops or intercrops, and the ground water depletion can also be checked. Efficiency of water use can be enhanced through a good combination of irrigation and mulching. Mulching is very important in arecanut plantation because the highly porous soils in which the crop is

Table 1.6 Comparative Merit of Drip Irrigation Over Surface Irrigation on Arecanut Yield

Details	**Yield (qtl/ha)**
Drip irrigation	24.93
Surface irrigation	19.90
Incremental yield (qtl)	5.03
Percentage increase	25.28

Note: qtl: quintal = 100 kg.

Table 1.7 Effect of Irrigation Levels on Photosynthetic Characters and Dry Nut of Arecanut

Treatment	Photosynthesis (μmol/m²/s)	Transpiration (μmol/m²/s)	Stomatal Conductance (μmol/m²/s)	Yield (kg/ha)
I 1 (10 l/day)	3.98	3.46	0.105	1.80
I 2 (20 l/day)	3.62	3.68	0.122	1.96
I 3 (30 l/day)	4.48	4.75	0.168	2.01
LSD (5%)	N.S.	0.96	0.041	N.S.

Note: 1, 2, and 3 = irrigation treatments; N.S., not significant; LSD, least significant difference.
Source: Balasimha et al. (1996).

planted can lead to water loss through seepage. Different plastic and organic mulches have been found effective (Abdul Khader and Havanagi, 1991; Shama Bhat, 1978).

Arecanut Nutrition

Available information on arecanut nutrition is primarily based on classical field experiments. Results obtained from several field trials indicate that each palm requires $N:P_2O_5:K_2O$ in the ratio of 100 g:40 g:140 g, respectively, clearly showing that of the three principal nutrients (N, P, and K), it is potassium that is required most. The requirements given here are annual per palm (Shama Bhat and Abdul Khader, 1982). These nutrients are, generally, supplemented by 12 kg of organic manure, either through cattle manure, compost, or green leaves like that of *Glyricidia maculata*, which can grow well on the boundaries of arecanut gardens. Recent fertilizer trials have shown that doubling the mentioned rates can substantially enhance yield (Sujatha et al., 1999). Doubling the rates of nutrients indicated here has not only resulted in higher yields, but increase in net income and "Benefit–Cost Ratio." Fertilizer nutrient requirements have been standardized for arecanut by Mohapatra and Bhat (1982).

Arecanut Physiology

Balasimha (1986) and Balasimha and Subramonian (1984) reported that approximately 30–50 percent of photosynthetically active radiation (PAR) is transmitted through arecanut canopy. This varies with season and time of the day. In the arecanut canopy, light environment is highly dynamic due to variation in cloud cover, solar angle, and canopy. The pattern of light transmission varies with the spacing of palms. Midday light profile shows that a spacing of 3.3 m × 3.3 m or 1.8 m × 5.4 m allowed maximum transmission. However, in the afternoon, 3.9 m × 3.9 m spacing allowed maximum transmission. On average, arecanut canopy intercepts 70 percent of incoming radiation. The interception of the remaining radiation depends on the nature of the intercrop canopy and leaf area index (LAI). For instance, a cocoa plant with a compact canopy and LAI of 3–5 intercepts nearly 90 percent of the incident light. Pepper, which is trained on arecanut stems, receives differential light depending upon directional and diurnal effects.

A fully grown arecanut palm, on average, bears 8–9 leaves. The canopy area and leaf area are about 11.2 m^2 and 22.0 m^2, respectively. The arecanut canopy covers a ground area of 9.1 m^2 with a LAI of 2.44. Photosynthesis in arecanut leaves ranged from 2.8 to 8.2 μmol CO_2/m/s depending on the cultivar and leaf position. The first fully open and third leaves showed the highest rate of photosynthesis. Total chlorophyll content also varied among the species of *Areca,* and the highest content was recorded in *Areca triandra* (Yadava and Mathai, 1972).

Mixed Cropping Systems in Areca

The interspace between arecanut palms can be put to good use by planting associate crops. This would bring in additional income to the farmer. The practice has been quite successful (Bavappa, 1961; Sannamarappa and Muralidharan, 1982). The long prebearing age of the areca palm has prompted farmers to grow different annual or semiperennial crops in the interspace to ensure economic sustainability of the areca plantations. The initial period of 5–6 years (the prebearing phase) is the ideal time to grow these intercrops, especially short duration ones. In later years of the palm, as the canopy enlarges in height, mixed cropping with other shade tolerant perennial crops can be practiced.

The choice of intercrops in arecanut plantations has been quite wide. A number of annual crops, such as rice, sorghum, beans (*Vigna unguiculata*), vegetables, and yams are grown as intercrops (Abdul Khader and Antony, 1968; Abraham, 1974; Muralidharan, 1980; Shama Bhat, 1974; Shama Bhat and Abdul Khader, 1970; Thomas, 1978). When these crops are cultivated, the cultural and nutritional practices followed are used for the pure stands of the respective intercrops (Sannamarappa and Muralidharan, 1982). Leaving 1 m radius around arecanut palm, the interspaces are prepared for cultivation of the intercrops during the premonsoon period. Rice, sorghum, corn, beans, groundnut, and sweet potato are sown in beds. For yam, banana, and pineapple, pits or trenches are dug. With the exception of banana and beans, biological productivity of the intercrops was found to be lower than when the crops were sown singly. In most studies in different regions, no deleterious effects on main crops due to intercropping were observed (Abraham, 1974; Muralidharan and Nayar, 1979; Sadanandan, 1974). Banana is the most preferred intercrop in all the arecanut gardens (Bavappa, 1961; Brahma, 1974; Shama Bhat, 1974). It also provides good shade during early growth of arecanut palms.

Abraham (1974) and Nair (1982) found black pepper (*Piper nigrum*) to be an excellent intercrop. The arecanut stems are used as live "standards" (prop on which the pepper twines). Of the pepper cultivars successfully experimented with, *Panniyur-1* and *Karimunda* were found to be best. It was also observed that arecanut yield was not depressed due to the presence of pepper.

Another important intercrop is cocoa. The microclimate, especially shade, soil moisture, and temperature, in the arecanut gardens was found to be ideal for cocoa growth. Shama Bhat and Leela (1968) and Shama Bhat and Bavappa (1972) have found an arecanut–cocoa mixture as good. A confounded asymmetrical factorial design, with different spacings and two fertilizer levels, showed significant influence on the number and weight of cocoa pods in the normal spacing of arecanut. A spacing of 2.7 m × 2.7 m or a spacing of 2.7 m × 5.4 m combination could be safely followed, although operational advantages are better in the latter spacing (Shama Bhat, 1983, 1988). High-density multispecie cropping in arecanut is typically composed of arecanut–black pepper, cocoa, coffee, or banana occupying different vertical air space levels. Bavappa et al. (1986) tried six crop species in a 17-year-old arecanut plantation. A steady increase in arecanut yield was observed, and the intercrops started to yield from the third year onwards after planting. The economic dry matter yield of intercrops accounted for about 27 percent of total economic yield. This intercropping system could accommodate 1300 areca palms with pepper, cocoa, clove, banana, and pineapple numbering 3700 in 1 ha. Investigations on the microflora in the rhizosphere of such combinations indicated that the population of bacteria, fungi, actinomycetes, N_2 fixers, and P solubilizers increased in the areca–cocoa or –pineapple combinations than in monocrop system of areca (Bopaiah, 1991). P solubilizers such as vesicular arbuscular mycorrhizae (VAM) were lowest in areca–banana combination. Microbial biomass was also found to be higher in the combinations, in general. The asybiotic N_2 fixers isolated from an arecanut-based, high-density, multiple-cropping system had an N fixing capacity in the range of 2.8–11.8 mg N/100 ml of medium (CPCRI, 1988a).

Arecanut Pathology

A wide range of diseases attack the areca palm. Of these, the yellow leaf disease (YLD), *Phytophthora* fruit rot, Anabe caused by *Ganoderma,* and inflorescence dieback cause considerable economic loss (Sampath Kumar and Saraswathy, 1994; Rawther et al., 1982). By far, the most serious disease is the YLD, though it is restricted only to certain areas such as Southern Kerala State and parts of Karnataka State. This is a highly debilitating disease and an effective control is still elusive. Phytoplasma is found constantly associated with the disease (Nayar and Selsikar, 1978). The plant hopper *Proutista moesta* acts as vector for the disease (Ponnamma et al., 1997). The diseased palm when treated with tetracycline showed remission of symptoms, which provides additional supporting evidence for phytoplasmic etiology (CPCRI, 1994). Characteristic symptoms are leaf yellowing followed by necrosis. Comparative physiology of healthy and diseased plants showed significantly higher stomatal resistance with increased water potential and turgor pressure in the latter (Chowdappa et al., 1993, 1995), while photosynthesis and transpiration were lower. Chlorophyll content was also low in diseased plants (Chowdappa and Balasimha, 1992). The altered values of chlorophyll fluorescence indices reflected in normal arrangement of antennae pigments of photosystem II (PS-II) and reduction in photosynthetic quantum yield (Chowdappa and Balasimha, 1995). Both nuts and roots were also affected. Different methods have been investigated to control the disease, and one such is intercropping and multiple cropping in YLD-affected areca plantations.

In arecanut, fruit rot is a major disease. In Karnataka State, it is locally known as "*Koleroga*" or "*Mahali*" and occurs in epidemic form, especially in the monsoon season. At first, water-soaked lesions appear on the surface of the nut, and infected nuts turn dark green after losing their natural color. A white mycelial mass then covers the entire nut. The causal organism is identified as *Phytophthora Mardi*, a fungus. The infection can cause up to 15 percent loss in yield (Coleman, 1910; Sampath Kumar and Saraswathy, 1994). Effective control of the disease is systematic and periodic spraying of 1 percent Bordeaux mixture on fruit bunches. Covering of the bunches with polyethylene paper gives complete protection (Chowdappa et al., 1999; CPCRI, 1983; Sastry and Hegde, 1985). The fungus can also cause bud rot. The spindle leaf is affected, which subsequently kills the plant. Early detection, removal of affected plant parts, and treatment with 10 percent Bordeaux paste can save affected plants.

Ganoderma lucidum fungus causes the foot rot, commonly known as "*Anabe disease*." This is a soil-borne fungus (Sampath Kumar and Nambiar, 1990). Poor drainage and a high water table are the predisposing factors for the onset of the disease. Yellowing of the leaves in the outer whorl are the initial symptoms. Stems show dull brown patches. Subsequently gummy exudation starts, and fruiting bodies of the fungus appear. To control the disease, checking the garden periodically is a must. Drenching of the root zone with 0.3 percent Calixin (15–201) or root feeding with the same chemical at the rate of 125 ml/palm is recommended.

Inflorescence dieback is another serious disease that causes button shed. Yellowing and subsequent drying of inflorescence rachillae are the usual symptoms. The disease can be observed through the year, but most severely in summer months (Saraswathy et al., 1977; Chandramohanan and Kaveriappa, 1985). The fungus *Colletotrichum gloeosporioides* causes the disease. Spraying Indofil M-45 at the rate of 3 g/l water at the time female flowers open controls the disease. Stem bleeding is a minor disease occurring in some parts of India caused by the fungus *Thielaviopsis paradoxa*. Small discolored depressions appear in the basal portion of the stem. Subsequently these patches coalesce and cracks appear, leading to the disintegration of fibrous tissues. Exudation of brown liquid out of the cracks follows. Removing the affected parts and applying coal tar or Bordeaux paste control the disease. Bacterial leaf blight caused by *Xanthomonas campestris* (Rao and Mohan, 1970; Sampath Kumar, 1981) causes the bacterial leaf blight, and the disease is confined to the "Maidan Region" of the State of Karnataka. Partial to complete blighting of leaves is caused by the infection. Spraying streptocycline or tetracycline (500 ppm) effectively controls the disease.

Arecanut Entomology

Mites, spindle bugs, inflorescence caterpillars, and root grubs are the insect pests of arecanut (Nair and Daniel, 1982). These are either seasonal or persistent pests on the crop. Two major foliage mites, *Oligonychus indicus* and *Raoiella indica*, occur abundantly during dry seasons. Dicofol, carbophenothion, or chlorobenzilate spray controls the disease. The spindle bug, *Carvalhoia arecae,* infests the leaf axils, and its highest population is found from August to September (Nair and Das, 1962).

The bugs suck the sap from the spindles and young leaves resulting in the leaflets getting linear necrotic lesions. This results in the drying up of the spindle, which fails to open. Spraying Thimet 10G or Sevin 4G at the rate of 10g/palm controls the disease.

Leucopholis burmeisteri and *Leucopholis lepidophora* are the most common root grubs. These feed voraciously on the roots causing serious damage. The pest damage is more in low lying areas. Leaves turn yellow and stems taper at the top. Rogor (dimethoate) and phorate, which are soil-applied insecticides, control the white grubs effectively (Prem Kumar and Daniel, 1981). Collection and destruction of adult beetles on emergence from soil is an effective management practice.

The pentatomid bug, *Halyomorpha marmorea*, causes tender nut drop (Vidyasagar and Shama Bhat, 1986). The infestation is prominent between March and August. Bugs pierce the nut, suck the sap, the kernel dries up, and the nut abscesses and drops. Spraying 0.05 percent endosulfan can effectively control the pest. The inflorescence caterpillar *Tirathaba mundella* damages areca inflorescence, feeding on tender rachillae. Malathion insecticide controls the pest.

Leaves, leaf sheaths, and nuts are colonized by scale insects (*Aonidiella orientalis*), which suck the sap from tissues. When these insects feed on nuts, they yellow prematurely and nut quality is lost. These pests are present throughout the year, but are most abundant between October and February. Although scale insects were once considered minor pests, there have been widespread outbreaks of the insect attack in some of the districts in coastal areas of the country. One region is coastal districts of Karnataka State. Control of scale insects is rather difficult, although spraying malathion (0.1 percent) and fenthion (0.1 percent) showed partial control. The most effective is biological control, and *Chilocorus* sp., the coccinellid beetle, has been found to be quite effective in controlling the pest.

Arecanut Nematology

Radopholus similis, the burrowing nematode, is the most commonly found pest of arecanut (Sundararaju and Koshy, 1988). The intercrops found to be infested with nematodes are banana, cardamom, and pepper. Intercultivating these crops in arecanut gardens increases the nematode infestation (Sundararaju and Koshy, 1988). Cocoa, pineapple, clove, cinnamon, and nutmeg were completely free of nematodes and, thus, are ideal species for mixed cropping with arecanut (CPCRI, 1988b). Nematode infestation, unlike that of other pests like diseases and insects, has not been found to seriously impact economy of arecanut production. Nematicides or neem (*Azadirachta indica*) cake keep the nematode population under control.

Harvesting and Processing of Arecanut

It is very important that harvest of the nuts is done at the correct stage, and this is determined by market demands. Harvesting at the most appropriate stage ensures nut quality. For instance, ripe nuts are harvested if the objective is to have dry nuts. On the other hand, green nuts of 6–7 months age are harvested for tender nut processing. Dried whole nuts, colloquially called *Chali*, are the most popular type of arecanut. Following harvest, ripe nuts are sun dried for 40–45 days. Drying is done by spreading the nuts in a single layer on a polyethylene sheet. Only proper drying will prevent fungal attack. Uniformity in drying can be ensured by turning the nuts in the evening each day. Dry nuts are dehusked manually or mechanically and then marketed. Good quality *Chali* is free from immature nuts which have no surface cracking, sticking husk, or, of course, fungal or pest (insects) infection. If the market requirement is tender-processed nuts, then the nuts are harvested at 6–7 months of maturity, when they are green and soft. Processing consists of dehusking, cutting nuts into halves, and boiling them with water or diluted extract of previous boiling. After boiling, arecanut pieces are coated with *Kali*, which is a concentrated extract after boiling three to four batches of arecanut, to obtain good-quality processed nuts. These nuts are then sundried, although ovendrying is also done. Dried nuts are broken into bits, blended with flavor, and packed for marketing. These dried bits of flavored nuts are known as *Supari,* and the flavor depends on specific demands of the region. Tender processed nuts are widely used in making scented *Supari*. Spices like cloves, cardamom, or cinnamon and sometimes synthetic flavors are added. Essence of rose and menthol is commonly used additives. The scented *Supari* is packed in aluminum foil to preserve its flavor and dryness or in butter paper pouches and eventually marketed under brand names. Also, small portable aluminum containers with the brand name are used for marketing. Recently, an arecanut-based product blended with bits of cashew nuts, known as "*Kaju Supari*" (*Kaju* meaning cashew nut in *Hindi*, the most widely spoken language of India) has become quite popular. No North Indian meal, be it on occasions of marriage or any other festivities, is complete without *Supari*. The habit is taking a firm hold in Southern India and in Pakistan as well.

A Look into Arecanut's Future

Like most other agricultural crops in India, arecanut also has taken a beating on the economic front ever since the globalization process set in. Prices have tumbled during the last decade to its lowest value compared to the preglobalization period. This has become a matter of great concern to the arecanut industry and also exporters of this valuable crop, not to mention the arecanut farmers. There is substantial scope to utilize most of the production within India, which is world's biggest producer of the crop (Bavappa, 1982). This is because of the internal demand. The National Commission on Agriculture, which was constituted by the Government of India in 2000, projected the internal requirement of the crop within India at 190,000 tons. During the last decade of the past century, production was about 330,000 tons, which means only about a third of the internal requirement was met by internal production. Despite high production, internal price prevailed at Rs 13,181 (approximately US$300) a quintal (100 kg) in 2000, which fell to less than Rs 4000 (approximately US$100 at current US$–Indian rupee exchange rate). This steep price fall has devastated areca farmers, and the situation has been compounded by the onset of many diseases that also decreased the crop yield. As a consequence of a World Trade Organization (WTO) agreement, 3022 quintals of arecanut was imported from neighboring Asian countries at a price of Rs 3122 (approximately US$70) a quintal in 2000–01, which was less than 25 percent of the price prevalent in 1999–2000. This was followed by a crash in price in 2001–02 of the local nut at 7893(approximately US$175 a quintal), nearly 100 percent less compared to prices prevalent at earlier times, which ruined the farmers and the industry. Although the price of imported nut was much lower than that of the local nut, the reduction in the local price was only about 60 percent. Despite the import, internal demand continued to be steady. Import was mainly from Sri Lanka (2426 tons) and Thailand (419 tons).

As has been stated, India is the largest producer and consumer of arecanut in the world, accounting for nearly 52 percent of world production. Other than in India, no other country cultivates the crop in a comparatively organized manner. The positive side of the globalization process is that external markets are opened to the country's crop. But unless newer competitive marketing strategies coupled with innovative production technology is in place, the crop's future is bleak. There are quite a few areas which need attention. One is production technology with specific reference to areca nutrition. Classical "textbook knowledge" still rules in devising fertilizer requirements of the crop. Recent techniques like "The Nutrient Buffer Power Concept," developed by the author, has been tried very successfully in other perennial crops, such as with black pepper, cardamom, and so on; it needs to be tried with arecanut as well.

2 Cashew Nut (*Anacardium occidentale* L.)

Cashew, which is a native of Brazil, is widely cultivated throughout the tropics for its very nutritious, free-of-cholesterol nuts. It is one of the first fruit trees from the New World to be widely distributed in the tropics by the Portuguese and Spanish adventurers (Purseglove, 1988), who had set out across the world on a sea route. The cashew plant has a checkered history. In both Asia and Africa, cashew nuts and its false fruit have been used in local small-scale entrepreneurial projects for more than three centuries. Large cashew-based enterprises were unknown until the early part of the 20th century, when international trade started to show keen interest in the nut. This is the period when export also began from India. The start was rather slow, but subsequently the momentum picked up and cashew nut plant became an important commercial crop. In Kerala State, in the central part of India, cashew processing and export became the centerpoint of commerce. It also brought in a lot of revenue through exports and provided a lot of employment to the local populace. Places like the Kollam District in central Kerala became the hub of cashew nut-related activities. In India, use of cashew apples (to distill alcohol) and nuts was adopted by the local population, and such accounts can also be observed in Africa, notably Eastern Africa, in countries such as Tanzania. In fact, in recent years Tanzania has turned out to be the biggest competitor to India in cashew export. India imports large quantities from there to meet the growing domestic demand. Making cashew wine appears to have been a common practice both in Asia and Africa (Johnson, 1973). The Maconde tribe of Mozambique calls cashew the "Devil's Nut." It was offered as a token of fertility at weddings, and research carried out at the University of Bologna has shown the presence of numerous vitamins including the old-age health

The Agronomy and Economy of Important Tree Crops of the Developing World. DOI: 10.1016/B978-0-12-384677-8.00002-3
Copyright © 2010 by Elsevier Inc. All rights of reproduction in any form reserved.

enhancer and an aphrodisiac vitamin E (Massari, 1994) in cashew kernel. Lately, enterprising Indian industrialists have started to distill alcohol from cashew apples as "green fuel" for automobiles. At the time of the first attempt to colonize India by the Portuguese, the name used by local population (*Tupi* natives of Brazil) for cashew was *acaju* (meaning *nut* in Portuguese), which later changed to *caju* (i.e., the name prevalent for cashew nut in north India, Pakistan, and Bangladesh), which became *cashew* in English. There is an interesting anecdote in Kerala State as to the origin of the word "cashew nut." When the British colonized India, the foreigners traveling to Kerala were fascinated by the nut for its palatability (it made a very good snack with a Scotch drink) and asked how much it cost. And the locals said, "Eight for pie" (pie being the lowest denomination of British currency in India, and in the local language it is called "Kashu." In a nut shell, the statement in the local language Malayalam was that one could get eight nuts for a pie—"Kashini ettu"—literally changed to *cashew nut*, as coined by the British). Most of the names for cashew in Indian languages are also derived from the Portuguese name *caju* (Johnson, 1973). However, the British incident stuck as the name for cashew nut globally.

Cashew is a versatile tree nut. The kernels contain a unique combination of fats, proteins, carbohydrates, minerals, and vitamins. The nut contains 47 percent fat, but 82 percent of this fat is unsaturated fatty acids, and thus can be considered not cholesterol enhancing. The unsaturated fat content not only eliminates the possibility of cholesterol enhancement, but also balances or reduces total cholesterol in blood. The nut contains 21 percent proteins and 22 percent carbohydrates and the right combination of amino acids, minerals, and vitamins. Therefore, nutritionally speaking, it can be considered on par with milk, meat, and eggs. Since the carbohydrate content in the nut is low, which has only 1 percent of soluble sugar, the sweet taste comes without excess calorie intake from sugary substances. The nut checks diabetes and its consumption does not lead to obesity. In summary, cashew nut is an excellent snack, a good appetizer, and an excellent nerve tonic and stimulant. Although the crop has its roots in Brazil, it is India which nourished it and brought it to international eminence. Today, India is the largest producer, processor, exporter, and second-largest cashew consumer in the world (Nayar, 1998).

Area and Global Production

The significant producers of cashew are India, Indonesia, Brazil, China, Mozambique, Tanzania, Sri Lanka, and Vietnam. Other countries on the Asian and African continents also grow the crop, on a smaller scale. The global scenario on cashew nut production is depicted in Table 2.1. Global cashew production is about 1.09 million tons (Table 2.3, Balasubramanian, 2000; Bhaskara Rao and Nagaraja, 2000). Between 1980 and 1995, cashew production increased by 108 percent, from 0.422 million tons to 0.878 million tons. However, between 1995 and 200, the growth declined to just 24 percent. Subsequent to 1995, world raw nut production was around 1.09 million tons. Global raw nut production is given in Table 2.2, while country-wise raw nut production data is given in Table 2.3. India's share of the world raw nut production

Table 2.1 Global Cashew Production

Indian Subcontinent	Southeast Asia	Africa	Latin America
India*	Vietnam*	Angola	Brazil*
Bangladesh	Thailand*	Benin	Barbados
Sri Lanka	Indonesia*	Burkina Faso	Dominicon Republic
	Malaysia*	Guinea-Bissau	El Salvador
	Philippines	Madagascar	Guadeloupe
	China	Mozambique*	Honduras
		Mali	
		Nigeria*	
		Kenya*	
		Senegal	
		Tanzania*	
		Togo	

*Major producers.

Table 2.2 Global Production of Raw Nuts (mt)

Year	Indian Subcontinent	Southeast Asia	Africa	Latin America	Total
1980	0.1483 (35.1)	0.0275 (6.5)	0.6120 (38.4)	0.0841 (19.9)	0.4219
1981	0.1651 (35.6)	0.0296 (6.4)	0.1856 (40.0)	0.0835 (17.9)	0.4638
1985	0.2309 (46.4)	0.0305 (6.1)	0.1114 (22.4)	0.1246 (25.0)	0.4974
1990	0.2957 (48.2)	0.0815 (13.3)	0.1119 (18.3)	0.1200 (19.6)	0.6127
1995	0.3820 (44.4)	0.1570 (17.9)	0.1397 (15.9)	0.1993 (22.7)	0.8780
1996	0.4328 (39.8)	0.1638 (15.1)	0.3205 (29.5)	0.1689 (15.6)	1.0861
1997	0.4450 (43.8)	0.1708 (16.8)	0.2917 (27.7)	0.1175 (11.6)	1.0149
1998	0.4450 (44.5)	0.1485 (14.8)	0.3690 (36.9)	0.0380 (3.8)	1.0005
2000	0.5200 (47.7)	0.1520 (13.9)	0.2000 (18.3)	0.1800 (16.5)	1.0900*

Source: Cashew Export Promotion Council, Government of India, 2003.
*Includes 0.38 mt under others. Figures in parentheses indicate percent of total raw nut production; mt = million tons.

is 47 percent, while the share of Southeast Asian countries has ranged from 14 to 16 percent. From 1980 to 2000, raw nut production in Southeast Asian countries increased by 45.2 percent. During the last two decades of the 20th century, Latin American countries registered a 114 percent increase.

Table 2.3 Cashew Raw Nut Production by Country (mt)

Country	Production
India	0.520
Indonesia	0.030
Vietnam	0.122
Brazil	0.180
Mozambique	0.030
Nigeria	0.040
Tanzania	0.130
Others	0.038
Total	1.090

Source: Cashew Export Promotion Council, Government of India. Data refer to the latest available figures (2003).

The Indian Scenario

In India, the crop is mainly grown in the states of Goa, Karnataka, Kerala, and Maharashtra along the west coast of the country, and in the states of Andhra Pradesh, Tamil Nadu, Orissa, and West Bengal along the east coast of the country. In Andaman and Nicobar Islands, and the states of Madhya Pradesh, Manipur, Meghalaya, and Tripura in northeastern India, the crop is grown to a limited extent. India's raw nut production has increased from 0.079 million tons in 1955 to 0.52 million tons by 2000, an increase of 558 percent in half a century—that is, more than an annual growth rate of 10 percent, which is, indeed, a remarkable increase. Data on raw nut production is given in Table 2.4. In the last decade of the century past, raw nut production almost doubled. In 1955 the total acreage in India was 0.11 million ha, which by 2000 had jumped to 0.683 million ha, a remarkable increase of 520 percent, or an annual growth of more than 10 percent. This shows that the jump in production during this period was mainly on account of increased acreage. During 1970–80, though the area increased there was a deceleration in production. It was during the last 5 years of the century past that the increase in both area and production was phenomenal.

If India is to maintain its preeminent place in the international market, the productivity has to necessarily increase. Up to 1970, productivity was around 630 kg/ha. Between 1975 and 1985, productivity was low at 430 kg/ha. Since 1985, productivity has been steadily increasing from 430 to 865 kg/ha in 2000 (Balasubramanian, 2000; Bhaskara Rao and Nagaraja, 2000). This is due mainly to improved technology of production, such as availability of high-yielding planting material, institutional research back up in providing nearly 10 million grafts annually, and an extensive replantation program. Private nurseries also provided good planting material. Statewise area on production and productivity is given in Table 2.5.

Table 2.4 Area, Production, and Productivity of Cashew Raw Nuts in India

Year	Area (million ha)	Percentage Increase in 5 Years	Production (million tons)	Percentage Increase in 5 Years	Productivity (kg/ha)
1955	0.110	–	0.079	–	720
1960	0.176	60.0	0.110	39.2	630
1965	0.232	31.8	0.141	21.9	610
1970	0.281	21.1	0.176	24.8	630
1975	0.358	26.7	0.166	−5.7	460
1980	0.451	25.9	0.142	−14.4	310
1985	0.509	11.4	0.221	55.6	430
1990	0.531	4.1	0.286	29.4	540
1995	0.577	8.7	0.371	22.9	640
1996	0.635	–	0.418	–	720
1997	0.650	–	0.430	–	835
1998	0.700	–	0.360	–	740
1999	0.730	–	0.460	–	800
2000	0.683	18.4	0.520	40.2	865

Source: Cashew Export Promotion Council, Government of India.

Table 2.5 Area, Production, and Productivity of Cashew in India by State (1999–2000)

State	Area ('000 ha)	Productive Area ('000 ha)	Production ('000 mt)	Productivity (mt/ha)
Maharashtra	121.20	85.00	125.00	1.47
Andhra Pradesh	100.00	90.00	100.00	1.10
Kerala	122.20	118.00	100.00	0.85
Karnataka	90.50	86.00	60.00	0.70
Goa	54.40	49.00	30.00	0.61
Tamil Nadu	85.20	84.00	45.00	0.54
Orissa	84.10	65.00	40.00	0.62
West Bengal	9.10	9.00	8.00	0.90
Others*	16.70	15.00	12.00	0.80
Total	683.40	601.00	520.00	0.865

Source: Cashew Export Promotion Council, Government of India.
*Madhya Pradesh, Manipur, Tripura, Meghalaya, and Andaman and Nicobar Islands.

World Trade in Cashew

For more than half a century, India has been in the cashew export trade. Over the years, both the quality and quantum of export have been on the rise. The established processing capacity of raw nuts is around 7 lakh tons. However, domestic production is around 5.2 lakh tons. Hence, India has been importing raw nuts from Africa, principally from Tanzania. This has been done to meet the demand of the cashew-processing industries. The export–import scene over the last half a century is presented in Table 2.6. Export earnings have been on the increase since 1955. In 2000, India had an all-time export earning of Rs 2500 crores (approximately US$625 million). Between 1980 and 1985, although export earnings increased, the quantity of cashew kernels exported decreased. The quantity of cashew kernels exported has steadily increased since 1985. It is estimated that the processing industries can absorb up to 10 lakh (1 million) tons of raw nuts for processing (Bhaskara Rao and Nagaraja, 2000).

India, which had been the pioneer in raw cashew nut production, has been pushed to third place (Table 2.7). However, India maintains its top position in processing and exports using imported nuts, mainly from Africa. FAO statistics clearly indicate that Vietnam is the biggest producer followed by Nigeria, pushing India to the third

Table 2.6 Import of Cashew Raw Nuts and Export of Cashew Kernels from India

Year	Import of Raw Nuts (tons)	Export of Kernels (tons)	Export Earning (million Rs)
1955	63000	31000	12.9
1960	95000	39000	16.1
1965	191000	56000	29.0
1970	163000	60000	57.4
1975	160000	65000	118.1
1980	24000	38000	118.0
1985	33000	32000	180.0
1990	59000	45000	3650.7
1995	222000	77000	12458.0
1996	222819	70334	12405.0
1997	192285	68663	12855.0
1998	224968	76593	13961.0
1999	181009	75026	16099.0
2000	199000	95000	25000.0

Source: Cashew Promotion Council of the Government of India.

Table 2.7 Global Cashew Nut Output ('000 tons) Scenario Vis-à-vis India

	1990	2005	Percentage Growth in 16 Years	Percentage Annual Average Growth
World	732	2,662	263	16.4
Vietnam	140	812	480	30.0
India	286	526	84	5.3

position. In 2005, Vietnam produced 9.61 lakh tons (100,000 tons equal 1 lakh ton), followed by Nigeria with 5.94 lakh tons, while Indian output stood at 5.44 lakh tons. The Indian cashew processing industry has an installed capacity to process around 1.2 million tons of raw cashew nuts. But the indigenous raw nut availability remains still around 50 percent of its annual requirement. As a result, imports have been on the upsurge. As against 2,49,315 tons valued at Rs 960.54 crores (approximately US$240 million at current exchange rate) in 2000–01, imports in 2006–07 were at 5,92,604 tons valued at Rs 1,911,62 crores (approximately US$485 million). When the country imports to meet the local industrial need, valuable foreign exchange is drained from the country.

The Tale of Cashew Trade in India Over the Years

The Indian cashew industry earns crores of rupees in foreign exchange and employs around 1 million laborers, mostly women. The nation has also to its credit the privilege of having pioneered cashew exports in 1945. It took the ingenuity of a British butler to develop the "Vita Packing," which provides an interesting sidelight as, in fact, cashew is the poor man's crop and the rich man's food. The Keralites are the pioneers of the cashew industry in India. It is believed that cashew was first discovered in Eastern Brazil by the Portuguese travelers. The Brazilians, while devouring the false fruits, discarded the nuts. The Portuguese brought the cashew to Goa and planted the seeds along the sea coast to check sea erosion. The country saw the processing and trading of cashew kernels take off in Kollam (in central Kerala State), Mangalore (in southern Karnataka State), and Vettapalem (in Andhra Pradesh). The kernels were first exported in oil cans. This method was not foolproof as the nuts were found to be infested with weevils at their destination. Subsequently, an English butler tried to store the kernels in cans infused with carbon dioxide by commissioning a soda maker and found the method to be successful. The industry is still using this method, which came to be known as "Vita Packing."

Nuts have been imported from African countries, to be processed and reexported to various global destinations. This helped develop a flourishing cottage industry around Kollam. The local women became adept at processing the nuts and, to date, they have been able to retain the skill with finesse. But, over a period of time, India lost its frontal position to Vietnam.

The Kerala Government, through its Cashew Development Corporation, is bringing out value-added cashew products. It has coordinated with the Central Food Technology Research Institute, the only one of its kind situated in the city of Mysore, Karnataka State, to bring out five value-added products, namely cashew powder, cashew soups, cashew bits, cashew vita, and cashew kernels. These value-added products will reach both the domestic and global markets. The Government of India is supporting the move by providing the corporation Rs 1 crore (approximately US$250,000) for international publicity. The target countries would be those in the Persian Gulf region, Russia, and China, which are the non-conventional consumers, in addition to the traditional markets like Japan, the European Union, and so on. It is in this context that the organization of a cashew export development authority (CEDA) becomes relevant to implement the plans of the Government of India to increase production of raw nuts in the country and decrease India's dependence on imported raw materials at a time when the nation has immense potential to become self-sufficient in production and further venturing into processing and export.

Against the background of all these, it is important to realize that of all the edible nuts, cashew nut remains world's most favorite nut. If a global perspective has to develop on the cashew nut as a "global food crop," it is imperative that a global effort must follow. It is in this light that world's important cashew producers, such as Brazil, India, and Vietnam, are coming together in the Global Cashew Alliance (GCA), which is developing with the assistance of the Cashew Export Promotion Council of the Government of India. A memorandum of understanding (MOU) between India, Brazil, and Vietnam has already been reached within the GCA. An initial grant of US$1.5–2 million is expected from the Government of India to get the GCA going. A similar contribution is expected from the other two partners. Finally, if the GCA comes into being, as is sure to, it would chart a new course in the history of this world's favorite nut.

The History, Evolutionary Origin, and Distribution of Cashew

It is from Southern Honduras to Parana, Brazil, and Eastern Paraguay that cashew, or *Anacardium*, is distributed naturally. It is not indigenous to South America west of the Andes, except Venezuela, Colombia, and Ecuador, where *Anacardium excelsum* is prevalent. *Anacardium occidentale* is cultivated throughout the Old and New World tropics. The genus has two centers of diversity, Central Amazonia and the Planalto of Brazil. This is illustrated by the occurrence of four species in the vicinity of Manaus and by three species occupying the same habitat in the Distrito Federal, Brazil. The following five distribution patterns are found in *Anacardium*.

1. *Anacardium excelsum* is isolated taxonomically and geographically from its congeners by the Andes. The uplift of the Andes was probably the driving force in the early differentiation of *Anacardium excelsum* from the rest of the genus.

2. *Anacardium giganteum* and *Anacardium spruceanum* have Amazonian-Guyanan distributions.
3. *Anacardium occidentale*, which is the most widespread species in the genus, has disjunct populations in the Planalto of Brazil, the restingas of eastern Brazil, the savannas of the Amazon basin, and the Ilanos of Colombia and Venezuela. It should be remembered that the natural distribution of this species is obscured by its widespread cultivation in both the Old World and New World.
4. Three closely related species, *Anacardium humile*, *Anacardium nanum*, and *Anacardium corymbosum* are restricted to the Planalto of central Brazil.
5. Two species of Anacardium are narrow endemics. *Anacardium corymbosum*, which is restricted to south-central Mato Grosso, is an allospecies of *Anacardium nanum*, and *Anacardium fruticosum* (a new species) is endemic to the upper Mazaruni River basin in Guyana. It is closely related to the Amazonian *Anacardium parvifolium*.

The eastern portion of the Amazon River figures prominently in distributions of many plants and animals, many of which are found either exclusively to the north or to the south of the river. However, in the case of *Anacardium*, all the Amazonian species are found on both sides of the Amazon. The reason for this is probably the ease with which bats, large birds, and water (in the case of *Anacardium microsepalum*) carry fruits across water barriers (Mitchell and Mori, 1987). *Anacardium occidentale* is cultivated throughout the Old and New World Tropics where the geographical limits of its cultivation are latitudes 27° north and 28° south, respectively (Nambiar, 1977). *Anacardium occidentale* is a native of tropical America where its natural distribution is unclear because of its long and intimate association with man. The questions concerning its origin and distribution have been investigated by Johnson (1973) who suggested that it originated in the *restinga* (low vegetation found in the sandy soil along the coast of eastern and northeastern Brazil). The author assumes that the cultivated form of *Anacardium occidentale* came from Brazil, because cashew trees cultivated in the Old and New Worlds are identical in appearance to native trees found in *restinga* vegetation. In particular, cultivated and wild populations of cashew species from eastern Brazil share chartaceous leaf blades and long petioles. *Anacardium occidentale* is probably an indigenous element of the savannahs of Colombia, Venezuela, and the Guyanas. It is clearly a native and occasionally a dominant feature of the *cerrados* (savannah-like vegetation) of central and Amazonian Brazil. The *cerrado* populations of *Anacardium occidentale* differ from the *restinga* populations by having undulated, thickly coriaceous leaves with short and stout petioles. The hypocarps (cashew apples) of *cerrado* trees are usually smaller and sometimes have a more acidic flavor than those of the *restinga*. The natural distribution of *Anacardium occidentale* extends from northern South America to south of Sao Paulo, Brazil. It is probably not native to Central America, the West Indies, or South America west of the Andes. It is believed that *Anacardium occidentale* originally evolved in the *cerrados* of Central Brazil and later colonized the more recent *restingas* of the coast. Central Brazil is a center of diversity for *Anacardium* where the distribution of *Anacardium occidentale* overlaps the ranges of *Anacardium humile*, *Anacardium nanum*, and *Anacardium corymbosum*. *Anacardium humile*, the closest relative of the cultivated cashew, is closer morphologically to the *cerrado* ecotype than it is to the *restinga* and cultivated populations of *Anacardium occidentale* (Mitchell and Mori, 1987). The earliest reports of cashew are from Brazil coming from the French, Portuguese, and Dutch observers (Johnson, 1973). The French naturalist and monk A. Thevet was the first to describe in 1558 a wild plant extremely common in Brazil: the cashew tree and its fruits. He recounted that the cashew apple and its juice was consumed, the nuts were roasted in fire, and kernels eaten by the natives. He also drew the picture of the natives harvesting the apple and squeezing by hand the juice into large jars (Johnson, 1973; NOMISMA, 1994). Indications that native Tupi Indians used cashew for centuries also exist. They took the crop across to the Brazilian northeastern coast during the migration, which is indicated by the considerable intraspecific variations (Ascenco, 1986). The entire cashew fruit, nut, and peduncle will float when mature. This explains the coastward dispersal of the species by rivers draining in the northern and eastern directions. Fruit bats also might have accelerated the dispersal process through the ingestion of the nuts. Fruit bats are the most effective seed dispersal agents within the Amazonian forests (Johnson, 1973). From its origin in northeastern Brazil, cashew spread into South and Central America (Van Eijnatten, 1991). Human intervention spread cashew to other continents (Johnson, 1973). The Portuguese discovered cashew first in Brazil and then spread it to Mozambique and later to India between the 16th and 17th centuries, possibly between 1563 and 1578 (De Castro, 1994). It first arrived in Africa on the east coast in the second half of the 16th century and then spread to the west coast and finally to the islands (Agnoloni and Giuliani, 1977). Though the possibility of the Portuguese bringing cashew to Africa can be reasonably surmised, there are no records to substantiate the specific dates of introduction. Dispersal to eastern part of Africa must be due to the elephants known for their love of fruits (Johnson, 1973). Possibly attracted by the vibrant color of the fruits, the elephants must have been swallowing the entire apple with the nut and since the nut is too difficult for digestion, the dung must have carried on this dispersal. This is how the cashew spread to African east coast along the Indian Ocean (Massari, 1994). The spread of cashew within the South American continent was gradual (NOMISMA, 1994). The plant was first found and described along the coast of Malabar in Kochi. Following introduction to southwestern India, the spread occurred through bats, birds, and most important, through human intervention. Kochi was the focal point from where the cashew spread to other parts of India, as well as to southeast Asia (Johnson, 1973). The plant was spread primarily to control soil erosion in coastal areas (Johnson, 1973). This interpretation may refer to a 20th-century concept in soil management, from a 16th-century phenomenon. It was the Portuguese who realized that the nuts had medicinal value and that the juice could be made into a good wine. This also led the early Portuguese colonizers of India to the realization of the potential economic value of cashew. Following introduction into India, the plant was taken to Southeast Asia (NOMISMA, 1994). Dispersal in Southeast Asia was aided by monkeys. It is uncertain whether the plant reached Philippines through India. It might have come directly from the New World on the Manila galleons (Johnson, 1973). Later, it spread to Australia and parts of North America, such as Florida. The present diffusion is between 31° north (latitude) and 31° south (longitude) both as a cultivated and wild species (NOMISMA, 1994). Presently cashew is cultivated in many tropical countries along the coast (Ascenco, 1986; Van Eijnatten, 1991). It was in the 19th century when the idea of growing cashew as a plantation crop took root and the idea spread to Africa, Asia, and Latin America (Massari, 1994).

Economic Botany of Cashew

Anacardium is one of the most important genera of the Anacardiaceae family. This importance is due to *Anacardium occidentale*, which is the cashew nut of commerce, a major export of the developing world. The false fruit (hypocarp) is both consumed locally and used to distill "Feni," an alcoholic beverage. In South America, especially Brazil, the juice from the cashew apple is marketed widely as a popular drink. The cashew nut shell liquid is used for industrial purposes and has medicinal value. Some of the other *Anacardium* species have economic potential, but they are currently underutilized. *Anacardium excelsum* is used in construction and also as a shade tree in coffee and cocoa plantations. *Anacardium giganteum* is a locally important timber in South America, and its hypocarps are very much relished by the local people. The spectacular white leaves of the inflorescence of *Anacardium spruceanum* make it a tree with excellent ornamental potential. *Anacardium humile*, a subshrub closely related to *Anacardium occidentale*, possesses edible hypocarps and seeds. Selective breeding for better quality hypocarps and seeds, as well as hybridization with *Anacardium occidentale*, could yield subshrubs with fruits that could be harvested mechanically. The economic potential of the other two subshrubs, *Anacardium nanum* and *Anacardium corymbosum*, also should be investigated (Mitchell and Mori, 1987).

Taxonomy of the Cashew Plant

The cashew plant belongs to the family Anacardiaceae, genus *Anacardium*, and species *occidentale*. The genus belongs to the Latin American genus of trees, shrubs, and geoxylic subshrubs, the taxonomic treatment of which is provided by Mitchell and Mori (1987). Anacardiaceae is a moderately large family consisting of 74 genera and 600 species. There are five tribes, namely Anacardieae, Spondiadeae, Semecarpeae, Rhoeae, and Dobineae. The tribe Anacardieae consists of eight genera: *Androtium*, *Buchanania*, *Bouea*, *Gluta*, *Swintonia*, *Mangifera*, *Fegimanra*, and *Anacardium* (Mitchell and Mori, 1987). Bailey (1958) suggests that *Anacardium* is a small genus of eight species indigenous to South America. However, Agnoloni and Giuliani (1977) recognize 11species, while Johnson (1973) recognizes 16 species. Valeriano (1972) names five species: *Anacardium occidentale* L., *Anacardium pumilum* St. Hilaire, *Anacardium giganteum* Hanca, *Anacardium rhinocarpus*, and *Anacardium spruceanum* Benth. This author suggests recognition of only two species: *Anacardium nanum* and *Anacardium* giganteum, which can be subdivided based on the color (yellow or red) and shape (round, pear-shaped, or elongated) of the pseudo fruit (apple).

This author also considers the division into dwarf and giant species to be the only way to classify cashew in a rational and practical manner. His arguments are based on the characteristics of the apple, the pseudo fruit. However, the description provided by Peixoto (1960) separates recognition of more than two species. It appears from published literature that *Anacardium occidentale* L. is the only species that has been introduced outside of the New World. As many as 20 species of *Anacardium* are known to exist within Central and South America (Table 2.8). Mitchell and Mori (1987) recognize 10 species of the genus *Anacardium*, one of which, *Anacardium fruticosum*, is described as new. The genus has a primary center of diversity in Amazonia and a secondary enter in the Planalto of Brazil. All known species of *Anacardium* genus can be found in the South American continent; only four of them, *Anacardium coracoli*, *Anacardium encardium*, *Anacardium excelsum*, and *Anacardium rhinocarpus*, are not

Table 2.8 The Different Species of *Anacardium* Linn.

Botanical Name	**Country of Origin**
Anacardium brasiliense Barb. Rodr.	Brazil
Anacardium curatellaefolium St.Hil (same as A.subcordatum Presl.)	Brazil
Anacardium encardium Noronha	Malaysia
Anacardium giganteum Hancock ex.Engl	Brazil
Anacardium humile St. Hil (Anacardium subterraneum Liais)	Brazil
Anacardium mediterraneum Vell. Fl. Flum	Brazil
Anacardium nanum St. Hil (same as A.*humile* Engl.,A.*pumilum* Walp)	Brazil
Anacardium occidentale L. (Cashewnut)	Brazil
Anacardium rhinocarpus D.C. Prod.	Brazil
Anacardium spruceaum Benth ex Engl.	Brazil
Anacardium microsepalum Loes.	Amazon region
Anacardium corymbosum Barb. Rodr.	Brazil
Anacardium excelsum Skeels (same as Rhinocarpus excelsa)	Brazil
Anacardium parvifolium Ducke	Amazon region
Anacardium amilcarianum Machado	Brazil
Anacardium Kuhlmannianum Machado	Brazil
Anacardium negrense Pires and Fro'es	Brazil
Anacardium rondonianum Machado	Brazil
Anacardium tenuifolium Ducke	Brazil
Anacardium microcarpum Ducke	Amazon region

Source: Index Kewensis 1996, Royal Botanical Gardens, Kew.

present in Brazil. There, the presence of the high number of wild species suggests that the northeast coast is the site where the *Anacardium* genus, *Anacardium occidentale* L., originated. In fact, here different forms of cashew can be found with high variability in local populations, mainly along the coast and the dune. Most species belonging to the genus *Anacardium* can be found anywhere in Brazil (NOMISMA, 1994). Ascenco (1986) reported that *Anacardium occidentale* L. is the only species in the genus that attained economic importance. The *Anacardium* genus appeared to have originated in the Amazon region of Brazil and hence speciation followed different geographic patterns.

Cytogenetics of the Cashew Plant

There is no detailed study of the cytology of *Anacardium occidentale* L. Only for *Anacardium occidentale* has been the chromosome number reported. This morphologically polymorphic species also exhibits chromosome polymorphism (Mitchell and Mori, 1987). In the literature, reported chromosome number varies from 2n = 24 (Goldblatt, 1984; Khosla et al., 1973), 2n = 30 (Machado, 1944), and 2n = 40 (Goldblatt, 1984; Simmonds, 1954) to 2n = 42 (Darlington and Janaki Ammal, 1945; Goldblatt, 1984; Khosla et al., 1973; Purseglove, 1988). In many domesticated trees, such chromosome polymorphism is well documented (Khosla et al., 1973).

Collection, Conservation, and Cataloging of Genetic Resources of the Cashew Plant

Records on the introduction of the cashew plant into the Malabar Coast of Kerala State, India, from where it spread to other parts of the country are imprecise. However, it is presumed that initially the introduction was a few *Anacardium occidentale* trees and that because of their hardy nature, they naturally spread to other parts of the country, especially the coastal region. Initially, the focus was on establishing plantations of seedlings. As cashew plant is crosspollinated and heterozygous, considerable segregations have resulted in the cashew population (Bhaskara Rao and Bhat, 1996).

Germplasm collection got an impetus following the establishment of the Indian Council of Agricultural Research (ICAR) controlled National Research Center for Cashew (NRCC) and State Agricultural University (SAU) administered research centers. Segregants or variants were collected as seed from different parts of the country. Wide variability has been observed in the collections (Bhaskara Rao and Swamy, 1994, Table 2.9). The NRCC established in Puttur, Karnataka State, has the National Cashew Gene Bank (NCGB). These have exclusive clonal accessions. The accessions are collected after a survey map is established during the fruiting season and the scions from the identified mother plant are collected during the propagation season (June–September). The grafts are produced and the clonal accessions are planted in the NCGB.

Table 2.9 A Selection of Cashew Germplasm Collections from Various Research Centers in India Released for Commercial Cultivation

State	Center	Varieties
Andhra Pradesh	Bapatla	BPP-3, BPP-4, BPP-5, BPP-6
Goa	ICAR, Research Center, Goa	Goa-1
Karnataka	National Research Center for Cashew, Puttur (NRCC)	NRCC Selection 1 and 2
	Chintamani	Chintamani-1
	Ullal	Ullal-1,Ullal-2, Ullal-3, Ullal-4,UN-50
Kerala	Anakkayam	Anakkayam-1 (BLA 139-1)
	Madakkathara	Madakkathara-1 (BLA 39-4) Madakkathara – 2 (NDR 2-1) K-22-1, Sulabha
Maharashtra	Vengurla	Vengurla-1,Vengurla-2
Orissa	Bhubaneswar	Bhubaneswar-1
Tamil Nadu	Vridhachalam	VR-1, VR-2, VR-3
West Bengal	Jhargram	Jhargram-1

Conservation field blocks are also established with clonal accessions. The efforts are coordinated by the All India Coordinated Research Project on Cashew (AICRPC). A total of 1490 accessions have been conserved (Bhat et al., 1999). At NRCC a total of 433 clonal accessions have been conserved in the NCGB. According to the International Plant Genetic Resources Institute (IPGRI) descriptors, 255 accessions have been characterized and cataloged after 6 annual harvests (10 years of planting), and the Catalogue of Minimum Descriptors of Cashew (*Anacardium occidentale* L.) Germplasm Accessions—I, II, and III have been published (Swamy et al., 1997, 1998, 2000).

The Genetic Improvement of the Cashew Plant

Breeding

Several plant attributes, such as number of inflorescence per unit area, number of nuts per inflorescence, and the mean weight per nut, decide the ultimate plant yield. These yield attributes, either directly or through their interaction with each other, decide the plant yield. Any attempt to improve the plant yield should precede a clear understanding of the various processes governing the physiology, nutrition, and so on of these yield components. The process of differentiation of reproductive shoot from vegetative shoot is an important aspect that needs to be investigated. Though data currently available inadequately explain the differentiation between vegetative and reproductive shoots, indications are that this could be governed by environmental variables, such as nutrition, soil moisture availability, and weather. Understanding the interaction between these will help breeding varieties having higher yield contributing factors, such as number of inflorescence per unit area, nuts per inflorescence, and fruit-to-nut ratio (Bhaskara Rao et al., 1998). Foltan and Ludders (1995) found that there were no significant differences in fruit set following selfing compared to crosspollination, except in one case where selfing H-3-13 resulted in significantly lower fruit set while the same variety when crossed with *Guntur* accessions gave maximum fruit set. The reciprocal combination of these parents resulted in a lower fruit set, indicating the need to understand the crosspollinated nature between preferential combinations of parents to realize higher yields. Therefore, studies on the compatibility relationship of cashew varieties and designing the models to establish orchards with polyclones to ensure highest compatibility and higher yields in cashew are a priority. An option to realize high yield is to go in for high-density planting ranging from normal spacing—that is, 8 m × 8 m (156 plants/ha) or 7.5 m × 7.5 m (175 plants/ha)—going up to 625 plants/ha, depending on soil fertility and canopy structure of the variety to be planted. High-density planting ranging from 200 plants/ha (10 m × 5 m) up to 625 plants/ha (4 m × 4 m) will only be possible with dwarf genotypes with compact canopy structure and intensive branching, with high proportion of flowering laterals per unit area. Therefore, breeding strategies must focus on dwarfing nature of the plant as well, an attribute that can be used as root stock or planting types as such. These can then be put through hybridization program to include other yield attributes to structure a genotype that can fit into a high-density planting program, in which case suitable canopy management should follow, such as proper pruning techniques, either conventional or through using chemicals, such as Paclobutrazol. Among the different *Anacardium* species, it is only *Anacardium microcarpum*, supposedly a dwarf genotype, that can be used as a root stock to multiply varieties that will result in compact canopies (Bhaskara Rao et al., 1998).

A very important aspect in breeding programs is to focus on the dietary aspects of cashew nuts. Tree nuts are generally considered highly nutritive and have been placed in the base of the Mediterranean Diet Pyramid developed by the World Health Organization (WHO), which recommends their consumption as part of a balanced diet. Differences with respect to neutral lipids and glycolipids without difference in phospholipids in varieties have been reported (Nataraja, 1987a, b). A quality index for cashew kernels was developed (Anonymous, 1994) based on protein, lysine, and sugar content. However, the recent emphasis is also on low fat content so that the misapprehension that consumption of cashew kernels is detrimental to health is not propagated among consumers. Some of the varieties having more than 35 percent protein and lysine, more than 50 μg/mg protein, and less than 14 percent sugar were identified. These can be used in breeding program to develop varieties with better nutritive value for the diet-conscious consumer.

One of the major production constraints in India as well as in other cashew-growing countries is the severe incidence of the tea mosquito bug (TMB) in the flushing and flowering season. Screening of the available germplasms in India showed little promise to encounter any with built-in resistance to TMB. However, in one accession, *Goa* 11/6, a phonological evasion has been noticed that enables the accession to escape severe TMB infestation (Sundararaju, 1999). Hybrids H-3-17, H-8-1, H-8-7,

H-8-8, H-15, and H-1600 have shown moderate tolerance to TMB. One must look for varieties whose flowering does not coincide with the onset of peak population in the TMB. Possibility of identifying tolerant types through screening of somaclonal variants is also a line worth pursuing in breeding. Since cashew yield structure is an integration of the different yield attributes enumerated earlier, an insight into breeding for yield enhancement can also be obtained through specific partitioning of these variables either through experimentation or through statistical analysis, such as multiple regression or path analysis. Through such an approach, high heritability components contributing to yield need to be identified and efforts made to integrate these attributes through hybridization. These approaches in breeding must also be coupled with yield-enhancing management practices (Bhaskara Rao et al., 1998).

Selection

Of the 40 cashew varieties released in India for commercial cultivation, 25 are selections from among the germplasm collections available in different cashew research centers in the country (Abdul Salam and Bhaskara Rao, 2001). These 25 varieties were identified and released based on the germplasm evaluation carried out at the different research centers (Table 2.9). As the cashew crop was initially propagated from "plus trees" for soil conservation and forestation, there was not much emphasis on the varietal concept described earlier in the section on breeding. This varietal concept is of recent origin. Initially, the focus was only on the total yield obtained per tree. This has resulted in the release of varieties with kernel grades of more than W 320. Important attributes, such as kernel weight, shelling percentage, and recovery percentage of whole kernels, received little attention. Because of the emphasis on quality of the nut, there is greater focus on identification of varieties with kernel weights over 2 g, which fall in the export grade of W 210 and W 240. To realize higher recovery of whole kernels, standards have been fixed for shelling percentage, which is not less than 30. This demands identification of donor parents that have these attributes which can be transmitted from parent to progeny (Bhaskara Rao et al., 1998).

Hybridization

Cashew hybridization is becoming increasingly popular because of its potential for enhancement of yield and several other desirable traits in the plant. Australia is currently placing great emphasis on hybridization. Parents of wide genetic variability obtained from different countries around the world are used for the hybridization work in Australia (Chacko, 1993; Chacko et al., 1990). This has necessitated the standardization of pollination techniques in cashew that is reliable, fool proof, and, most important, simple to follow. A simple technique of pollination in cashew has been developed at the NRCC (Bhat et al., 1998). The new pollination/crossing technique involves the use of butter paper rolls or pantographic paper rolls. It is as follows:

- Panicles with flower buds that will open the following day are selected on male and female trees. All opened flowers and nuts, if any, are removed from the selected panicles on the female parental tree.

- Each morning between 8 a.m. and 9:30 a.m., all the opened male flowers from the selected panicles on female parental trees are removed. Then anthers are removed (emasculated) using ordinary pins prior to anther dehiscence from freshly opened hermaphrodite flowers of the panicles. The stigma along with the style is enclosed with a butter paper roll or pantographic paper roll, which is prepared using a small piece of butter paper sheet (2.5 cm × 1.5 cm in size) by rolling it between the fingers.
- Freshly opened male flowers with undehisced anthers are collected in a Petri dish from selected male parents between 8 a.m. and 9:30 a.m., and the anthers are allowed to dehisce under partial shade.
- The butter paper roll from the emasculated flower is then removed, and the sigma is pollinated with the pollen from freshly dehisced anthers of the male parent collected in a Petri dish.
- The pollinated stigma along with style is reenclosed in the butter paper roll.
- Each panicle is labeled indicating the names of the male and female parents of the cross as well as the panicle number. Each panicle is used only for one cross combination.
- The procedure is repeated until 8–10 hermaphrodite flowers are pollinated in each of the selected panicle.
- All opened hermaphrodite flowers that are unused for pollination are removed each day.
- The remaining flower buds are removed from the panicle on the last day of pollination for that specific panicle.
- Each panicle with developing hybrid nuts is enclosed in a cloth bag in order to collect the nuts on maturity. Details of the crosses should be clearly written on the cloth bags.
- The hybrid nuts obtained as per the outlined procedure are grown in polyethylene bags for subsequent planting in the field.

The procedure described here gives higher percentage of hybrid nuts when compared to procedures of pollination currently used on account of the low physical injury to the delicate cashew flowers.

Fruit Set and Retention of Fruits

Investigations carried out at Darwin, Australia, on flowering, fruiting, and genotype compatibility (Foltan and Ludders, 1995) indicated that among the five cultivars used in the investigation, only one, H-3-13, performed differently compared to others. No significant differences were observed in the fruit set following selfing compared to crosspollination in all the combinations except in the cultivar named previously, where selfing resulted in significantly lower fruit set. When H-3-13 was crossed with *Guntur*, a fruit set of 51.7 percent (maximum) was obtained, while in the reciprocal combination of *Guntur* × H-3-13, only 38 percent fruit set was obtained. This clearly shows that very careful parental selection is a prerequisite in cashew to obtain good fruit set (Bhaskara Rao, 1996). In selfed progenies compared to crossed ones, owing to post-zygotic mechanism responsible for sterility, general yield reduction was observed (Wunnachit et al., 1992). Premature preferential flower shedding of selfed fruits has also been noticed in avocado (Degani et al., 1989). These investigations point to the crucial fact that compatibility of genotypes must be clearly understood to contain premature fruit drop, which is a major problem in cashew production (Bhaskara Rao, 1996).

Performance of Cashew Hybrids

Bhaskara

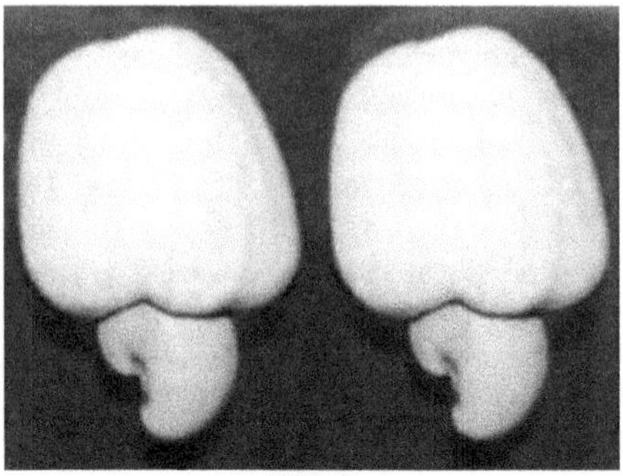

BPP-8

Dhana

Ullal-3

Vengurla-4

Vengurla-7

Vridhachalam-3

Priyanka

NRCC-Sel-2

The review of performance of the 34 cashew varieties showed that in the Indian states where the selections and hybrids were released for commercial cultivation, the hybrids performed better than the selections. On account of the fact that the cashew plant is amenable for vegetative propagation, it is possible to exploit hybrid vigor in the plant. The technique of softwood grafting has been standardized and is the best method suitable for commercial multiplication of cashew varieties and clones. It was in Kerala State, in Kottarakkara, that hybridization work was first initiated in 1963 and later continued at the Cashew Research Station in Anakkayam in the same state. Currently it is being pursued at Cashew Research Station at Madakkathara in the same state. In the initial breeding programs, three parents with prolific bearing habit (T.No. 12A, 30, and 30A) and three bold nut type parents (T.No. 27, 8A, and Brazil-18) were used in hybridization at the Cashew Research Station in Anakkayam (Damodaran, 1977). The reports on the evaluation of these hybrids indicated marked variation in the progenies derived from the same parental combinations. Where Brazil-18, an exotic bold nut accession, was used in hybridization, the percentage of progenies with high yield (more than 8 kg raw nuts/tree) was higher by 35 percent compared to those involving the accessions that were collected within the country (9.1 percent increase). Of the 28 parental combinations evaluated at the Cashew Research Station at Anakkayam (191 hybrid progenies) and at Vellanikkra (114 progenies), two hybrids, H-3-17 and H-4-7, were found to be superior than all the other combinations (Damodaran et al., 1978). It must be noted here that for both these hybrids, Brazil-18 accession parent was the exotic male parent. Research results on hybridization from other cashew research centers such as at Vengurla in Maharashtra State and Bapatla in Andhra Pradesh indicate that when a prolific bearer is crossed with a bold nut type, chances of realizing a hybrid with better nut weight are far greater (Nagabhushanam et al., 1977; Salvi, 1979).

Based on these results, varieties with smaller nut size but high yield were crossed with bold nut types, namely Vetore-56 and Brazil-711 at the Cashew Research Stations in Vengurla, Maharashtra State, Bapatla in Andhra Pradesh, and Madakkathara in Kerala State. Results indicated that Vetore-56 possessed high transmitibility of bold nut type trait to the progenies (Nawale and Salvi, 1990). Seven hybrids, which were released by the Kerala Agricultural University, namely *Dhana*, *Kanaka*, *Priyanka*, *Dhanashree*, *Amrutha*, *Akshaya*, and *Anagha*, have at least one parent with bold nut trait (Table 2.10). *Dhana* is a cross between ALGD-1 and K30/1 and *Priyanka* is a cross between BLA 139-1 and K30/1, which has a good nut weight of over 8 g/nut. *Kanaka* is a cross between BLA 139-and H3-13, which itself is a cross between two parents, one of which is the bold nut type Brazil-18. Among the 15 cashew hybrids released in India, three hybrids, BPP-1, BPP-2, and Vegurla-5, have small nuts (4–5 g/nut) with kernel grade between W 400 and W 450, whereas the remaining 12 have kernel grade between W 180 and W 240 (Abdul Salam and Bhaskara Rao, 2001, Table 2.10). These 12 hybrids have had at least one of the parents with a bold nut type and so have derived this advantage in the progeny, and this greatly helps in international trade. In addition, a higher shelling percentage in one parent is also a decided advantage that will lead to higher kernel output. Among the hybrids released so far, *Kanaka* and *Priyanka*, for which the common parent is BLA-139-1, have a short flowering phase (Bhaskara Rao and Bhat, 1996). Future thrust in cashew improvement must not only focus on higher yield, but other important attributes such as export-grade kernels, higher shelling percentage, and high nutritive value of kernels. For commercial cultivation, a desirable hybrid can be multiplied by soft wood grafting. Current research strategy is to have high-density hybrid progeny planting at closer spacing for preliminary cultivation and subsequently multiply the identified hybrids with desirable attributes, which takes 6–7 years to evaluate, through soft wood grafting for final testing at different locations. This method of evaluation could be modified to reduce the time lag between production of hybrid combinations and final testing (Bhaskara Rao et al., 1998).

Biotechnology

For rapid multiplication of elite lines, a biotechnology route can be accessed via micropropagation. It is also a tool to produce clonal root stocks. The procedure to use seedling explants has been standardized (D'Silva and D'Souza 1992; Lievens et al., 1989; Thimmappaiah and Shirly, 1996, 1999). Owing to high contamination, browning, slow growth, and poorly rooting microshoots, regeneration from mature tree explants has been quite difficult. However, micrografting in cashew is being attempted to rejuvenate mature cashew tree explants. Micrografting as a means for germplasm exchange has earlier been attempted (Mantell et al., 1997). Ramanayake and Kovoor (1999) reported success in micrografting using a scion of seedling origin on *in vitro* root stock. Somatic embryogenesis from maternal tissue such as the nucellus and leaf is an alternative for micropropagation and development of synthetic seeds. Somatic embryos can be used as a target organ for transformation investigation and also as organs to conserve germplasm. Hegde et al. (1993) reported obtaining somatic embryos from cotyledon leaves. Thimmappaiah (1997) observed embryogenesis from both cotyledons and nucelli. But the germination of embryoids in all cases was far from satisfactory. Somaclonal variation induced through culture can be exploited to select useful variants, but regeneration from callus (direct organogenesis) is yet to be

Table 2.10 Cashew Hybrids Released in India and Their Salient Features

Center	Hybrid	Parentage	Yield Potential (kg/tree)	Nut Weight (g)	Kernel Weight (g)	Shelling %	Kernel Grade
Bapatla	BPP-1	T.No.1 × T.No.273	10.0	5.0	1.3	27.5	W-400
	BPP-2	T.No.1 × T.No.273	11.0	4.0	1.0	25.7	W-450
	BPP-8	T.No.1 × T.No. 39	14.5	8.2	2.3	29.0	W-210
Madakkathara	Dhana (H1608)	ALGD-1 × K30-1	17.5	9.5	2.2	28.0	W-210
	Kanaka (H1598)	BLA 139-1 × H3-13	19.0	6.8	2.1	31.0	W-210
	Priyanka (H1591)	BLA 139-1 × K-30-1	16.9	10.8	2.8	26.5	W-180
	Dhanashree (H3-17)	T30 × Brazil	15.0	7.8	2.4	30.5	W-240
	Amrutha (H 1597)	BLA 139-1 × H3-13	18.3	7.1	2.2	31.5	W-210
	Akshaya (H 7-6)	H4-7 × K30-1	11.7	11.0	3.1	28.3	W-180
	Anagha (H 8-1)	T 20 × K30-1	13.7	10.0	2.9	29.0	W-180
Vengurla	Vengurla-3	Ansur-1 × Vetore-56	14.4	9.1	2.4	27.0	W-210
	Vengurla-4	Midnapore Red × Vetore-56	17.2	7.7	2.4	31.0	W-210
	Vengurla-5	Ansur Early × Mysore Kotekar-1/61	16.6	4.5	1.3	30.0	W-400
	Vengurla-6	Vetore-56 × Ansur-1	13.8	8.0	2.2	28.0	W-210
	Vengurla-7	Vengurla-3 × M 10/4	18.5	10.0	2.9	30.5	W-180

Source: Abdul Salam and Bhaskara Rao (2001).

demonstrated in exploiting as a tool for breeding. *In vitro* multiplication by culturing callus induced at the base of the microcuttings repeatedly over a period of time has been demonstrated by Bessa and Sardinha (1994). Because immature embryos can be regenerated to a complete plant (Das et al., 1996), embryo rescue techniques can be used to retrieve and regenerate inviable hybrids. Anther culture can be used to produce haploids and dihaploids, which in turn can be used in genetic studies and to produce homozygous lines (inbreds). Cell or protoplast culture is useful in making somatic hybrids to transfer beneficial characters from alien sources. Thimmappaaiah (1997) has reported protoplast isolation in cashew. However, protoplast cultures and regeneration are yet to be reported in cashew. Protoplasts can be used as target organs for transformation, provided they are made regenerative to a complete plantlet.

Employing micropropagation techniques, it is possible to establish international germplasm exchange, clonal propagation of elite lines, and *in vitro* conservation. Molecular markers, such as DNA markers (random amplified polymorphic DNA [RAPD]), random fragment length polymorphism [RFLP], amplified fragment length polymorphism [AFLP], and bio chemical markers [isozyme, protein]) can be employed to characterize germplasm and soma clonal variants. DNA fingerprinting of varieties using RAPD markers is being done at the National Research Center (NRC) for DNA fingerprinting (NRCDNAF) in New Delhi, and Department of Horticulture, University of Agricultural Sciences (UAS), Bangalore, in collaboration with NRCC, Puttur. RAPD profiles of 20 Tanzanian cashew accessions have been reported by Mneney et al. (1997). Also, RAPD profiles of 19 accessions were done at NRCDNAF, New Delhi. DNA fingerprints of 34 released varieties and one TMB-resistant accession were done by Murali Raghavendra Rao (1999). These techniques can be successfully employed to correlate markers with economically important plant attributes, which will aid in marker-assisted selection. Genetic transformation techniques, such as *Agrobacterium*-mediated gene transfers, can be used in cashew for transfer of genes for biotic (TMB/CSRB—cashew stem and root borer resistant genes) and abiotic (drought related) stress.

Establishing and Managing a Cashew Orchard

Soil Requirement

A prevalent myth around the world about cashew cultivation is that it is the most suitable crop to conserve soil, wasteland development, and afforestation. In India, the plant was first introduced by the Portuguese seafarers, turning later into colonialists, along the Malabar sea coast in the State of Kerala, from where it spread to other coastal areas of the country. This is also the reason why the crop has been relegated to soils of poor productivity, along hilly slopes and attracting little attention, which has led to poor productivity. This is not only true of India, but also in many other Asian and African countries. Cashew can thrive well in a variety of soils, such as hard degraded laterites, red sandy loam, and coastal sands. A rating chart for land selection has been suggested by Mohapatra and Bhujan (1974, Table 2.11).

Table 2.11 Guidelines for Selection of Land to Establish Cashew Orchard

	Very Good Class I	**Good Class II**	**Fair Class III**	**Poor Class IV**	**Unsuitable Class V**
Soil characteristics					
Soil depth	>1.5 m	90 cm-1 m	45–90 cm	23–45 cm	<23 cm
Texture	Loam Sandy loam	Loamy sand Silty loam Coastal sand	Clay loam Silty clay loam Sandy clay loam Loamy skeletal	Gravelly clay loam Gravelly silty loam Gravelly sandy loam	Gravelly clay Sandy clay Silty clay Clay
Soil reaction	Very slightly acidic to neutral (pH 6.3–7.3)	Slightly acidic (pH 6–6.3)	Medium acidic (pH 5.6–5.9)	Strongly acidic (pH 5.1–5.5) or mildly alkaline (pH7.4–7.8)	Very strongly acidic (pH <5) or alkaline (pH > 7.8)
Land features					
a. Slope (%)	<3	3–5	5–15	15–25	>25
b. Water table (m)	2–5	1.5–2 (coastal belt)	8–10	10–13	>13
c. Erosion	None to slight (eo)	Slight(e1) (sheet erosion)	Moderate(e2) (rill and sheet erosion)	Severe(e3) (gully erosion)	Very severe(e4) (gully and ravine erosion)
d. Drainage	Well drained	Well drained to somewhat excessively drained	Moderately well drained	Excessively and imperfectly drained	Poorly drained
e. Physiography	Coastal plains	Alluvial plain	Plateaus	Denuded hill slopes with shallow soils	Swamps
	Delta reaches	Natural levees	Hills	Ridges	Valley bottoms

(*Continued*)

Table 2.11 *(Continued)*

	Very Good Class I	Good Class II	Fair Class III	Poor Class IV	Unsuitable Class V
	Shield plains	Upland plains	Domes, mounds	Steeply undulating terrain with severe erosion	Escarpments
	Inland lateritic region adjoining coastal plain	Coastal ridges			Steeply sloping mountains Creek plain
Climate and environmental factors					
a. Altitude(m)	<20	20–120	120–450	450–750	750
b. Rainfall(cm/year)	150–250	130–150	110–130	90–110	<250
c. Proximity to sea (km)	<80	80–160	160–240	240–320	<320
d. Temperature (°F) Maximum in summer	90–100	100–103	103–106	106–110	<110
e. Minimum in winter	60	57–60	53–56	48–52	<48
f. Humidity (%)	70–80	65–70	60–65	50–60	<50 or >80
g. Occurrence of frost	None	None	Very rare	Occasional	Very often
	(once in 20 years)	(once in 15 years)	(once in 10 years)	(once in 5 years) to frequent (every year)	

Source: Mohapatra and Bhujan (1974).

The authors suggest that instead of considering the soil type only, the class of soil with a grading from Class I to V should be adopted while selecting a site to develop a cashew orchard. Class I to III types of soil in the medium acidic to neutral range (pH 6.3 7.3), with a slope of 0 15°, and water table up to 10 m were recommended to be the best for cashew orchards. However, in many countries, other plantation crops such as rubber compete for such soil types, and as such currently Class IV and V types of soil are put to cashew cultivation. However, while Class IV types require good soil amelioration, Class V types are best avoided, if high production is targeted.

Water Requirements

Cashew is principally a rainfed crop; irrigation is uncommon. In most of the cases where cashew plantations exist, surface water sources are nonexistent. However, new plantations are being raised where supplementary irrigation during summer months is possible by tapping underground water sources. Experimental results in India indicate that supplemental irrigation at the rate of 200 l/tree at fortnightly intervals in summer months (November–March) can enhance fruit retention leading to doubling the yield. This irrigation schedule would require ten irrigations during these months and this led to 44 percent fruit retention, while those plots where no irrigation was provided retained only 30 percent fruit set. Table 2.12 shows the results (Yadukumar and Mandal, 1994). Such a practice can be followed in homestead gardens, which are quite common in the State of Kerala. The positive effect of supplemental irrigation has been reported by Nawale et al. (1985). In China, supplementary irrigation is only provided during the early stages of the orchard establishment. The monocrop orchards or the adult orchards rarely receive supplementary irrigation, whereas the practice is seen in some gardens where intercrops are also cultivated. Other Asian countries rarely practice supplementary irrigation. Trials have been conducted on the efficacy of drip irrigation coupled with graded doses of nitrogen (250–750 g/tree), phosphorus, and potassium (62.5–187.5 g P_2O_5, K_2O/tree) at the National Research Center for Cashew. It was observed that irrigation alone at the rate of 60–80 l/tree without fertilizers increased yield by 60–70 percent when compared to trees receiving no fertilizers or irrigation. When the same level of irrigation was provided once in 4 days during dry summer months along with the highest dose of fertilizers like nitrogen (750 g/tree), phosphorus, and potassium (187.5 g each/tree as P_2O_5 and K_2O), nut yield increased up to 117 percent over the plots that received neither fertilizers nor irrigation (NRCC, 1998, Table 2.13).

Manuringa Cashew Orchard

Fertilizer application for cashew varies from one country to another, even so within the same country. This has been the experience in India as well. A cashew tree bearing 24 kg nuts and 155 kg apples removes annually 2.85 kg N, 0.35 kg P_2O_5, and 1.26 kg K_2O through uptake by root, stem, nut, and apple (Mohapatra et al., 1973). Beena et al. (1995) have estimated that every kg of nut harvested along with apples requires 64.1 g N, 2.05 g P, 25.7 g K, 4.19 g Ca, and 1.57 g S. Almost all reports on

Table 2.12 Retention of Fruits as Affected by Irrigation Treatments

Treatments	Fruit Set (Average of 5 Panicles)	Number of Fruits Harvested	Fruit Retention (%)	Yield (kg/tree)
Inrrigation once In 15 days @ 200/tree From November to January	27	9	33	4.53
January to March	16	6	37.5	4.93
November to March	25	11	44.0	7.32
Control		31	4.0	3.54
LSD (5%)				1.47

Source: Yadukumar and Mandal (1994).

Table 2.13 Effect of Drip Irrigation and NPK Doses on Cumulative Nut Yield (kg/tree) 8 Years Post Planting

Treatments	Fertilizers (g/tree)				
	M1	M2	M3	M4	Mean
Irrigation (l/tree)					
0	7.6	10.2	9.5	9.9	9.3
20	10.4	12.4	14.4	14.5	12.9
40	10.8	12.4	13.3	14.5	12.8
60	12.9	12.9	15.0	16.5	14.3
80	12.3	14.0	16.4	16.6	14.8
Mean	10.8	12.3	13.7	14.4	
M1 = 0:0:0	M2 = 250:62.5:62.5		M3 = 500:125:125		M4 = 750:187.5:187:5

Source: NRCC (1998).
Note: The numbers in M2, M3, and M4 refer to N, P_2O_5, and K_2O, respectively.

fertilizer use by cashew from different countries indicate a marked response to the application of nitrogen. But, for balanced application, addition of both phosphorus and potassium is required. In China, cashew trees are fertilized twice a year, during July–September. In the second year, when the plants are about 40–50 cm tall, 0.25 kg urea is applied, which is subsequently increased to 0.5 and 1 kg, respectively, in the third and fourth year. In addition, 0.5 kg calcium phosphate, 0.3 kg muriate of potash, or 20–30 kg organic manure is also added. The current recommendation of fertilizer application in India is 500 g N, 125 g each of P_2O_5 and K_2O per tree annually. In the case of high-yielding varieties, response to N was noticed up to an application rate of 750 g/tree. In India, fertilizer mixtures are commonly used in plantation crops. Because these mixtures do not conform to the ratios mentioned here, straight fertilizers to supply these quantities of nutrients are used. Fertilizers are only applied when the monsoon ceases into a shallow trench at the drip line of the tree. It is also recommended that fertilizers are applied in split doses during the pre-monsoon phase (May–June) and post-monsoon phase (September–October). However, if a single application is preferred, the post-monsoon period is preferred when there is still enough moisture in the soil. But the application has to, more or less, commence with the cessation of rains and should not be delayed much thereafter when soil moisture will deplete. In the first year, one-third of the recommended dose is applied, which is subsequently increased to two-thirds, and the full dose is applied in the second and third years, respectively. Rainfall on the eastern coast of India is scanty, while on the west coast, heavy rainfall is received. In high rainfall regions, fertilizer application in circular trenches of about 25 cm width and 15 cm depth at a distance of 1.5 m away from the trunk is recommended. In low rainfall regions, fertilizer is applied at the soil surface and raked into the soil. In Indonesia, the practice is to apply graded doses from the first year through the third (Abdullah, 1994, Table 2.14). In Myanmar, cashew farmers seldom apply chemical fertilizers to their crop. This is primarily because of the paucity of chemical fertilizers in the country and, hence, application of green manures and organic manures is used. Growing *subabul*, which is the locally available green manure crop, in the interspaces between trees and cutting them and incorporating the biomass into the soil is the routine practice (Bhaskara Rao, 1994). In Sri Lanka, chemical fertilizers are very rarely applied to the cashew crop, and only about 3.8 percent of the crop grown in the country receive chemical fertilizers. Locally available fertilizer mixtures in the ratio of 3:2:1 (N, P, and K, respectively) at the rate of 2.5 kg/ha are applied. Table 2.15 summarizes fertilizer application schedules to establish a cashew garden in India.

Table 2.14 General Recommendations of Fertilizer Application in Cashew Plantations (g/tree)

Age of Plantation (years)	Nitrogen	Phosphorus	Potassium
1	100	80	–
2	200	80	60
3	400	120	120
4	500	130	130
5	700	250	420
6	900	250	420
7	1000	500	300

Source: Abdullah (1994).

Table 2.15 General Recommended Fertilizer Quantities on an All-India Basis to Establish Cashew Orchard (g/tree)

Years After Planting	Urea	Rock Phosphate	Muriate of Potash
I year	300	125	40
II year	660	250	80
III year	990	375	120
IV year	1230	500	160
V year and onwards	1650	625	200

The Relevance of "The Nutrient Buffer Power Concept" in Cashew Nutrition

The fertilizer management of cashew is still carried out on the basis of "textbook knowledge." The soil analysis for nutrient bioavailability continues to be based on routine soil tests. It is in this context that the role of "The Nutrient Buffer Power Concept" has to be examined in cashew nutrition. The fundamental concept has been explained by Nair (1984), and experimental evidence validating the concept provided by Nair and Mengel (1984) and Nair (1996) has presented an extensive review on the concept in *Advances in Agronomy*, where experimental evidence supporting the concept on the nutrition of black pepper (*Piper nigrum* L.) and cardamom (*Elettaria cardamomum* Maton.) has also been included. Cashew is a perennial crop like black pepper and cardamom, and there is a very strong case to examine the validity of this globally accepted concept in cashew nutrition.

Some Salient Aspects of Raising Soft Wood Grafting

The following is the standardized technique of softwood grafting:

Step 1: About 40- to 45-day-old seedlings raised in polyethylene bags (25 cm × 15 cm, 300 G thickness) are utilized as root stocks.

Step 2: For a selected variety, lateral shoots of current season's growth (nonflowered, 35 months old, pencil thick with prominent terminal bud) are selected and precured on the mother tree by clipping of leaf blades, leaving behind petiole stubs. After 10–15 days, these precured shoots are collected and utilized as scions for grafting.

Step 3: The root stock is prepared by removing all the leaves except the two pairs of bottom leaves. The terminal growth (soft wood portion) at a height of 15 cm from ground level is decapitated, and a cleft 5–6 m deep is made on the stem.

Step 4: The precured scion stick is mended into a wedge shape 5–6 cm long by chopping off the bark and a little portion of wood from the two opposite sides taking care to retain some bark on the remaining two sides.

Step 5: The wedge of the scion is inserted carefully into the cleft of the root stock, taking care that the cambium layers of both the root stock and scion come in perfect contact with each other.
Step 6: Then the graft joint is secured firmly with a polyethylene strip (15 cm wide, 30 cm long of 100 G thickness).
Step 7: A long and narrow white polyethylene cap (20 cm × 4 cm size of 20 G thickness) is inserted on the grafted plant. This protects the apical bud from drying up and enhances sprouting.
Step 8: The grafting should be done under shade in the nursery shed, and the grafts are to be kept in the shade for about 10–15 days and later on shifted to the open area in the nursery. Alternatively, if silpaulin sheet roofing is made, instead of shifting the grafts, it is convenient to remove the temporary roof to allow all of the sunlight to fall on the grafts.
Step 9: Before shifting the grafts to the nursery, or dismantling the silpaulin roof, the polyethylene caps should be removed. These grafts are to be maintained in the nursery until the following planting season. The grafts will be ready for planting 5–6 months following grafting.

In a few comparisons that have been made, grafted trees have grown better and fruited earlier than the seedlings of similar age. However, once the soft wood grafts are planted in the field, it is also necessary to provide adequate care to establish a proper orchard to derive benefits of planting vegetatively propagated materials of high-yielding varieties.

Planting Technology

Planting of soft wood grafts is usually done during the monsoon season (July–August), on both the west and east coasts of India. Therefore, land preparation, such as clearing of bushes and other wild growth and digging pits to plant, should be done during the pre-monsoon season (May–June). A spacing of 7.5 m × 7.5 m or 8 m × 8 m is recommended for cashew; this will ensure about 156–175 plants/ha. A closer spacing of 4 m × 4 m in the beginning and thinning out in stages and thereby maintaining a spacing of 8 m × 8 m by the 10th year after planting can be followed. This leads to higher returns during initial years and, as the trees grow in volume, final thinning is done. However, where the land is flat, it is advantageous to adopt a spacing of 10 m × 5 m accommodating about 200 plants/ha, which leaves enough space for intercrops to be planted and may yield additional income to the farmer.

Cashew grafts are normally planted in pits of 60 cm × 60 cm size. The size of the pits can be 1 m × 1 m if hard lateritic substratum is encountered in the subsoil. Pits should be dug at least 15–20 days prior to planting; this exposes the lower soil to sun, which in essence is a biological sterilization process. Pits should be completely filled with a mixture of top soil, 5 kg of compost or 2 kg of poultry manure, and 200 g of rock phosphate. This will provide a good organic medium to obtain optimal plant growth. Grafts are best planted in July–August. Normally 5- to 12-month-old grafts are supplied by research stations and private nurseries in polyethylene bags. Since the cashew plant is normally grown along slopes, arresting soil erosion and runoff water during monsoon can be achieved by planting on terraces along the contour and opening pits to catch running water at the lower end. Therefore, before the onset of the southwest monsoon (May–June), terraces of 2 m radius should be made before the pits are dug. This helps in soil and moisture conservation, resulting in good plant growth in the first year of planting itself. Terraces are made first by removing soil from the top of the slope, spreading the soil to the lower side, and then making a flat basin of 2 m radius. Terraces may be crescent-shaped with the terrace slope facing the elevated side of the land so that top soil that is washed off from above due to rain is deposited in the basin of the plant. A catch pit across the slope, 200 cm long, 30 cm wide, and 45 cm deep, at the peripheral end of the terrace is made to withhold water during pre- and post-monsoon showers in sloping areas. A small channel to connect the catch pit sideways is made to drain out excess water during rains (Bhaskara Rao and Swamy, 2000).

High-Density Planting

Conventionally, cashew is planted in a square or triangular system, spaced 7.5–8 m. Field trials with plant density ranging from 156 to 2500 plants/ha have been conducted by the Central Plantation Crops Research Institute, Kasaragod, in Kerala State, also under the administrative control of ICAR and NRCC, Puttur, in Karnataka State. Plant density of 625 plants/ha, spaced 4 m × 4 m during the first 11 years and later thinning to obtain 312 plants/ha spaced 8 m × 5.7 m × 5.7 m gave maximum cumulative yield of nuts. Higher production resulting from high-density planting has also been reported from West Bengal State from the research center at Jhargram. Two benefits of high-density planting are effective soil conservation and checking weed growth, especially in forest lands. Judicial training and pruning in the initial stages after planting is a must (Bhaskara Rao and Swamy, 2000).

Covercropping

For covercrops, leguminous plants are used. The primary function of covercrops is to protect the surface soil from erosion, check the adverse effects of water run off, and smother weed growth. But when leguminous plants are used, they have the additional benefit of enriching soil fertility by adding fixed nitrogen, in addition to building up the carbon base through biomass incorporation. Crops such as *Pueraria javanica*, *Calapagonium muconoides*, and *Centrosema pubescens* enrich the soil with organic matter, add plant nutrients, check soil erosion, and also help conserve soil moisture. A seed rate of 7 kg/ha is used and sowing must be done at the commencement of the monsoon. Seeds must be soaked for 6 hours prior to sowing and are then sown in 30 cm × 30 cm beds in the interspaces of the cashew plants. Before harvest of the nuts, the cashew basins must be fully cleared of the covercrops to ensure easy harvest to gather all the fallen fruits with nuts intact. In degraded laterites, it is very difficult to establish covercrops because of very poor soil structure and absence of soil moisture. In China, natural grass and leguminous crops are usually maintained at the time of land clearance to conserve soil. During initial years after planting, green manure crops are also grown. Creeping covercrops, such as *Pueraria phaseoloides* and *Centrosema pubescens*, and bush covercrops such as *Glyricidia maculata* and *Leucaena leucocephala*, and nitrogen fixing trees such as *Acacia mangium* are the principal covercrops grown in cashew plantations in Sri Lanka.

Intercropping

Until systematic cashew planting started, intercropping received very little attention in cashew production. The practice of intercropping was picked up after the realization that as a sole crop cashew was not very remunerative, especially in small homestead gardens; much of interspaces between plants remained vacant, which could be put to profitable use. Intercrop must be established early in the plantation as delay would lead to a smothering effect due to the spreading cashew canopy. Further, heavy leaf fall is not conducive to the growth of any normally grown field crop. Field investigations with fruit crops, such as pineapple, *sapota*, and forest species such as casuarina and acacia, and green manure and cover crops such as *subabul* and mucuna have shown good promise. Pineapple, when grown in trenches across hilly slopes, helps check soil and water erosion. Casuarina and acacia were found unsuitable due to adverse effects on soil structure and soil moisture, as the former is a heavy depletory of soil moisture and its root spread almost menacing. In pineapple interplanted plantation, cumulative cashew yield was 61.41 kg compared to 37.74 kg in plot where cashew was grown as a monocrop (Table 2.16). Cashew with casuarina or acacia gave the least cashew yield.

From practical experience, it seems cashew with pineapple is the best combination, not only to generate more income, but also because of the ability of the latter to arrest soil erosion. Pineapple should be preferred to tree spices. An alternative to achieve high production is to plant cashew with high-density population at spacing of 5 m × 5 m in the square system and 4 m × 4 m in the hedge grow system and adopting judicious pruning to realize higher yields in the initial years (NRCC, 1998). Medicinal and aromatic plants can also be planted in the interspace. In Indonesia, sweet potato and peanut are popular intercrops. Recently, watermelon and sweet melon and chilli or hot pepper have also been tried as intercrops. Vegetables can only be grown as intercrops when facilities for supplemental irrigation are available. When melons are cultivated, a lot of biomass after the harvest of the fruits is available for incorporation into the soil, which will help build up the organic carbon content. In Myanmar, several intercrops, predominantly annuals, such as sweet potato, sesame, peanut, maize, cassava, pigeon pea, and so on, are grown. In Sri Lanka, banana is a popular intercrop. Pineapple, papaya, pomegranate, and coconut are also grown as intercrops. In Sri Lanka, the common annuals grown in cashew plantations are legumes (cowpea, black and green gram), oil crops (sesame, ground nut), and condiments such as hot pepper and onion. Field investigations conducted at Si Sa Ket Horticultural Research Center in Thailand have indicated that sweet corn,

Table 2.16 Yield of Cashew 2 Years after the Removal of Tree Species and Cumulative Yield (kg/plot of 384 m^2 area)

Cropping System	5 Years After Planting Before Removal of Intercrops (Tree Crops)	6 Years After Planting and 1 Year After Removal of Tree Crops	7 Years After Planting and 2 Years After Removal of Tree Crops	Cumulative Yield for the Past 6 Years
Cashew monocrop	5.60	7.78	14.42	34.74
Cashew + Pineapple	8.80	14.37	28.34	61.41
Cashew + Casuarina	4.06	6.75	12.12	26.16
Cashew + Acacia	2.03	2.15	10.32	16.23
Cashew + Subabul	4.12	5.44	13.15	26.65
Cashew + Mucuna	4.46	8.43	15.32	35.40
Cashew + Guava	5.30	5.94	13.55	31.07
LSD (95%) for treatments		3.27		7.07
			S.E of mean ± 2.29	

Source: NRCC (1995).
Note: LSD = least significant difference.

ground nut, and vegetables can be grown profitably as intercrops in the initial years of planting.

The principal constraint in growing intercrops is the availability of water for supplemental irrigation, as without water it is impossible to sustain intercrops. But because of the fact that in most of the situations cashew is grown in land where water is very scarce, growing intercrops becomes a problematic issue. Nevertheless, intercropping is essential in small homestead gardens where extra income can be profitably utilized to follow improved farming technologies.

Controlling Pests and Diseases in Cashew Plantations

Between diseases and insect pests, the latter are the most significant constraint to cashew production. In Asia, the pests attack cashew inflorescence and foliage. Many farmers are unaware of the initial symptoms of pest attack and so fail to take remedial measures. Subsequently, even if they adopt curative measures, damage that has already been caused cannot be undone. More than 194 species of insects and mites have been listed as pests occurring in different cashew-growing countries in the world (Nair et al., 1979). In China, more than 40 pests have been identified that attack the cashew stems, branches, leaves, tender shoots, flowers, and fruits in the Hainan plantations (Liu Kangde et al., 1998). In India, more than 84 species have been reported to attack the cashew plant (Pillai, 1979), of which 79 are insects and 5 are mites. Another 26 species of pests (17 insects and 9 vertebrate species) were added to the list of pests that damage cashew. Sundararaju (1993) compiled a list of 70 species, in addition to the previous reports, which cause damage. In total, 151 insects, 8 mites, and 21 vertebrate species damage cashew. In Indonesia, the main attack leading to low yields is caused by *Helopeltis* sp. and *Cricula*. In Myanmar, stem borer is the major pest, mainly due to poor phytosanitation measures. Sporadic attacks by shot tip caterpillar and leaf webber are also noticed (Maung Maung Lay, 1998). Hence, the current level of infestation is not cause to excessive economic loss in cashew. Control of stem and root borers is essential to save high-yielding trees in plantations. In the Philippines, the most significant pests are termites, leaf miners, shoot and root borers, and TMB, while in Sri Lanka, major pest attack is from stem and root borers and TMB. Sporadically, leaf miners and leaf and blossom webbers also cause damage. In Thailand, the major problem is the TMB, which causes heavy damage. Thrips (*Haplothrips* species) also cause heavy loss by attacking inflorescence and shoots, causing dieback. Although a large number of pests are reported to attack cashew (Nair et al., 1979; Rai, 1984; Sundararaju, 1993), the most significant that limit production are the cashew stem and root borer and TMB. Leaf miners (*Acrocercops syngramma*) and leaf and blossom webbers (*Lamida moncusalis*) are also major pests in certain areas. In addition to these, there are some pests of minor significance, in general, but in certain endemic areas they become a very serious problem. Such pests are defoliating caterpillars, leaf beetles, shot-tip caterpillars, foliage thrips, flower thrips, and apple and nut borers.

Pest Control

In most Asian countries, recommendations are available to control insect pests. In China, recommendations include pesticidal spray of 20 percent Fenvalerate (an insecticide) with a dilution of 1 ml in 200 ml water and a mixture of 40 percent Dimethoate and 80 percent Diarotophos(1:2 ratio) with a dilution of 1 ml in 200 ml water, applied as a low-volume spray. To control fruit borers, the recommendation is spraying 20 percent Fenvalerate or 2.5 percent Deltamethrin (1 ml in 200 ml water dilution). In India, three sprays are recommended to control foliage and inflorescence pests with monocrotophos and carbaryl during flushing, flowering, and fruit setting. A 0.05 percent spray of endosulfan or monocrotophos for the first and second rounds and 0.15 percent carbaryl for the third round is recommended. In Myanmar, plant protection is hardly practiced. In view of the level of infestation currently encountered, a general recommendation of spray is not warranted. In Thailand, thrips (*Haplothrips* sp.) are controlled by spraying 30 ml carbosulfan in 20 l of water or 50 g carbaryl in 20 l water. TMB is controlled by spraying 20 g carbaryl in 20 l water or cyhalothrin (10 ml in 20 l water). A major constraint in pest control is the farmers' general lack of awareness.

Control of Foliage and Inflorescence Pests

*Tea Mosquito Bug (*Helopeltis *sp.)*

Shoot drying due to TMB

TMB damage on shoots and panicles

TMB damage on tree

TMB damage on tender shoots

The tea mosquito bug (TMB) is the most serious cashew pest all over. The adult and immature stages of this mirid bug suck sap from tender shoots, leaves, floral branches, developing nuts, and apples. The injury made by the sucking parts of the insect's mouth causes tender shoots to exude resinous, gummy substances. Tissues around this point of entry of the stylets become necrotic and form brown or black scabs, presumably due to the action of the phytotoxin present in the saliva of the insect injected into the plant tissue at the time of feeding. Finally, the adjoining lesions coalesce and the affected portion of shoot/panicle dries up. Severe infestation on the floral branches may also attract fungal infestation, which will result in the inflorescence blight. Immature nuts infested by this pest develop characteristic eruptive spots and finally shrivel and drop off. Prophylactic sprays, detailed previously, at flushing, flowering, and fruiting can minimize losses, though complete eradication is still elusive.

Leaf Miner (Acrocercops syngramma *M.*)

The prophylactic spray schedule for TMB can also be effective in controlling leaf miner attack. However, if a serious outbreak is noticed, 0.05 percent spray of phosphomidon, fenitrothion, or monocrotophos has been found effective.

Leaf and Blossom Webber (Lamida moncusalis *Walker and* Orthaga exvinacea *Hamps.*)

Application of carbaryl (0.15 percent) has been found effective to control this pest.

Shoot-Tip Caterpillar (Hypatima haligramma *M.*)

The tiny yellowish or greenish brown caterpillars of the moth damage shoot tips and inflorescence. Systemic insecticides such as monocrotophos (0.05 percent) have been found to be effective to control the pest.

Foliage thrips (Selenothrips rubrocinctus *Giard,* Rhipiphorothrips cruentatus *Hood, and* Retithrips syriacus *M.*) *and flower thrips* (Rhynchothrips raoensis *G.,* Scirtothrips dorsalis *H.,* Haplothrips ganglabaueri [*Schmutz*], Thrips hawaiiensis [*Morgn*], H. ceylonicus *Schmutz, and* Frankliniella schultzei [*Tryborn*])

Cashew plantations, especially those raised on grafts that flush continuously, are prone to damage by foliage thrips. There are three species: *Selenothrips rubrocinctus* Giard, *Rhipiphorothrips cruentatus* Hood, and *Retithrips syriacus* M.

Flower thrips cause premature flower shedding and scabs on floral branches, apples, and nuts. Infestation on developing nuts results in formation of corky layers on the affected part. Malformation of nut and even immature fruit and nut drop are noticed. Endosulfan, monocrotophos, or quinalphos 0.05 percent spray controls both the foliage and flower thrips.

Apple and Nut Borer (Thylocoptila panrosema *M. and* Nephopterix *sp.*)

Apple and nut borers cause heavy economic loss (Dharmaraju et al., 1974). The caterpillars attack the fruit at all stages and the infested fruits shrivel and fall off with the nut. Spraying 0.1 percent carbaryl or 0.05 percent endosulfan effectively controls the pest attack.

Stem and Root Borer

Timely control of the pest is a must to preempt later damage that will lead to the complete succumbing of the plant. The pre-disposing factor to the attack is the lack of proper phytosanitation in the cashew orchards. Up to 35 percent loss can be expected, especially in plantations raised by forest departments. The primary species infesting the plant is *Plocaederus ferrugineus* L. Two other species, *Plocaederus obesus* Gahan and *Batocera rufomaculata* De G., also infest cashew. Small holes in the collar region of the plant gummosis extrusion of frass through holes, leaf-yellowing, twig drying, and finally the entire plant succumbing are the symptoms of the pest attack (Pillai, 1975; Pillai et al., 1976). The adult is a reddish-brown, medium-sized longicorn beetle, the head and thorax of which are dark brown or almost black. The eggs of the beetle are laid on the crevices of the tree bark as well as on the exposed parts of roots. When eggs hatch, grubs make irregular tunnels into fresh tissue and bark and feed on the subepidermal tissues and sap wood. This leads to injury of cells, and a resinous material oozes out when the vascular tissues are damaged. The ascent of plant sap is arrested, leaves turn yellow, and are subsequently shed. Several pest management techniques incorporating mechanical, chemical, cultural, and biological methods were tried against this pest. Removal of eggs, grub, and pupae from infested trees and swabbing the trunk after removal of the grubs from infested tree with 0.2 percent carbaryl or lindane, or painting with a mixture of coal tar and kerosene in the ratio of 1:2 will revive the tree. Whenever an infested tree is noticed in the plantation, it should be treated early in the infestation stage and all the adjoining trees must also be swabbed with coal tar and kerosene as a prophylactic measure.

Biological Pest Control

A number of natural predators from both Asia and Africa have been found to be effective against TMB, recorded by Simmonds (1970). Sundararaju (1993) has reported *Telenomus* sp. and *Chaetostricha* minor as natural predators of *Helopeltis* in India. Devasahayam and Radhakrishnan Nair (1986) reported that *Erythmelus helopeltidis* paratizes on the eggs of *Helopeltis antonii*. However, efforts to multiply egg parasitoids met only with little success as these are specialized parasitoids. *Crematogaster wroughtonii* Forel (Formicidae) was reported as a predator of the nymphs of the pest (Ambika and Abraham, 1979). Spiders, *Hyllus* sp. (Salticidae), *Oxyopes schirato*, *P. hidippes* Patch, and *Matidia* sp. have been found to predate *Helopeltis antonii* (Sundararaju, 1984; Devasahayam and Radhakrishnan Nair, 1986). Three species of reduvid bugs (*Sycanus collaris* [Fab.], *S. phadanolastas signatus* Dist., and *Endochus inornatus* Stil.) have also been noticed as predators of TMB (Sundararaju, 1984). Recently, Rickson and Rickson (1998) observed that a number of ants regularly visit the cashew tree and indicated ants as a possible defense against TMB. These authors have surveyed the plantations in Sri Lanka, India, and Malaysia and indicated the possibility that *Oceophyllas maragdina* is a promising ant species in this connection. These authors have recorded *Crematogaster* sp., *Monomorium latinode*, and *Tapinoma indicum* as possible predators, entering the open flowers and preying upon the flower thrips or mites. These ants did not appear to interfere with other pollinators.

About 80 percent of a cashew tree's current vegetative growth is destroyed by TMB (Rickson and Rickson, 1998). The outbreak is normally patchy and it is best that a tree-by-tree spraying program schedule is undertaken (Van der mere and Andow, 1986; Greathead, 1995). This schedule demands a keen surveillance to spot localized attack to be followed by pesticide spray. To control stem and root borers, *Bacillus thuringiensis*, *Bacillus popillae*, and fungi such as *Metarhizium anisopliae* and *Beauveria bassiana* have been recommended (Pillai et al., 1976). Presence of nymphal and adult endoparasitoid and mermithid parasite nematodes was detected in adult populations of TMB for the first time in cashew (NRCC, 1998).

Cashew Diseases and Their Control

When one compares the intensity and the consequential economic loss by pest attack on cashew, the ravages of diseases are lesser in intensity compared to that of insect pests. More than four dozen fungi attack cashew, but the intensity of their attack and its consequence is almost negligible. The disease problem is mainly in the nursery. In China, root rot, stem rot, and dieback are observed mainly in the nursery, while gummosis, defoliation, and root rot have been observed in the adult orchard. In India, dieback or the "pink disease" caused by *Corticium salmonicolor*, damping off of seedlings (Kamaraj and Bhide, 1962), and anthracnose disease (Singh et al., 1967) were found to be of significance. The other diseases are shoot rot and leaf fall (Thankamma, 1974), cashew nut decline (Ramakrishnan, 1955), and yellow leaf spot (Subbiah et al., 1986). *Oidium* sp. causes powdery mildew on the west coast of India in Maharashtra State (Phadnis and Elijah, 1968). Gummosis is also a problem in certain endemic areas, but none of the diseases reported in Asia are of major economic concern. Some of the important diseases and their control are described here.

Anthracnose

Colletotrichum gloeosporioides is the causal fungus not only in cashew, but many other fruit trees, such as mango, papaya, avocado, citrus, and so on. The pathogen continues to grow on the dead parts of the host tissues and perpetuates itself even in unfavorable conditions. Bordeaux mixture or copper oxychlorate spray effectively controls the disease. Fungal infection is preceded by the TMB attack (Nambiar, 1974). It is desirable to remove the infected plant parts and burn them rather than resort to chemical control. This is especially so in small orchards, while in large ones chemical spray of Bordeaux mixture or copper oxychlorate is preferable.

Inflorescence Blight

As the term implies, the disease is characterized by drying up of floral branches, and a gummy exudate can be observed at the site where the infection made a lesion by *Helopeltis antonii*. The causal fungi are *Gloeosporium mangiferae* and *Phomopsis anacardii* (Nambiar et al., 1973), which are secondary saprophytic colonizers and are not the pathogens.

Dieback

Terminal drying of twigs in cashew is caused by several fungi. These diseases are also called "pink disease" caused by *Pellicularia salmonicolor* or *Corticium salmonicolor* (Anonymous, 1950). A 1 percent spray of Bordeaux mixture during post-monsoon period is an effective prophylactic measure against the disease (Nambiar, 1974).

Leaf Spot

There are several leaf spot diseases in cashew. *Pestalotia microspora* causes the gray blight, *Phyllosticta* sp. causes the red leaf spot, and brown leaf spot is caused by *Colletotrichum gloeosporioides*. Spraying 1 percent Bordeaux mixture or 0.3 percent Benlate is recommended to control the disease. For a long time, yellow leaf spot was considered a disease of unknown etiology. However, Subbaiah et al. (1986) associated the disease with low (4.5–5.0) soil pH. Also, the affected leaves were found to contain excess Mg and low amounts of Mo. Spraying molybdenum salt was found to control the disease.

Powdery Mildew

Oidium sp. causes the powdery mildew disease, which is very rare in Asian countries, but very severe in Africa, and said to infect cashew blossoms in Maharashtra State in India on cloudy days (Phadnis and Elijah, 1968). Sulfur dusting controls the disease.

Effect of Inclemental Weather Conditions

Untimely rains result in late flowering. Many days with bright sunshine hours lead to bud break. Rise in night temperature to about 20°C together with fewer dewy nights, which coincide with the flowering phase, is detrimental to flowering.

Excessive cloudiness during flowering impedes opening of hermaphrodite flowers. However, the precise mechanism or any physiological causes for this is yet to be clearly understood. This is particularly so, even when infestation by TMB infection is not an adverse effect during such weather conditions. The only way the impact of adverse weather conditions can be mitigated to a limited extent is through a mix of cultivars with early or differential flowering characteristics and implementing such a strategy in the planting program (Rao et al., 1999).

Cashew End Products

Cashew Kernel

Cashew kernel is the most widely used nut in confectionary. There are 33 different grades, of which 26 are commercially available for domestic consumption and export. The Indian Standards provide precise specifications for the various grades. Broadly, the kernels can be classified as white wholes, scorched wholes, dessert wholes, white pieces, scorched pieces, and dessert pieces. The kernels as of now are mainly used as snacks in the roasted and salted forms. Bakery, confectionary, and chocolate industries use the broken kernels. Of late, different recipes with cashew have been developed in different parts of the world, primarily in Southeast and West Asia, where consumption of nut is very popular.

In the emerging global food market, quality and not price has come to rule supreme. This is because consumers, especially in the developed world, are highly conscious of the excessive use of chemicals in modern agriculture. Food safety is the most important criterion. To gain an entry into global markets, cashew has to conform to internationally stipulated standards inasmuch as quality is concerned. In this regard, even the type of packing materials used comes under scrutiny. These packing regulations relate to lead-free solder in tin containers, avoiding toxic/carcinogenic chemicals in preservation, and use of environmentally friendly and recyclable materials for packaging, storage, and so on (Nayar, 1998).

Cashew Kernel Peel

Rich in tannins (25 percent), kernel peels have great industrial use, especially in leather industry (Nair et al., 1979; Nayudamma and Koteswara Rao, 1967), and the peels with adhering pieces of cashew kernel form an excellent poultry feed (Nair et al., 1979).

Products from the False Fruit

The false fruit, also known as the cashew apple, is invariably discarded in many countries. Because of its astringent taste, many people do not consume it in Asia, while it is widely consumed in Brazil, the place of origin of the crop. In many small villages in Kerala, poor people cut the fruit into small pieces and consume it by dipping it in salt, along with the locally brewed alcohol. The more enterprising among them distill the fruits and make alcohol out of it for local sale or domestic consumption. It is a very juicy fruit, rich in vitamin C—a fivefold increase compared to citrus fruit—and contains 10–30 percent sugar. The apple is sucked and when it is empty of the juice the fibrous mass is left behind. The astringent and/or active acid is primarily due to the tannins in the fruit (0.35 percent). Steaming the fruit is the most efficient way to remove the astringent and acidic principles. Steam pressure varies 2–5 kg and exposure time is 5–15 minutes. This depends on the quality of the fruit and the end product to be made. The astringent principle can also be removed by boiling the fruit in 2 percent salt solution for 4–5 minutes. Alternatively, the fruit juice can be treated with gelatin (0.25–0.4 percent) and pectin (0.35 percent) or simple lime juice (25 percent). A number of cashew apple beverages, such as clarified juice, cloudy juice, apple syrup, or juice concentrate, can be made by following the above procedure in which the astringent content is removed. Other popular products are cashew vinegar, cashew apple candy and jam, canned apple, cashew apple chutney, and cashew pickle. As cashew is a seasonal plant and most of these products have only short shelf life, they are yet to become popular on a large scale. Post-harvest technology in the area of fruit preservation in cashew is still in a primitive stage. The most popular of all the apple products is the *Feni*, an alcoholic beverage produced in the Goa State in India (the original colony of the Portuguese) along the coast line of Maharashtra State, which has turned now as a global tourism hot spot because of its beautiful beaches. The drink has become very popular within and outside Maharashtra. There is a colonial legacy to the place, because the Portuguese seafarers landed there. They are the ones who also introduced the cashew crop to the island state.

Cashew apple residues left behind after extraction of the juice, constituting 30–40 percent of the fruit, are nutritious. It contains 9 percent protein, 4 percent fat, 8 percent crude fiber, and 10 percent pectin. This residue can also be made use of in manufacturing various products, such as drinks, jam, chutney, or as a preservative (Joshi et al., 1993). Also it finds its use as cattle feed after drying or can be utilized for the recovery of low methoxy pectin (Nanjundaswamy, 1984). A comparative picture of cashew apple and other tropical fruits' nutritive contents is given in Table 2.17. The chemical composition of cashew apple is given in Table 2.18, while Table 2.19 summarizes the composition of cashew kernel vis-à-vis other important nuts.

Cashew Nut Shell Liquid (CNSL)

One of the most important by-products of cashew nut processing is the cashew nut shell liquid (CNSL). Motor-driven expellers extract shell oil. Following extraction, the shell liquid is heated, filtered, and sealed into metal drums for export. Depending on the shell weight, 33–38 percent oil is extracted by weight (Johnson, 1982). CNSL is an excellent raw material for the manufacture of unsaturated phenol. It is a versatile industrial raw material and has innumerable uses in polymer-based industries, such as in friction lining, paints, varnishes, laminating resins, rubber-compounding resins, cashew cements,

Table 2.17 Vitamin and Mineral Contents of Various Tropical Fruits

Content/ 100 g	Cashew Apple	Pine-apple	Avocado	Banana	Lime	Grape Fruit	Mandarin	Orange
Thiamine (μg)	90	120	90	10	40	70	90	
Riboflavin (μg)	99	20	150	60	Traces	20	30	30
Vitamin C (mg)	240	24	16	10	45	40	31	49
Ca (mg)	41	16	10	8	14	–	33	33
P (mg)	11	11	38	28	10	–	23	23
Fe (mg)	3	0.3	0.3	0.6	0.1	–	0.4	0.1

Table 2.18 Composition of Cashew Apples

Moisture (g/100 g)	86.1
Proteins (g/100 g)	0.8
Fat (g/100 g)	0.2
Carbohydrate (g/100 g)	12.6
Fiber (g/100 g)	0.6
Ash (g/100 g)	0.3
Ca (mg/100 g)	0.2
P (mg/100 g)	19.0
Fe (mg/100 g)	0.4
Vitamin B1 (thiamine mg/100 g)	0.2
Vitamin B2 (riboflavin mg/100 g)	0.2
Niacin (mg/100 g)	0.5
Vitamin C (mg/100 g)	200

polyurethane-based polymers, surfactants, epoxtresins, foundry chemicals, and intermediates in chemical industry (Aggarwal, 1954; Aggarwal, 1973; Anonymous, 1993).

Cashew Shell Cake

The residual shell cake after extraction of shell liquid is currently used as fuel in the processing factories and in CNSL extraction plants. This oil cake could also serve as a raw material in the manufacture of plastics and container boards (Johnson, 1982).

Table 2.19 Kernel Composition of Different Tree Nuts (%)

Constituents	Almond	Hazelnut	Walnut	Macadamia Nut	Cashew Nut
Moisture	5.2	–	4.5	1.5–2.5	–
Protein	20.8	12.7	15.6	9.2	21.0
Fat (Ether extract)	59.9	60.9	64.5	78.2	47.0
Carbohydrate	10.5	17.7	11.0	10.0	22.0
Fiber	1.7	–	2.6	–	1.3
Mineral matter	2.9	–	1.89	–	2.4

Value-Added Products

Several products are made from raw cashew kernels. Product development leads to diversified uses of the original raw kernel. Value addition can substantially enhance the economy of cashew production as explained below.

1. *Cashew Kernel Flour*: The low grade kernels, which can neither be exported nor sold in the domestic market, are made into cashew flour, which is highly proteinaceous and is easily digested (Johnson, 1982). It is not used in baking bread or *chapathis*—the Indian bread—but, if a concerted effort is made, it may turn out to be an excellent supplement to usual wheat flour.
2. *Cashew Kernel Oil ("Caribbean Oil")*: As in the case of cashew kernel flour, lower grade kernels are also used to extract kernel oil, which is a highly nutritious edible oil, and in terms of quality compares quite well with the healthy and nutritious olive oil (Johnson, 1982). The kernel contains 35–40 percent oil (Van Eijnatten, 1991).
3. *Cashew Kernel Butter*: The residue of kernel after oil extraction is used to produce cashew kernel butter, which is similar to peanut butter (Nair et al., 1979). The oil-expelled kernel can also be processed into cashew nut cake, which is an excellent animal feed, especially for dairy cows (Van Eijnatten, 1991). The Central Food Technological Research Institute in Mysore, India, under the administrative control of the Council for Scientific and Industrial Research (CSIR), New Delhi, the premier research institute into food processing in the country, has perfected a procedure to extract cashew butter from the cashew kernels for export and the technique is available for commercial exploitation and has been passed on to the Cashew Export Promotion Council in Kochi, Kerala State, India.
4. *Coated Cashew Kernels*: Sugar-, honey-, and salt-coated cashew "baby bits" are available in the market and the technology of manufacture has been developed at the NRCC. Baby bits are the lowest grade kernels that are commercially marketed (Bhaskara Rao and Swamy, 2000).
5. *Cashew Kernel Milk*: Sweetened and flavored cashew milk can be prepared from cashew baby bits, and the procedure has been developed at NRCC (Bhaskara Rao and Swamy 2000).
6. *Cashew Spread*: The NRCC has developed a procedure to prepare cashew spread from baby bits (Bhaskara Rao and Swamy, 2000). There are different varieties, but the most popular is the sweetened vanilla-flavored spread. The salted spread is also popular.

There is great scope to generate value-added products in cashew. Following are some of the avenues:

1. Commercial exploitation of cashew butter and oil in the cosmetic industry, especially for the production of cold creams.
2. The 2–3 percent of rejected cashew kernels, which are rich in fat, can be used in which oil is extracted and the refined oil is used to develop various oil-based by-products. The cashew kernel is rich in vitamin E, which is a particularly useful vitamin in slowing the aging process and has wide use in the pharmaceutical industry.
3. The fiber from the cashew apple can be extracted to blend with other food materials to manufacture fiber-rich foods, which are particularly useful for healthy digestion. The NRCC has carried out investigations on cashew fiber after extraction of juice from the apples and characterized its physical and chemical properties. Quite possibly, the cashew fiber could have antidiabetic properties, and these need to be investigated.
4. Cashew kernels are rich in proteins and these proteins are known to contain amino acids. At NRCC, attempts have been made to investigate the functional properties of defatted cashew kernel flour and compare the same with other standard flours like wheat and maize flour, and the cashew flour compares quite well. It has also been compared with soybean and almond flours. It has also been shown that stable foam could be produced from cashew kernel flour over a wide range of pH. It would be a good idea to blend cashew flour with cereal and pulse flours to produce a nutritionally rich flour. Fortification of lower grade flours with cashew flour is an important avenue to explore regarding value addition.
5. The tannin-removed kernel testa has been shown to contain considerable quantities of carbohydrates and protein. Efforts are to be made to develop food and animal feed blends from tannin-free testa.
6. An enterprising industrialist in Kerala State, India, has succeeded in extracting ethanol, the "green fuel" from cashew apple, which has been tried on a pilot-scale to run automobiles. Commercial production holds much promise, especially in areas where supply of apples is not a constraint.

Organization of Cashew Research in India and Overseas

In 1950, India initiated systematic research in cashew, further strengthened by the establishment of the Central Plantation Crops Research Institute (CPCRI) in Kasaragod, Kerala State, under the administrative control of the Indian Council of Agricultural Research (ICAR) in New Delhi. In 1986, the independent research organization, the National Research Center for Cashew, was established in Puttur, Karnataka State. In recent years, considerable research has also been carried out in Australia under the administrative control of the Commonwealth Scientific and Industrial Research Organization (CSIRO) in Darwin.

There are other facilities, such as Cashew Training Research Center in Binh Duang, Vietnam; National Research Center on Cashew, Fortaleza, Brazil; Tanzanian Agricultural Research Organization controlled Research Institute in Naliendele, Mtwara, Tanzania; and Hainan Cashew High Yield Research Center in Hainan, People's Republic of China (Bhaskara Rao, 1996; NOMISMA, 1994). At present in India, there are nine research centers in eight cashew-growing states, all of them under the administrative control of ICAR, which coordinates the All India Co-ordinated Research Project (AICRIP) on cashew as shown in Table 2.20.

Table 2.20 Details of AICRIP Research Centers on Cashew in India

Institution	Location	State	Year Established
CRS, Andhra Pradesh Agricultural University	Bapatla	Andhra Pradesh	1971
ARS, University of Agricultural Sciences	Chintamani	Karnataka	1980
CRS, Kerala Agricultural University	Madakkathara	Kerala	1972
RARS, Kerala Agricultural University	Pilicode	Kerala	1993
ZARS, Indira Gandhi Krishi Vishwa Vidyalaya	Jagadalpur	Madhya Pradesh	1993
RFRS, Konkan Krishi Vidyapeeth	Vengurla	Maharashtra	1970
CRS, Orissa University for Agriculture and Technology	Bhubaneshwar	Orissa	1975
RRS, Tamil Nadu Agricultural University	Vridhachalam	Tamil Nadu	1970
RRS, Bidhan Chandra Krishi Vishwa Vidyalaya	Jhargram	West Bengal	1982

Note: CRS, Cashew Research Station; ARS, Agricultural Research Station; RARS, Regional Agricultural Research Station; ZARS, Zonal Agricultural Research Station; RFRS, Regional Fruit Research Station; RRS, Regional Research Station.

A Look into Cashew's Future

Since the start of systematic research of the crop in India in 1950, considerable progress has been made in evolving high-yielding varieties, calibrating nutrient requirements, standardizing vegetative propagation techniques, chemical control of major insect pests and diseases, utilization of cashew apple, production of a very large number of saplings for planting purposes, and replanting declining plantations. In view of the emerging global perspective, where cashew is most likely to play a major part as an important fruit nut, the research priorities have to be redrawn. Following are the areas that demand focused attention.

Genetic Resources

- The different germplasms that have been conserved in NCGB at NRCC in Puttur need to be consolidated and the preparation of a district-wise collection map of the cashew germplasms for the country has to be prepared.
- Germplasms should be collected from other cashew research centers, such as ARS Ullal, RRS, Brahmawar, CRS, Anakkayam, and CRS, Kavali, all of which are not affiliated to the AICRIP under the ICAR set up.
- Germplasms should also be collected from nontraditional areas, such as the hills of Garo in Meghalaya (Northeast India) and Bastar in Chattisgarh State (East India).

- The accessions in NCGB have to be characterized to support the cashew breeding program, focusing on processing quality of raw nuts, cashew apples for better fiber quality, and tolerance or resistance to major insect pests and diseases.

Varietal Improvement

- Genetic investigations on dwarfing traits and cluster bearing with a view to have high-density plantations and breeding for bold nut characters through molecular markers and isozyme banding pattern in different cashew genotypes.
- Investigations on the reciprocal differences in hybrids, polyclonal interactions, pollen viability, incompatibility, fertility, and post-zygotic abortions.

Biotechnological Interventions

- Through DNA fingerprinting in collaboration with the University of Agricultural Sciences in Bangalore, Karnataka State, and the National Research Center for DNA Finger Printing (NRCDNAFP), in New Delhi, the released cashew varieties and accessions should be characterized.
- Independent development of DNA fingerprinting facilities at NRCC, Puttur.
- Evaluation of biochemical and physiological basis of variations for observed responses in mature tree plants.
- Comparison of micrografts with normal grafts for their performance with regard to various attributes.

Crop Management Techniques

- Yield targeting and working out the production requirements with regard to optimum planting time, fertilizer requirements, irrigation if needed, onsite operations like weed and pest control, and so on.
- Soil fertility management, taking into consideration land use planning and all other state-of-the-art techniques.
- Evaluation of less vigorous varieties, such as Ullal-1, H-2, BLA39-4, and other varieties for high-density planting to obtain maximum yield per unit area.
- A critical evaluation of drip irrigation to enhance water use efficiency.
- A complete assessment of micronutrients in the soil, plant, plant parts, and defining critical levels in soil and plant to target high yields.
- Field evaluation of (1) performance of a scion variety on its own root stock and on different root stock varieties and (2) performance of a root stock variety with its own scion and scions from different varieties.
- Basic studies on the use of hormones, growth-regulating chemicals, growth retardants, and inhibitors on cashew yield.
- The role of "The Nutrient Buffer Power Concept" in the nutrition of cashew.

Crop Protection

- Standardization of the revival techniques in cashew stem and root borer (CSRB)—infested trees through induction of bark regrowth and root development.
- Design and development of traps for CSRB based on kairomones or utilizing infected trees as a bait.

- Approaches for reducing latent or residual CSRB inoculum.
- Developing models to identify the most vulnerable trees for CSRB infestation in the plantation.
- Investigations on the bioecological aspects of the CSRB infestation.
- The role of entomophilic nematodes on CSRB infestation.
- Management of cashew pests' infestation using the least chemicals, including the use of pheromones.
- Biology and bionomics of flower pests (thrips, apple and nut borers, and shoot-tip caterpillars).
- Monitoring the fauna in the cashew ecosystem. This is especially important when the crop is grown in the forest.
- Investigations on the pest complex of nuts during the post-harvest and processing stages.
- Development of a forecasting system of inflorescence pests.
- Investigations on weeds and pesticide residues in soil and water in the cashew plantations and adjoining area.

Post-Harvest Technology

- Development of appropriate protocols for extension of shelf life of the cashew apple for consuming fresh using proper package and storage methods.
- Development of cost-effective end products from cashew apples, specially to augment employment potential for rural men in the neighborhood of cashew plantations.

Technology Transfer

- Assessment of the impact of production technology recommended to the cashew farmers.
- Further refinement of the various aspects of cashew production using feedback from farmers and first-hand experience gained from working in farmers' fields.
- Recommendations of appropriate varieties depending on the ecological conditions.
- Production of quality planting materials.
- Establishing intimate institutional professional relationships with farmers' forums for constant updates on developments in cashew production technology.

Biodynamic Cashew

In 2005, a group of 31 farmers in Goa, Maharashtra State, came out with "Biodynamic Cashew." The production technique is based on the concept of dynamic agriculture propounded by Rudolf Steiner in Germany in 1924 to a small group of farmers in Koberwitz, East Germany. Biodynamic agriculture is a method of farming that aims to treat a farm as a living system that interacts with the environment, to build healthy living soil, and to produce food that nourishes and develops humanity. Steiner introduced the practice of making preparations based on cow manure, silica, and various herbal plants to be used in order to open up the soil to cosmic influence. He advocated discontinuing the use of chemical fertilizer altogether. Because of their inherent lack of life (inorganic nature), he felt that chemicals could not maintain life or increase soil fertility. The cashew produced in this fashion has a very high export value.

3 The Coconut Palm (*Cocos nucifera* L.)

Origin and Evolution

Coconut palm is a unique plant and stands apart from all other palms because of its high degree of consistency and continuity in flowering and fruit production, month after month, year after year, for decades. In *Sanskrit*, the ancient Indian language, it is called "*Kalpavriksha*," or that which provides humans with everything. In India, in the state of Kerala, the life of the people is woven around the coconut palm. The endocarp is everyday fare in the kitchen—scrapped or ground to extract the milk and used in all of the culinary specialities of the people. Coconut oil is used for both body massage and for the hair, water from the tender coconut is used as a very nourishing drink, leaves are used for thatching homes, and when dried, coconut is used as firewood. In fact, even in the case of the poorest peoples, there may be a lone coconut palm in front of the home, even if it is a thatched hut. It is at the sight of the coconut palm the poor wake up to a new day.

The above mentioned capability of the palm suggests evolution in an environment free from severe seasonal or episodic constraints on growth. Such an environment would most likely have resembled those few isolated niches where coconut presently thrives, unaided by human intervention or management. For instance, in southwest Java, on the island of North Keeling in the Indian Ocean, coconut dominates a dense woodland growing on low sand cays underlain by fresh water. On the east coast of Cape York Peninsula in Australia, where coconut was introduced by a European settlement in the nineteenth century, pockets of "coconut woodland" have also formed. Seeds cannot naturally move landward beyond this zone because of their large size, which makes it impossible for birds and most other mammals to carry them. Hence, the only avenue for dissemination is via ocean to other coasts with fringing sand cays. The coconut has been put to great use along the coast, also at great distances inland. The palm thrives well where rainfall is plentiful (>200 cm annually) and well distributed. In summer months, the palm needs good irrigation.

The Evolution of Coconut along the Drifting Coastlines

The general course of evolution of coconut, the modern prolific palm, which emerged from the prima ancestral palms of Mesozoic Gondwana, remains obscure. This has

The Agronomy and Economy of Important Tree Crops of the Developing World. DOI: 10.1016/B978-0-12-384677-8.00003-5
Copyright © 2010 by Elsevier Inc. All rights of reproduction in any form reserved.

been the theme of much speculation and limited molecular exploration by scientists (Lebrun et al., 1999). Remarkably, no fossil records exist, which can be attributed, in part at least, to the instability of "migration" of the coastal fringe environment. Even within the past 120,000 years, there have been no fewer than six cycles of glaciation when sea levels have fallen between a few tens of meters and more than 100m in one case around 20,000 years ago (Chappell, 1983; Veeh and Veevers, 1970). The corresponding movements of the coastal zone by many kilo meters, in some cases, would leave the coconut zone "high and dry" during the phase of falling sea level and progressively inundated during the rising phase. Potential fossil material would be destroyed by oxidation through exposure to aerobic conditions in the first phase and disintegrated by wave action in the second. Quite likely, the only "recent" coconut fossil to be found are those where humans planted the nut adjacent to a swampy environment, where it would not have been established by natural dispersal (Spriggs, 1984). In this case, the fossil remains were found to be slightly more than 5000 years old.

The ancestors of coconut possibly began to emerge from the palm branch of the tree of plant evolution around 100 million years ago (–10 Myr). According to Harries (1990) the ancestral coconut inhabited the north coast of Gondwana as that great land mass of the southern hemisphere began to break up around –80 Myr. Huge crustal plates carrying exposed or partly submerged land surfaces, which now comprise Australia, India, the Arab Peninsula, and smaller fragments such as New Zealand and Madagascar, began to drift northward.

The expanse of ocean between these wandering land masses and associated islands is referred to as the Tethys Sea (Harries, 1990). This sea is likely to have been warm and stormy, delivering high rainfall and periodic cyclonic wind gales onto the neighboring coastlines. Recent experience of extremes of weather suggests that a warmer ocean spawns more intense cyclones. The suggestion is that natural selection occurred in the palm population for traits resistant to extreme wind, which is described in the next section.

Development of Wind-Resistance

The previously described condition believed to have been in the Tethys Sea would favor the emergence, on the shore line, of a tropically adapted palm with a flexible, wind-resistant trunk that could flex rather than break up during episodes of violent wind. The modern palm has a "tubular" trunk structure elated to the density of its cortex and interspersed vascular fibers, forming a thick outer "wall" surrounding a softer core that undergoes compression without fracture when the trunk flexes itself in the wind. The coconut palm further exhibits an adaptation trait to survive strong wind, which is the capacity to progressively shed older fronds. This brings about a reduction of wind pressure, reducing the risk of damage to the heart of the crown of the palm. The coconut palm has also the ability to continue to grow even after the trunk has fallen flat on the ground, with new trunk growth resuming a vertical attitude (Marty et al., 1986).

The "Swimming" Coconut Fruit

The ancestor fossil remains of the coconut found in New Zealand include a small nut, less than 50cm in diameter and most unlikely to contain any endosperm liquid providing nourishment to a long swimming episode (Ashburner, 1994). Dispersal, by means of a hardy floating seed with a thick husk that allowed it to float high in the water, was a subsequent adaptation that allowed the nut to move on the ocean and occupy coastal niches throughout the favorable climatic zone. The coconut seed found in such wild places is still capable, in our time, of being picked up in large numbers by a tidal surge or in less numbers by falling from seaward—leaning palms, which survive 3–4 months at sea swim. Such a period is long enough to travel distances up to thousands of kilometers, depending on the wind current. This open "pooling" of genetic diversity would have counteracted the natural tendency toward separate paths or pockets of evolution fostered by isolation. Natural diversity in coconut appears to be limited, indeed, in contrast to most land-borne species in which genuine physical isolation of sub populations has occurred. For example, this has given rise, in the case of Macadamia populations, to easily distinguished species, sub species, and varieties determined by both morphological traits and molecular methods (Aradhya et al., 1998). The paradox of coconut dispersion naturally through its "swimming" trait has been the subject of intense speculation, much of it failing to appreciate the prolonged time frame during which dispersal has taken place. The major tropical land masses and islands of the Indian and Pacific oceans and the South China Sea have been in place for some million years. Ample time must have been there for the nut to travel far and wide, though frequent glaciation and rising and falling sea levels would have both aided and also disrupted in the sea "swim." Narrowed ocean barriers between neighboring islands and land masses would at least favor colonization and further mixing of populations at a greater regional level. When sea levels fall, the nut would have become widespread, not only along the coastlines as such, but also across the "temporary" landscape between the coast at the original sea level and the coast at the eventual low extreme. It has the capacity to compete quite strongly with other vegetation in the short term, but would eventually have been overshadowed by tall forest species. When the great northern hemisphere ice sheets receded, sea levels rose once again, and the nut in low-lying areas would have succumbed to inundation. A phase of rising sea level would most likely have been conducive to mobilization of seeds onto the ocean. In common with the fringing coral reefs, the fringing vegetation of the strand, including the coconut, would have followed the fluctuating shorelines of the tropics back and forth.

True Palm Traits

Though the coconut must have "branched off" in its early evolution, it retains the essential physical traits of a true palm (Ashburner, 1994). As with all mature palm trunks, diametrical expansion of the trunk is fixed once the attached fronds have developed, confining trunk growth to axial (length or height-wise) dimension. A "unit" of growth in the mature palm comprises a trunk section, supporting an attached frond and inflorescence. In a liberally irrigated, warm environment, the coconut sustains a most remarkably uniform and stable growth pattern through time, giving rise to a trunk increment, a new frond, and a new bunch about every 25 days. A frond generally persists for 2–3 years and detaches cleanly from the trunk (except in especially dry atmospheric environments) along with the remnant of the fruit bunch.

Human Influence on Coconut Evolution

As human (proto-Melanesian people) settlement increased along the southern and southeast Asian coastlines, coconuts provided a great bounty and a most convenient and welcome food source. Evidence for the settlement of the proto-Melanesian people is entirely obscure, although migration as far as Papua New Guinea and Australia can be located between 60,000 and 100,000 years ago. These migrants appear to have preferred the refuge of the secure and isolated mountainous hide-aways, far from the convenience but also the danger of the coast. Quite possibly, *Homo erectus*, the "**Java Man**" could have made use of the coconut 1.2 million years ago. It is only recently, between 5000 and 10,000 years ago that the sea-faring people, known as *Austronesians*, made use of the coconut palm in a more interdependent manner. By contrast, the earlier hunter-gatherer inhabitant had no knowledge of farming and there emerged the farming people who began to domesticate plants, such as taro and banana as well as coconut. Not only did coconut provide a source of food and drink for day-to-day living, but, it also became a staple commodity during travel on sea, including long-range exploration as it does to this day among many island communities. Probably the greatest age in maritime discovery in human history is the age of Polynesian exploration and colonization reaching out from Southeast Asia through Melanesia to Hawaii, the Marquesas, and Easter Island. By canoe, the domesticated coconut reached the Pacific coast of Central America, accompanied or not by surviving people, who might have perished during the tough journey or following arrival in a new environment or else been absorbed by the local inhabitants without trace. This era of exploration is believed to have begun around 4000 and 5000 years BC and extended well into the second millennium of the modern era, culminating with the arrival of Polynesian people in New Zealand less than a thousand years ago. The strong linguistic ties between Sumatra, Borneo, and Madagascar indicate that some *Austronesians* traveled westward. These voyagers might have arrived 3000 years BC, taking with them food crops such as coconut and banana (Simmonds, 1976).

The fact that coconut palm was found wherever Polynesian colonization took place is evidence to its ability to provide food and nourishing drink. The domesticated nut had a large size, thin husk, and came to be used as highly efficient and convenient water containers with a long shelf life. Such fruits taken during voyage would eventually germinate and sprout, but would remain as a vital source of food and drink. Whenever a voyaging Polynesian party colonized a new place, some of the remaining coconut supplies from the journey would have been available for planting among the native wild coconut palms. Most modern Polynesian communities in the tropics have inherited various forms of domesticated coconut, well-preserved through ongoing selection. Recent random fragment length polymorphism (RFLP) molecular analysis has shown a close affinity of coconut identity from Malaysia to Panama (Lebrun et al., 1999).

In India, coconut dates back to post-Vedic period—that is, about 3000 years BC, (Thampan, 1982) and Sri Lanka, for about 2300 years. That a distinct population of coconut has developed in South Asian coastal region is attested by the clear differences in DNA detected by RFLP analysis (Lebrun et al., 1999). There appears, however, to be scarce evidence in India of selection for large nuts with thin husks and high water content, although the variety *Kappadam* has thin husk, which is an exception, perhaps because a good supply of husk has been vital as a source of household fiber and fuel in southern India since ancient times. The coconut had probably spread naturally to some of the islands of the Indian Ocean, and through trade or migration to Seychelles, Madagascar, and the coast of East Africa. DNA analysis has also revealed that Southeast Asian germplasm entered the Indian Ocean via Madagascar, giving rise to intermediate forms that also extended to East Africa (Lebrun et al., 1999).

The spread of coconut from India to the outside world was triggered by the arrival of the Portuguese explorer Vasco da Gama who landed on the Malabar Coast at Calicut (now Kozhikode) in 1497/1498, from where the coconuts were taken for planting in tropical lands of the Atlantic Ocean, beginning at the Cape Verde Islands. From there, it spread to the West African mainland and also in the mid-sixteenth century to the Caribbean Islands, and from thereon to all coasts of Central America and tropical South America (Harries, 1978). Thus, from the obscure and untraceable early origins of the coconut on the coasts of migrating continents and Southeast Asian islands, the coconut distribution finally encircled the globe about four centuries ago. Current distribution is best described by summarizing production data (Table 3.1).

Botany of Coconut

In the genus *Cocos* of the palm family Arecaceae, coconut stands alone and has no close surviving relatives. In short, the coconut palm is considered to have evolved on the strand of the ever-changing coastline of islands and land masses fringing the Tethys Sea as they drifted north from Gondwana (Harries, 1990). The "melting pot" of the ocean provided frequent (in geographical time) opportunities for mixing of the diversifying progeny of many small populations of palms, occupying many contrasting strand environments, which evolved finally into the modern coconut *Cocos nucifera* L.

Table 3.1 Coconut Producing Countries Grouped into Eight Categories Based on Production

Category	Indian Ocean, Southeast Asia, and Pacific	West Africa, Caribbean, and Americas
<1 kt	Angola, Cocos Islands (Australia), Mauritius, Nauru, Niue, Oman, Seychelles, Singapore, Somalia, Tokelau, Tutuila (USA), Tuvalu, Wallis, and Futuna	Barbados, Benin, Cameroon, Cape Verde, Central African Republic, Gabon, Guadaloupe, Martinique, Puerto Rico, Senegal, St. Kitts, Democratic Republic of Congo
1–5 kt	Caroline Island, Cook Island, Maldives, New Caledonia	Belize, Costa Rica, Cuba, Dominica, Equatorial Guinea, Grenada, Guinea, Guinea-Bissau, Honduras, Liberia, Nigeria, Panama, Peru, Sao Tome and Principe, Sierra Leone, St. Lucia, St. Vincent, Suriname and Togo
5–10 kt	China, Comoros, Guam, Kenya, Palau, Tonga	Ecuador, Guyana, Haiti, Trinidad and Tobago
10–50 kt	Bangladesh, Fiji, French Polynesia, Kiribati, Madagascar, Mozambque, Myanmar, Solomon Islands, Vanuatu, Western Samoa	Colombia, Ghana, Ivory Coast, Jamaica, Nicaragua, Venezuela
50–100 kt	Papua New Guinea, Tanzania	
100–500 kt	Malaysia, Sri Lanka, Thailand, Vietnam	Brazil, Mexico
500 kt–1 Mt	India	
>1 Mt	Indonesia, Philippines	

Note: Oil equivalent – 1 kt = 1,000 tons; 1 Mt = 1,000,000 tons.
Source: BUROTROP (1992).

Two distinct forms of coconut are recognized: the tall and the dwarf. The primary difference is in the rate of trunk elongation, which is at the least twice as rapid in the tall form compared to the short. The diameter of the trunk of the tall form is also generally one and a half times to two times greater when compared to the dwarf form, which gives a cross-section area two- to fourfolds more. There is one important subgroup of the dwarf form, "*Niu Leka*" or "*Fiji Dwarf*," which, however, has a similar trunk diameter as that of the tall form but a trunk extension rate even less than that of the other dwarf forms. The frond length of the tall form is around 6 m, compared to the 4 m of the dwarf, resulting in a much larger crown. Another major difference is that the tall form is predominantly self-pollinated and therefore largely homozygous.

Morphology

The Trunk

The coconut trunk is composed of an outer dense zone or tube surrounding a central "rod" of much lower density, although both zones become more dense with age, with the outer zone reaching a maximum of around 1.1 ton/m^3. Except for the zone immediately below the crown, which is a low density trunk formed in the last year or two, the trunk is very tough and relatively flexible. The upper portion of a trunk 15 m long is capable of bending almost parallel to the ground, which allows critical relief of wind-induced mechanical shear. The crown adopts a position presenting a streamlined shape to the wind, minimizing pressure and potential damage. The trunk of the tall form typically has a bole (basal region of large diameter tapering from about 1 m height to the standard diameter at 2 m height) that develops between 3 years and the initiation of the first inflorescence. It can be understood in terms of the assimilate produced by the canopy that is available to support development of the "sink." During this period, prior to any demand by the reproductive system, the early growth of the tall trunk is very well supported by photosynthetic assimilate supply. As the palm diverts and assimilates to meet the demand from its developing inflorescence and later the fruit bearing bunches, the trunk diameter gradually tapers. In the dwarf form, however, flowering is initiated much earlier and there is practically no development of a bole.

During early fruiting, the trunks of both tall and dwarf form elongate quite rapidly. However, when yield increases to a high degree, competition for assimilate gradually reduces the rate of trunk elongation in the tall form from its early peak of greater than 1 m/year to 50 mm/year at 60 years of age, whereas the diameter diminishes only by 30 percent over that period (Foale, unpublished data). Restriction of the diameter and trunk growth is triggered by water and nutrient deficiency, and the same is restored to the original when these deficiencies no longer persist.

Properties of the wood in a mature trunk are such that valuable timber can be milled out for use in construction and manufacture of ornamental materials. Special sawing equipment, with stellite or tungsten teeth, and special technique, such as injection of cold water into the active cutting zone, are adopted to prevent heat damage to the saw blade and help clear fibers that are released by the blade during sawing. The outer zone of the trunk is milled separately from the inner zone, which is lower in density.

The Root System

As with other monocots, coconut has an adventitious root system. Attached to the base of the trunk, there are many primary roots that radiate in an ideal light-textured soil to form a hemispherical root zone. Where the soil is shallow or has a compact B horizon, with the water table close to the surface, downward extension of the root is restricted. Roots of 6–10 mm diameter can extend 5–7 m outwards from the base of the palm (Thampan, 1982). Vertical extension down the profile is normally

1–15 m, but this is found more in sandy soil. In a plantation, significant overlap of the root systems exist, which shows that fertilizer applied anywhere across the interrows can be accessed for absorption. In a soil with 30 percent clay, cored to a sandy one, root depth is much reduced (Pomier and Bonneau, 1984). Primary roots have first order (major) root branches of 4–5 mm diameter. Second (2 mm) and third (0.5–1 mm) order branches fill the role as feeder roots. Coconut roots have no root hairs and no nutrient-scavenging mycorrhiza. The resilience the coconut palm shows while growing under water and nutrient stress points to the experimentally unestablished fact that there might be a positive interaction between soil microorganisms and the coconut roots. Extending a few mm from the upper surface of primary roots can be found small, whitish, pointed organs known as pneumatophores, which evidently evolved to maintain oxygen supply to the root tip during diurnal submersion in the water table.

The Frond

As in most plants, coconut's spear-shaped frond emerges vertically from the single terminal growing point. While tightly packed together, the leaflets lack chlorophyll, but are transformed rapidly, becoming green as they unfold. The angle of phyllotaxis (arrangement around the axis of the stem) of the coconut is close to 140°, either clockwise or counter clockwise. The fifth frond above or below any chosen frond subtends at an angle of 20–30° with that frond. That is, each frond is only slightly aside from being placed in vertical alignment with fronds that are five positions above or below. This relationship enables very rapid counting of the number of frond in a palm crown, which is convenient in research investigations on coconut. Collectively, the fronds form the crown of the palm. Some important aspects of the dimensions of the frond base and the interval in the height of the trunk between successive fronds influence the shape and behavior of the crown at different stages in the life of the palm. These effects are crucial to forming a realistic expectation of productivity as the palms get older.

The Crown

There is a sideways overlap between neighboring fronds of the thick wad of tissue, referred to henceforth as the "base pad" (Foale et al., 1994). This section of the base of a frond adds mechanical strength to resist the tendency of the long axis of the frond to rotate around the pivotal zone of firm attachment to the trunk. The attached surface is actually wrapped one-third of the way around the circumference of the trunk, providing a powerful "grip" to resist any pivoting action or "lowering" of the frond from its almost vertical position, hugging the trunk. On the lower trunk of the palm, up to 5 m in height, the average vertical spacing (interval between leaf scars) of fronds has a maximum value of 7 cm. The spacing diminishes to around 4 cm in 25 years, causing the upsweeping attitude seen in all the fronds of a young palm. Gradually, over 10–20 years or so, the behavior of the frond changes. The frond, pivoting on its attachment to the trunk, droops or descends more rapidly from its early vertical position. In time, half of the fronds hang at angles below the horizontal plane and the other half are above the horizontal plane, giving the crown a distinctly spherical shape—the iconic shape so much liked and loved by the tourists to a coconut grove. This change can be explained in terms of mechanical pressure exerted between the base pads of neighboring fronds. The rearrangement of the "angle of repose" of the frond is simply a response to internal pressure on its base pad as adjacent younger fronds expand. Angular movement or drooping of the frond axis of the older (outer) frond about its point of attachment relieves this pressure. When the average interval on the trunk between fronds is 3 to 5 cm, there is firm pressure due to contact between the base pads of any frond and its neighbors two positions up and two positions down from it. The frond has one-third overlap with each of these neighbors. The pressure increases as trunk extension diminishes so that eventually the base pads of fronds 1 and 4, which have a one-half overlap, are also in firm contact. It appears that the resulting sustained outward pressure weakens the attachment of older fronds so that these fronds progressively give way to as shown by their drooping. The base of the inflorescence would also contribute to this overcrowding at the surface of the trunk, generating even more outward pressure. In extremely old coconut palms, the overcrowding is so severe that the frond base is almost wrenched from the trunk, remaining loosely attached by elongated, fibrous-looking xylem bundles, leaving a sunken "scar" on the trunk. Other palms of economic importance, such as the oil palm and date palm, differ from coconut palm in that the trunk is of greater diameter and the width of basal frond attachment is much less than that of coconut, so that inter-frond pressure does not come about until the palms are very old.

There are two important consequences of this evolution of the shape of coconut crown. Palms of 25 years age have a crown that evolves during that period from an inverted cone shape to hemispherical. This range of shapes achieves maximum light interception and the incident energy is fairly well distributed over all fronds, ensuring high phytosynthetic efficiency (Foale, 1993a). As the crown takes on a spherical shape, however, with a decreasing proportion of the fronds angled upward, light interception falls until when the fronds are "half up and half down." Only 50 percent solar radiation is intercepted at a standard plantation density. Progressive reduction in length of the frond with age serves to reduce interception of light even further. The obvious consequence of the declining rate of capture of solar energy is that potential biomass production falls in proportion, following a downward spiral of reducing photosynthetic infrastructure and also increasing the intensity of competition for resources between developing frond, inflorescence, and trunk. The second consequence of loosening of the older frond in the coconut crown is that these fronds can be shed more readily in a devastating wind. The coconut minimizes risk of such devastation by its ability to reduce the size of the crown, which complements its ability to adopt a streamlined position, already mentioned in the previously discussed section. A young palm is more prone to wind damage because the crown is more robust and also because there is insufficient length of the trunk to bend down on the wind direction. Yet, it may survive. A further consequence of falling light interception is that in a natural coconut woodland, more light reaches the young palms below that are struggling to capture sufficient energy to become productive.

The Inflorescence

Each mature and healthy frond has an inflorescence in its axil, emerging about 1 year after the frond expands. First, the encasing spathe (a somewhat leathery sheath) appears and extends its full length, then its lower side splits open and the multibranched "flower bunch" is released. Several thousand small male flowers (4–6 cm long) are borne on the 20–40 branches and begin to shed pollen progressively from the distal ends for about 6 weeks. The rare spicata form of coconut has only one or two short branches bearing just a few dozen male flowers. The female flowers, which expand to about 25 mm diameter during the period of opening of male flowers, are borne toward the proximal end of the mid and lower branches, normally singly or in pairs. The male phase starts with pollen shedding on the day the inflorescence opens and continues for about 3 weeks. However, the duration of the female phase generally lasts only 5–8 days, and the individual stigma will remain receptive during 1–3 days. The female flowers of the tall form become receptive for pollination after pollen shedding is complete, although this depends on the genotype and prevalent environmental conditions. However, on the dwarf varieties, pollen is still abundant within the inflorescence during female receptivity. Thus, the tall varieties are generally out breeding and heterozygous (diverse traits between palms), whereas the dwarf verities are generally inbreeding and homozygous (uniform traits). There is some inbreeding in the tall varieties, as well, when a favorable environment (which is seasonal when rainfall is plentiful in many coconut regions) speeds up the rate of emergence of fronds and inflorescences. This results in the next youngest inflorescence being released while some female flowers of the older one are still receptive. Some outcrossing occurs in the dwarf varieties through pollen from other palms, brought about by wind and insects, which in fact turns out to be in competition with the native pollen.

The Fruit

The number of female flowers vary from season to season, is subject to the environmental stresses the palm undergoes, and is usually well in excess of the number of fruits developed. A full 12 months elapse between pollination and fruit maturity, less in warm regions, but as much as 15 months at higher altitudes with cold background. At higher altitudes within the tropics, the delay rate in maturity will be in proportion to the drop in mean temperature below the coastal mean of 28°C. After 6 months, the fruit reaches its full size and its vacuole is filled with water, which is the liquid endosperm. Thereafter, the volume of the vacuole diminishes as the 10–15 cm of layer of kernel (endosperm) is formed on the inner surface of the shell. Maturity of the fruit is usually indicated by the entry of air into the vacuole, so that the water "splashes" audibly when physically disturbed. In an environment with a large atmospheric vapor pressure deficit, early loss of nut water takes place, and the splash in the vacuole can be generated before maturity. In that case, one should avoid interpreting the splash as an indicator of maturity. Both seasonal temperature and soil water deficit appear to influence the shape of the shell within the fruit. A common response is the decrease in diameter of the nut, giving it an elongated appearance in contrast to the dominant spherical shape in most varieties. The size of the fruit differs greatly with genotype, and the weight of mature fruit ranges between 1 and 3 kg, although fruits outside this range are encountered. Fruit size varies with the number borne on the palm at a time, and the contrast is especially notable between the bunches carrying few and many fruits respectively. Coconut fruit is the very largest seed in the plant kingdom after coco de mer (*Loidecea seychellarum* L.). The seed provides an almost unique long-term source of energy for the emerging seedling, which enables it to cope with both water and nutrient deficiencies during the first year of the life of the palm (Foale, 1968b). This ability was an important contributor to the survival and competitiveness of the coconut in newly colonized habitats, and has earned it the "weed status" in some environmental reserves and semi-urban beach environments where it has proliferated and suppressed other strand vegetation (Foale, unpublished).

The Seed and Seedling

The shell with its contents constitute the seed where the husk performs the function of protection and dispersal. The entire fruit is normally left intact, except to trim some husk from the region where the shoot is expected to emerge, when selected for seedling production. Following maturity, the single embryo present in the nut begins to develop by first emerging through the germ pore. In domesticated Pacific coconut types, germination is rapid, often taking place prior to fruit fall, whereas in others there may be several weeks' delay or the need for added moisture to stimulate the embryo, prior to germination. The presence of the husk prevents the observation of true germination, as there is about 2 months' delay before a sprout develops sufficiently to emerge through the husk. By dehusking and storage, unbagged and free of moisture, germination of the nut can be controlled. A nut in this state will not sprout until the atmosphere is vapor saturated by placing it under moist mulch or in a sealed bag with enclosed moisture (Foale, 1993b). The embryo expands through the germ pore, developing within 10 days into a rounded mass of soft white tissue. The internal end of the embryo expands to form the haustorium, an enzyme-secreting organ that breaks down and absorbs the kernel and progressively fills the vacuole. Absorption of the energy-rich tissue of the kernel supports the continuing expansion of seedling tissue and its differentiation, which commences after about 14 days, eventually forming a shoot on the upper side and a root initial below.

Growth of the coconut seedling could continue for months in the dark as absorption of the kernel proceeds, but at about 3 months from true germination, the first small leaf appears and photosynthesis begins. Over the next several months, as leaf area expands, a gradual transition from total dependence on the kernel to complete independence occurs around 12 months of age, when uptake of the kernel is complete (Foale, 1968b). During the early months, the endosperm-assisted growth rate of the seedling is very high compared to the small-seeded plants. If microorganisms invade the endosperm within the first 6 months of germination, the growth rate of the seedling falls sharply, usually leading to rejection of the seedling.

Cytogenetics of Coconut

There was early interest in descriptive research of the cytology of coconut (Abraham et al., 1961; Nambiar and Swaminathan, 1960; Ninan et al., 1960), but interest waned. However, Louis and Rethinakumar (1988) presented a useful report entitled "Genetic Load in Coconut Palm," which identified mechanisms, arising from the heterozygous nature of the coconut palm, that resulted in the elimination of undesirable recessive genes. Developments in unraveling the genetic code and identification of specific genes and linkages has led to diminished interest in the discipline of cytogenetics.

Genetic Improvement of Coconut

Source of Diversity

That coconut improvement has been an important objective over a long period in human history is testified by the fact that the diversity of coconut populations and its association of readily identifiable variants with specific dispersed ethnic groups, especially across Southeast Asia and the Pacific. Large-fruited populations, for instance, are widespread wherever there are Polynesian people, who have detectable linguistic and cultural links with island people close to Southeast Asian landmass. The large fruit fulfills the primary human objective of gathering a convenient source of food and water for long sea voyages. A variant of the large fruit has especially long fibers, valuable in the manufacture of rigging for sea-going vessels. Since 1950, there has been an industrial initiative for genetic improvement of the coconut, with the primary objective of raising productivity of oil output and economic profitability of the coconut plantations. All genetic improvement is based on the diversity present in the species. However, in order to be useful to plant breeders, this diversity needs to be characterized reliably.

Characterization of the Genome of the Coconut Palm

On the basis of appearance and utility, coconut populations have been distinguished since antiquity. Variations in fruit size, shape, and color have been seen in tall forms, and humans have shown a preference for larger fruit, especially for water and supply of food at sea, but also for convenience in processing the kernel for domestic use. Dwarf palms were preferred for their ease of harvest to provide drinking water to the household, but their kernel is less palatable. Tall palms, on the other hand, are more difficult to manage because of outbreeding and heterozygosity, the consequence of which is that tall forms provide greater challenges in their characterization to establish the fact whether or not real and heritable differences exist (Liyanage and Sakai, 1960).

Fruit Component Analysis

Pieris (1935) was the first to employ this method as a tool to characterize coconut palm. He confined his measurements to the dehusked nut. Subsequently, Whitehead (1966) and Harries (1978) continued the work. Later researchers used the method to much advantage. Fruit component analysis allowed identification of similarities and differences between populations with a fair degree of confidence, based on the low sensitivity of fruit composition to most changes in the environment. Harries (1978) made use of differences in fruit components in developing a general theory of the evolution and dissemination of coconut around the world. The basic approach is to weigh the whole newly mature fruit and then its dissected components. The fruit is selected only when a trace of the fresh color remains on the husk. Moisture content of the husk would be expected to be similar in fruits from different populations, at around 20 percent, and water content of both shell and kernel would be fairly stable. Broad categories were established, named *domestic* and *wild*, at the extremes, with low and high proportion of husk respectively. Many populations were of intermediate proportions, and these were described as introgressed. It is impossible to determine true genetic affinity or difference between diverse populations that are identified relying solely on fruit component analysis. Such distinctions, based on molecular parameters, were impossible to be made earlier.

Use of Molecular Markers

Today a variety of tools is available, based on price labels and efficacy, to characterize the coconut genome. The basic entity in this approach is called a "molecular marker," which is defined as "*an inherited chemical trait that can be used to distinguish between individuals, groups of individuals, or positions on chromosome*" (Ashburner, 1999). Until the early 1990s, attention was focused on "gene products," such as isozymes, to detect genetic difference, but a direct assay on the DNA itself has proven to be far more effective. A review of the application of molecular markers on coconut improvement was published by Ashburner (1999), but scientific advances in this field are being made very rapidly. A comprehensive investigation of a very wide range of germplasm encompassing global variations in coconut germplasms was undertaken by the Centre de cooperation internationale en recherché agronomique pour le development (CIRAD) in France during the 1990s, which has been reported by Lebrun et al. (1999). Their report reveals the groupings of coconut genotypes based on RFLP analysis. The investigation showed, for instance, that all germplasms in populations extending from Southeast Asia eastward right across the Pacific, including the Pacific coast of Central America, could be grouped together as quite distinct from that of Southern Asia and the Caribbean. An intermediate group was apparent that included Madagascar and the east coast of Africa. Dwarf types were found to be genetically similar to the tall types of their geographic region.

There are a variety of uses in taking recourse to molecular markers. They are specially useful in many resource management applications, including diversity analysis based on at least 20 individuals per population. This will provide data on genetic distance both within and in between populations, genetic finger printing, such as establishing true crosses from inter population hybrids, and outcrossing analysis, which can establish with certainty the mating system of a coconut population (Ashburner, 1999). Many interesting hybrid combinations have been tested in coconut, selected in the early days using a mix of both intuition and geographical separation. Molecular characterization of coconut populations will make possible the choice of future combinations with a higher probability of significant improvement, although a great deal more research would be needed to validate this strategy.

Early Breeding Work

While coconut provided food for the inhabitants of the tropics for millennia, it was the oil that was cherished by the Europeans and Americans in late nineteenth and early twentieth centuries, facing little competition from other oil sources. Following the Second World War (WWII), there was a concerted effort to enhance genetic improvement of coconut as it was found to be an important oil-bearing crop. The genetic improvement of the coconut palm has been occurring during thousands of years, rather unknowingly. However, it was during the twentieth century that the crop assumed a plantation status. Though the effort was widespread, the earliest attempt was initiated in India, with the establishment of four research institutes in 1916. The coconut research institute at Kasaragod (Kerala State), which was enlarged to become the leading coconut research institute of Asia called the Central Plantation Crops Research Institute (CPCRI, which also brought under its ambit arecanut), and another at Nileshwar (Kerala State) carried out pioneering work.

Hybrid Vigor in Coconut

It was Patel (1937) who first discovered the existence of hybrid vigor in coconut, between West Coast Tall × Chawghat Green Dwarf, which was a landmark in the history of genetic improvement of coconut. Since then, a number of hybrids, both D (Dwarf) × T (Tall) and T × D were evolved and released in India. Some of the important hybrids are Lakshaganga (Laccadive Ordinary × Gangabondam), Anandaganga (Andaman Ordinary × Gangabondam), Keraganga (West Coast Tall × Gangabondam), Kerasree (West Coast Tall × Malayan Yellow Dwarf), Kerasawbhagya (West Coast Tall × Straight Settelelemt Apricot), Kerasankara (West Coast Tall × Chawghat Orange Dwarf), Chandrasankara (Chawghat Orange Dwarf × West Coast Tall), Chandralaksha (Laccadive Ordinary × Chawghat Orange Dwarf), VHC-1 (East Coast Tall × Malayan Green Dwarf), and VHC-2 (East Coast Tall × Malayan Yellow Dwarf). Drought tolerance investigations at CPCRI, Kasaragod, showed the possibility of identifying the desirable traits of drought-tolerant cultivars under field conditions (Rajagopal et al., 1988). The promising drought-tolerant varieties/hybrids were West Coast Tall × West Coast Tall, FMS (Federal Malay State), Java Giant, Fiji, Andaman, and Laccadive ordinary × Chawghat Orange Dwarf. In Fiji, a hybrid was evolved by crossing the indigenous dwarf known as Niu Leka with Red Malayan Dwarf in the 1920s (Marechal, 1928). Worldwide depression and war halted the further progress of this pioneering research. The Coconut Research Institute in what was then Ceylon (established in 1929 as the Coconut Research Scheme, becoming a full-fledged Institute in 1950), contributed much to early investigations (Pieris, 1935). A broad foundation of understanding was developed in those early decades of the opportunities and the constraints presented to those seeking to improve coconut yield.

Constraints in Coconut Breeding

Compared to most other field crops, constraints in genetic improvement of coconut are formidable. Height is the most important of all the constraints as this renders pollination extremely difficult. Another is the large size of the palm, which requires at least 8 ha to conduct a properly laid-out field trial (Bourdeix et al., 1993). Yet another is the long delay in developing a successful technique for harvesting a manageable quantity of pollen from the inflorescence, which was finally tackled in the 1960s (Whitehead, 1963). Additionally, there are biological constraints of few seeds per palm and per bunch, requiring multiple visits for pollination, as well as the long turnaround time of 6–10 years between generations of tall varieties. Genetic constraints include a high degree of heterozygosity within any population of tall palms and different levels of combining ability between genotypes. There is also the constraint of lack of a clonal propagation method for outstanding selected or bred genotypes, which are otherwise constrained by the low number of seeds generated (Santos, 1999).

Selection and Its Progress

Tall populations of coconut have been distinguished mostly by fruit characters locally, while globally or regionally by place of origin. The latter approach seemed somewhat arbitrary until collections of different populations were assembled for research. It was observed that the introduced germplasms were attacked by one or more insect pests that were of mild consequence locally or by diseases that were unknown in the area previously. Examples are the attack on Malaysian Tall in Solomon Islands by the leaf beetle *Brontispa* sp., the outbreak on the leaves of Polynesian Tall in Solomon Islands by the fungus *Drechslera* sp., and the attack on hybrids that had a West African Tall parent, planted in Indonesia and Philippines, by a *Phytophthora* strain that differed from strains that were tolerated in West Africa. In India, the "root wilt" is devastating, and elsewhere in Kerala State, they experience a coconut "decline," of which the exact etiology remains unknown even to date. A most unusual experience in Vanuatu in Sri Lanka is the appearance of foliar decay by a virus attack. Although it was presumed that the virus was introduced along with new varieties, later it transpired that it had existed locally all along in Vanuatu. The indigenous population possessed tolerance, which allowed it to remain symptom free; when the pathogen when transmitted to the nontolerant introductions, the result was, indeed, devastating. These examples point out that many coconut plantations have adopted to local environment with particular pests, diseases, and aberrant climate, as in cold episodes in Hainan Island (Zushum, 1986). This is an aspect of population identity to which molecular markers are beginning to make an important contribution. Breeding for yield improvement or other positive traits, while retaining "invisible" traits, such as pest resistance or disease tolerance, will be feasible only with the aid of molecular tools. Owing to difficulties in demonstrability of positive traits through heritability (Liyanage and Sakai, 1960) and because of a negative correlation between the number of fruits borne and the copra content, researchers were slow to develop confidence inbreeding for yield improvement. Beginning 1950, yield improvement became a major objective inbreeding in India, Sri Lanka, Philippines, Ivory Coast, and several Pacific countries, whereas in Jamaica, the main objective was to overcome the ravages of the devastating yellowing disease. In India and Sri Lanka, emphasis was at first on mass selection within a population, even though the first report of dwarf × tall hybrid vigor was made in India in 1937. Other researchers concentrated on inter population hybrids, especially in the 1960s.

Hybrids and Their Future

The enormous diversity in the F2 generation from hybrids produced in the 1920s in Fiji (Marechal, 1928) between Malayan Red Dwarf and Niu Leka (Fijian Dwarf) and in the 1960s in Solomon Islands evoked much research interest. On a parallel, employing hybrids in the Pacific and in Ivory Coast also has yielded rapid progress for some combinations, such as Malayan Yellow Dwarf × West African Tall and the Malayan Red Dwarf × Rennell Tall, which became commercial cultivars in the 1970s. In each case the first generation hybrids produced a yield advantage of 30 percent over the best performing tall population (Foale, 1993b). Outstanding yield increase has also been demonstrated for Tall × Dwarf hybrids in India (Nair and Nampoothiri, 1993) and in Sri Lanka (Peries, 1993) in the recent past. On account of the risk of a specific parent lacking tolerance to a potential natural enemy in locations far from the source of the parent genotypes, local industries have lately moved to include at least one local parent in any Tall × Dwarf hybrids tested. Apart from general combining ability between Tall and Dwarf populations, subsequent testing of many combinations in Ivory Coast has revealed that some are significantly more productive than others (Baudouin, 1999). This author has reported that from 15 Rennell Tall (RLT) palms combined with each of the three "testers"—West African Tall (WAT), Malayan Red Dwarf (MRD), and Cameroon Red Dwarf (CRD)—produced very different yields of copra, ranging from 15 to 27 kg/year. The mean value (kg) of 15 palms in each group was (by tester): MRD: 19, WAT: 21, and CRD: 24. Other combinations of "genetically distant" tall parents have proved to be very interesting. Geographic separation, supported by isozyme and lately RFLP markers (Lebrun et al., 1999), has enabled three major groups to be defined within the global coconut population. Crosses between sub populations from Group I (Southeast Asia/Pacific) and Group II (Southern Asia/West Africa/Caribbean) gave higher yield than any combinations within these two groups (Baudouin, 1999). An interesting observation was that an intermediate group (Group III) could be identified from Madagascar and East Africa from which the Mozambique Tall hybrid with WAT equaled the best combinations between the geographically more separated Groups I and II. It can be concluded that molecular markers will turn out to be the most important tool in identifying outstanding coconut germplasms in the future in hybridization research.

Commercial Production of Hybrid Seeds

To produce commercially viable seeds, interplanting of the two parents is involved. A Dwarf × Tall cross is most readily done with the Dwarf as female parent, requiring emasculation or removal of all male flowers before pollen is shed. The dwarf palms planted in two or three rows per row of the pollen parent must be checked daily. Any inflorescences that have begun to emerge from the opening spathe are dealt with by complete removal entirely from the field so that the only possible source of pollen is the tall parents nearby. Multiple male parents can be included in such seed gardens to provide the possibility of producing different hybrids as required, but only one hybrid is available by open pollination at any given time. Controlled pollination of bagged inflorescences to produce small lots of seed is also possible but obviously more costly and less successful. Where a smaller scale crossing is to be achieved, with few selected parents, the female parent is bagged after emasculation and the chosen pollen blown into the bag during several days in a row. The procedure is expensive, but suitable for research only.

Commercially viable hybrid seed gardens have been set up in many countries. For example, 2 million hybrid seeds are produced each year in India, and in Sri Lanka, the capacity to produce hybrids is rising rapidly. Nursery managers know that both yellow and red colors of widely used female parents is due to a recessive gene, which if expressed in a progeny from seed garden is evidence of self pollination of the mother palm. A seedling that shows yellow or red color on the petiole is therefore rejected.

In vitro *Propagation*

The method has produced only one commercially viable application—the propagation of embryos from *makapuno* nuts. *In vitro* propagation is still a challenge to coconut breeders. A *makapuno* nut is filled with a jelly-like substance that is rich in coconut flavor, but incapable of stimulating and supporting the embryo germination. The raw *makapuno* product is sufficiently valuable that there is commercial support for culture of embryos to ensure production of *makapuno* nuts in the progeny. Though considered promising and presented for funding (Santos, 1999), *in vitro* propagation's viability is still elusive (Harries, 1999). There has to be other objectives besides oil yield to evoke commercial interest in *in vitro* propagation. Limited propagation of outstanding elite palms to form the basis of a seed garden might possibly be justified, although technically is not feasible. The long-term possibility that genetic transformation can be performed on coconut callus, which is then differentiated into propagules, such as has been achieved in many other agricultural crops, still interests researchers in the quest to streamline this technique. No major changes in the present uncertain prospects of *in vitro* propagated coconut palms contributing to yield or quality improvement can be expected in the near future. To a limited extent, embryo culture is useful in germplasm exchange, although even here the technique is not yet well advanced for use by technicians of moderate scientific skill.

Agronomy of Coconut

Soils

The coconut palm must have been subjected to diverse environmental conditions in the course of its evolution on the tropical edge of the free land masses that drifted from the Gondwana. The basic requirement of the palm for its water could have been met from diverse sources—highly alkaline sands derived from erosion of coral, silica sands, deltaic silts and loams, black and red clays formed from volcanic ash, and highly acidic lava sands. Tall populations adapt best to such diverse soils, which reinforces the idea that a great deal of natural mixing of genotypes between sub populations from these diverse environments has taken place over geological time. The result of this phenomenon is that the coconut palm has thrived in the tropics far from its shoreline home, wherever its need for a regular supply of available water has been met. The extremes of soil environment are the coral atoll with sand and gravel at pH 8.3, where some critical nutrients are rendered difficult to be absorbed as, for instance, in the organic peat of Sumatra where the pH is lower than 4.5. This decline is because drainage oxidizes the elemental sulfur in the soil to sulfuric acid, releasing Al and some Mn in the process from clay minerals, which turn out to be toxic.

Soil Water

The coconut palm accesses the life-giving freshwater "lens" or reservoir found under atoll soil or the water table of larger land masses that commonly flows to the sea under coastal sand dunes. In other situations the soil must be capable of retaining sufficient water for the coconut palm to survive the longest seasonal intervals between significant rainfall events. Deep, well-structured clay or clay loam soils, which hold up to 250 mm of plant available water, could from a fully saturated starting condition sustain growth for 2 dry months. Such soils are found on the uplifted coralline "benches" of many South Pacific islands and on river plains and deltas worldwide. The coconut palm is also planted widely on banks between irrigated fields, where the water level is controlled by the farmers irrigating other crops, thereby providing an accessible water table. Although the coconut palm appears to thrive in the vicinity of the swamp, which often occupies the swale adjacent to a coastal dune, the palm actually is very sensitive to waterlogging. Where the water level is static and low in dissolved oxygen, coconut roots become inactive. On the other hand, the sort of diurnal, tide-induced vertical oscillation of the water level in the freshwater lens of an atoll soil or under a coastal dune is ideal. The roots have pneumatophores (a sort of snorkel), attached to short vertical branchlets which supply oxygen to the main root, while its physiologically active root-tip region is temporarily submerged. It has been found necessary to provide good drainage for coconut on heavy-textured lowland plains to avoid waterlogging during sustained rainfall. While the coconut palm is susceptible to waterlogging, it has some tolerance to water deficit as well. Whereas in most island environments the coconut palm was rarely moved far from the coast prior to the plantation era of the late nineteenth century, it has evidently been grown in the sub coastal zone and further inland in India since antiquity. This provided an opportunity for adaptation to more severe episodes of water deficit than would be experienced elsewhere. This development is supported by evidence that the modern West Coast Tall and related populations in East and West Africa, and the Caribbean Islands are more drought-tolerant than most other populations from Southeast Asia and the Pacific. An exception may exist in the Pacific on some of the dryer atolls such as those of northern Kiribati, where the level of salinity in the freshwater lens rises sharply toward the end of long periods of low rainfall. The local tall population generally shows less stressful signs (collapsing lower fronds, premature nut fall, failure of inflorescence emergence) than material introduced from better-watered environments.

In this connection, it would be interesting to know about the effect of drip irrigation. Coconut palm is almost always rainfed and sometimes irrigated. And when it is irrigated, it is invariably basin irrigated. The importance of irrigating coconut for sustained yield has been emphasized. Among the irrigation systems, drip irrigation is gaining importance as it maintains soil moisture availability and air balance in the root zone of coconut near field capacity throughout the dry season and saves irrigation water (Vidhana Arachchi, 1998). The experiment was carried out during a 6-year period (1993–1999) during the summer months, when water requirement is very severe, at the field experiment station of Central Plantation Crops Research Institute (CPCRI). The results revealed that annual leaf production and leaf nutrient status (N and K) of coconut palm was significantly higher in the irrigated treatments compared to the rainfed control treatment. Female flower production and nut yield with 66 percent of open pan evaporation (Eo) was on par with 100 percent Eo through drip irrigation and 100 percent of Eo through basin irrigation. Drip irrigation equal to 66 percent of open pan evaporation (Eo) proved to be an economically efficient method of irrigation with water saving up to 34 percent compared to 100 percent Eo through basin and drip methods. Results are summarized in Table 3.2.

Table 3.2 Nut Yield and Female Flower Production (Number/Palm) in West Coast Tall (WCT) as Influenced by Irrigation in a Laterite Soil

Treatment	Pre-experimental Period (1991–1993)		Average of (1993–1999)	
	Female Flowers	Nut Yield	Female Flowers	Nut Yield
T_1	79.9	28.2	184.5	68.2
T_2	73.2	30.1	214.7	96.5
T_3	66.0	24.9	200.8	89.8
T_4	78.6	31.6	225.5	98.2
T_5	79.5	30.8	157.4	52.6
LSD (P = 0.05)	N.S.	N.S	25.6	9.5

Note: T1= Drip irrigation at 33% Eo (Open pan evaporation) daily. T_2= Drip irrigation at 66% Eo daily. T_3= Drip irrigation at 100% Eo daily. T_4= Basin irrigation at 100% Eo applied once in 4 days through hose pipe. T_5= Rainfed control. N.S.= Not significant.

Plant Nutrients

Like most other plants, coconut requires all the essential elements, and unlike others, need much chlorine (von Uexkull, 1972). Legends have existed in coconut-based cultures that when grown away from the sea coast, the palm needs to be sprinkled with saltwater. This folklore was finally shown to have scientific substance, when substantial need for chlorine was scientifically proved. Chlorine is deficient most commonly beyond the range of cyclic salt, delivered near coastlines in rainfall, and where soil is readily leached during intense seasonal rainfall. Since 1950 extensive research has been carried out to understand why in a given local situation the crop yield does not reach the potential yield and the implication of specific fertilizer nutrients has been identified through classical fertilizer experiments. Numerous case studies of the association of a specific nutrient deficiency with specific soil characters or landscape history have enabled some findings to be extrapolated to identical locations. For example, on atolls it is found that when nitrogen is not limiting, Fe and Mn become important, rendered poorly available in soils of high pH. Another case concerns sulfur, which has been found deficient in many areas of grassland plantation. Frequent natural or human-induced burning of the grass has led to a cumulative loss of sulfur to the atmosphere and, in time, exhaustion of the sulfur reserve in the soil (Southern, 1969). N deficiency is most common in drier environments, especially on sandy soil, which leaches readily during the wet seasons. Where rainfall is high and the dry season is short or absent, N is rarely in short supply as the rapid recycling of nutrients from the considerable body of organic residue being returned from the palms and the under growth meets the needs of the entire plant community. Exported N is replaced by the accession of nitrate from rainfall and from any N-fixing herbs or shrubs. In this well-watered environment, K commonly becomes limiting because of the large amount that is exported to the fruit, as well as what is lost through leaching. When rainwater is high in cyclic salt (the mineral salts present in seawater, usually present near the coast), Na displaces much of K from the clay particles, which is easily leached out of root zone, thereby starving the coconut palm of the much needed K. A low available P level is quite common in many coconut-growing soils, especially the laterites of Kerala State, where the soluble Fe and Al bind the P into insoluble forms. Hence, care must be taken in applying P fertilizers, and it is best done in bands so as to minimize soil contact and thereby minimizing interaction at the clay colloidal level. Mg has been found limiting in some soils, especially after other limiting factors have been met. On some clay soils situated in uplifted coral benches, low Mg was evident due to loss induced by both high Ca and Na.

Tissue Analysis

Manciot et al. (1979a, b, 1980) published a comprehensive review of the mineral nutrition of coconut, where the tissue analysis form an important guide to the characterization of nutrient deficiencies and thereby lead to appropriate fertilizer recommendations. The coconut palm renders itself particularly well to nutrient investigation based on tissue analysis because of the regular production of foliage and fruit throughout the year. The leaf analysis data lead to determining "critical levels" of nutrients, which can be calibrated to arrive at appropriate fertilizer recommendations to tap the potential yield of the coconut palm. Leaf analysis is used most commonly as material of a similar stage of maturity, such as from the 14th frond, which can be used as a standard source for sample collection. Table 3.3 summarizes data on critical levels of important nutrients (Manciot et al., 1979a, b; de Taffin, 1993). The prescription of fertilizer application based on either nutrient diagnosis or fertilizer experiments is an exercise in economics depending on the relativity of input costs to the increased market return from increased yield. Coconut palm is rather slow to respond to fertilizer application. But its response is relatively quick (in about 6 months from the date of application) to K and Cl, which is reflected in increased kernel per nut (Foale, 1965; Manciot et al., 1979b). Clearly, early detection of limiting nutrients is vital to achieving the most economical yield potential in the local environment.

Table 3.3 Critical Values for Concentration of Mineral Elements in the Leaf Tissue (14th, the Youngest Frond) of Adult Tall Type

Major Nutrients	Percentage Dry Matter	Remarks
N	1.8–2.0	In Tall × Dwarf hybrid—2.2
P	0.12	
K	0.8–1.0	In Tall × Dwarf hybrid—1.4
Mg	0.20–0.24	Strong inverse sensitivity to extremes of K
Ca	0.30–0.40	Some inverse sensitivity to extremes of K
Na	Not essential	Substitutes for K in case of deficiency
Cl	0.5–0.6	
S	0.15–0.20	
Trace elements	Parts per million (ppm)	Remarks
B	10	
Mn	>30	Difficult to fix values as very interactive with Fe in strongly alkaline soil; potentially toxic in extreme acid soils
Fe	50	Deficient only in strongly alkaline soils
Cu	5–7	Deficiency very rare, uncertain
Mo	0.15	Common value, no response observed, yet
Zn	20	Common value, no response observed, yet
Al	>38	Nonessential element, but, always present; potentially toxic at values well in excess of this common level

Source: [132,133,134] Maciot et al. (1979a, b, 1980); de Taffin (1993).

Coconut-Based Mixed Cropping Systems and Their Management

Coconut plantations are, invariably, dual cropping systems, where in the field with coconut as main crop, a number of others, such as diverse fruit trees, pineapple, cacao, coffee, root crops, banana, pastures, and several others are also grown. The mixed cropping systems help capture better the solar radiation. For reasons outlined in the section on coconut botany, the coconut canopy follows an evolutionary trend in its capacity to intercept light. In a healthy crop, with a plant density of 180 palms/ha, light interception ranges from low values at pre flowering stage to around 90 percent at 10 years of age, continuing up to 25 years of age (D. Friend, unpublished data), and then gradually declining to around 45 percent at 50 years of age. The brief pre-production phase is often quite productive for short-term crops, especially if the land has been newly cleared, providing plentiful supply of nutrients released from the great bulk of residue generated by the destruction of the previous vegetative cover. In some cases coconut density may be lowered, either uniformly or by "hedge-grow" planting, in order to allow continuous intercropping, but the potential yield of the coconut stand will be proportionately reduced. Intercrops under old and so-called "declining" coconut plantations frequently become more productive with any increase in the amount of solar energy that bypasses the coconut canopy.

Seed and Seedling Management

Both challenges and opportunities are provided by the big coconut seed. There has been much interest in seedling vigor since recognition in India and Sri Lanka of its correlation with yield potential of the adult palm (Liyanage and Abeywardena, 1958). One challenge in recognizing relative seedling vigor is to provide a clearly defined starting point for a batch of newly germinated seeds. Some populations, particularly in Southeast Asia and the Pacific, have seeds that germinate very quickly after maturity, some tending to sprout well before the fruit falls naturally. Most other populations have a brief dormancy after maturity before being ready to germinate when the dry husk is moistened. If a batch of the latter comprises fruit that has been harvested at a comparable stage of maturity, a firm starting point is achieved. On the other hand, the fast germinating seeds may contain a mix of maturity when harvested, with some embryos already emerged from the germ pore, but remaining hidden for many weeks within the husk. Such seeds are subject to the risk of misorientation in the seed bed when the still-hidden shoot is pointed downwards. This can lead to protracted delay or even fatal entrapment as the shoot changes its direction of growth upward once more.

Germination Rate

The time lag between expansion of the embryo through the germ pore, the index of true germination, to sprouting (emergence of the sprout from the husk) depends on both the vigor of the seedling and on the thickness of the husk. As these traits are quite variable, in a Tall population, the time lapse could be 6–10 weeks (Foale, unpublished data). A reasonable comparison of the vigor, at least within small batches of seedlings, can be achieved by taking all the nuts that sprout within 1 week, for example, and keeping them together for a within-the-group comparison in the nursery. Checks done at time intervals thereafter will allow the seedlings to be sorted into fast, medium, and slow growers on the basis of leaf length and area, robustness of the collar, and general appearance. Another potential influence on the growth rate of the seedling is the amount of kernel present within the nut. Most of the seedlings continue to draw energy from the kernel for about 12 months and, interestingly, smaller nuts within a population support a slightly higher early growth rate (Foale, 1968b). The haustorium, obviously, makes more rapid and effective contact with the kernel when the vacuole is smaller. As both genetic and environmental factors influence the amount of kernel in the nut, it would appear that the response of the seedling growth rate to nut size would introduce a small bias into early seedling selection.

Poly Bag Seedlings

The ability of the seedlings to recover from "bare root planting," aided as it is by the energy supply from the kernel, has resulted in that method being still in use in many traditional coconut cultures. However, raising seedlings in poly bags (made of polyethylene, of size 40 cm, height and 25 cm radius, for 8-month planting out, while larger ones for older ones) has become common. While transplanting, no damage occurs to the root system and the growth continues unchecked (Foale, 1968a). Widespread use is made of poly bag seedlings in many areas, in spite of greater logistic challenge of carrying into the field the seedling in the poly bag, weighing 20–25 kg. Comparative health and vigor can be checked when the seedling is 8 months old, but, the poly bag can hold the seedling up to 12 months. A 50 percent increase in the size of the poly bag used would be advisable in this case, thereby saving a few months of more expensive field maintenance. It is important, however, to spread older seedlings further apart in the nursery to avoid slowing the growth rate due to mutual shading.

Seedling Selection

It is best to select the 50–60 percent of the vigorous seedlings from an open-pollinated population, but one would expect to discard very few seedlings from a batch of hybrid seeds. Where a high proportion of seedlings is to be discarded, this must be done quite early so that selected seedlings can be transferred from an open nursery bed to polythene bags without a severe setback. On the basis of correlation studies for 1-year old seedlings, selection criteria in India include the collar girth (10–12 cm), number of leaves (6–8), and early splitting of leaves. Nuts that do not germinate within 6 months of sowing should be removed from the nursery (Iyer and Damodaran, 1994). To guard against insect pests, rigorous protection of seedlings in the nursery is required, employing appropriate chemicals. The genotype chosen determines the plant density, with Talls at 160–180/ha and hybrids at 180–220/ha for sole coconut plantation. While intercropping, different densities and palm arrangements are chosen other than isometric (triangular) planting.

Field Management

The early management is very important in a seasonally dry environment where it is common to plant the seedlings with the nut slightly below the soil surface. On atolls, where the water table is 1 m or lower than at the surface, sometimes the seedlings are placed quite deep in order that the roots reach the most soil zone quickly. Weeds must be kept in check in the initial stages to preempt competition for water and nutrients. A common practice is to establish a weed-free circle by manual weeding or herbicide application and subsequently enlarging the spread of the circle as the palm grows. A few South Pacific Talls (e.g., Rennell Tall) flower within 4 years, and their hybrids with Red Malayan Dwarf flower almost a year earlier, especially in the case of poly bag seedlings. In Africa and India, the populations must endure the summer water deficit. Talls tend to flower after about 7 years, but the Tall × Dwarf hybrids flower 2 years earlier.

Productive Palms

The nature of the end product determines once palms start bearing. Where high human population density occurs, every part of the palm is processed to make value-added products, like coconut milk, cream, milk powder, desiccated coconut, oil, coconut drink (fresh or bottled), fuel, charcoal, fiber, and coco peat from coir dust, in addition to the number of utensils and curios made out of coconut shell.

In the case of large plantations, the only end product is the oil via the copra pathway. Increasingly, the oil is separated for export from the country of production, and the residual cake is either exported or fed to local livestock.

Control of pests and diseases is done where a fatal attack from these is expected. For example, phosphoric acid treatment can arrest the incidence of *Phytophthora* bud rot, and oxytetracycline is used to protect against the lethal yellowing. Bourgoing (1991) published the recommendations to control insect pests, including the hygienic practices, like removing old coconut logs where pests harbor and breed.

Adaptation to Biotic Factors

Throughout the evolution of the coconut palm, it has been exposed to diverse insect and disease pests, and this is particularly so in the case of those populations established next to diverse rainforests on the coastline of large islands or continental landmasses. Establishment of the palms along such coasts, such as on the Australian coast, might not have been always successful. An indigenous rat species, the white-tailed rat (*Uromyces caudimaculatus*), is capable of chewing through the husk and shell to feed on the kernel. In the past millennia, probably in a period of about 60,000 years, colonizing hunter-gatherer human populations would have contributed to failure of the coconut as they assiduously collected seeds or seedlings on the beach for food. The tribal language of the Cook-town and Lockhart River regions in the far northeast coast of Australia contains separate words for a pail nut (on the beach) and a nut with an haustorium and emerged seedling (Tucker, 1983; Diana Wood, personal communication). The examples mentioned in this section add to those mentioned earlier in this chapter of adaptation in the form of tolerance or resistance to attack or invasion by insects and microorganisms. A range of types of organisms is presented below to convey the apparent broad attractiveness of coconut tissues and organs as food or habitat.

The Range of Coconut Pests

Some biota are rapidly fatal, as with the infestation by the palm weevil, *Rhynchophorus* sp., which lives and multiplies within the upper, softer part of the trunk and growing part, consuming soft tissues until the palm perishes. Similarly, the red ring nematode, *Rhadinaphelencus cocophilus*, enters the phloem of the coconut trunk, causing slow death of the palm caused by clogging the internal tissue. The Polynesisan rat, *Rattus exulans*, a native of Indochina, is widespread on many coconut islands, feeding on immature nuts that fall, once damaged. In the western Pacific, a highly specialized coconut crab, *Birgus latro*, may have evolved along with the coconut. It possesses a very powerful claw capable of tearing off the husk and smashing the shell in order to feed on the kernel. The termite also attacks the trunk of the palm consuming tissue inward from beneath the outer layer and eventually proving to be fatal to the palm. Among the microorganism pests causing fatal diseases, there is a range, starting from the heart rot caused by trypanasome (Dollet, 1999), through fungal *Phytophthora* (bud rot, Dollet, 1999), phytoplasma causing lethal yellowing and many related variants (Harrison et al., 1999), and virus (foliar decay, Randles et al., 1999) to a viroid (cadang–cadang, Rodriguez, 1999).

Insect Pests

Evidence exists that some populations have adapted and developed resistance to insect attack. The different intensity of the attack of the leaf beetle *Bronstispa* sp. was discussed earlier in this chapter. This pest is endemic to Papua New Guinea, Solomon Islands, and Vanuatu and is currently spreading to northern Australia. Any exotic genotype brought to Papua New Guinea requires concerted protection, at least during the first 2–3 years in the field after planting. Local populations, on the other hand, while needing protection in the intense environment of the nursery, are rarely attacked in the field. The Papua New Guinea Rhinoceros beetle *Scapanes australis* also shows a preference for exotic germplasms, but is still inclined to inflict serious damage on indigenous populations, especially when no choice is available. No record exists of resistance to many other insect pests, such as *Oryctes rhinoceros*, which has an almost global distribution, and various locusts and stick insects. Both species of Rhinoceros beetle often disfigure the coconut canopy without fatal results, but their damage commonly opens the door to *Rhynchophorus* sp., which infests with fatal consequences. There are many other insect pests of coconut, such as white fly, mite, scale insects, locust, stick insect, leaf miner, and hemipterous nut-fall bugs, for which no observed adaptation has been observed.

Disease Pathogens

When the coconut palm adapts to a disease, it normally takes the form of tolerance, but the practical situation is quite often hard to evaluate. The plant pathologist starts an investigation first by concentrating on the question why the palm is not doing well, often identified by foliage discoloration or inflorescence distortion (Rodriguez, 1999). By the time such symptoms are visible the palm would have been heavily infected by the pathogen, which can be identified positively using appropriate microscopy and molecular probing, and the investigation rests there. Other palms in the population might also have been infected, but it might be quite possible that the symptoms are either not showing or the palm has the capacity to withstand the pathogenic onslaught and keep its spread in check, or the infection is of a recent origin and so does not show any classical symptoms. When there is actual tolerance to the infection, some loss of vigor of the palm is observed. If such a situation is observed, it is quite possible the effect is environmental, such as the limiting effect of a specific plant nutrient. Molecular tools for the detection of sub clinical pathogens would be useful to dispel any uncertainty in cases of nonlethal or even nonexistent reaction to an invading organism. Tolerant individuals are identified, whereas previously, all survivors of disease onslaught would have been lumped together as "escapes."

Adaptation in Coconut Palm

The question of adaptation in coconut palm is only beginning to be understood scientifically. With the availability of affordable molecular tools, which are widely applied, the uncertainties surrounding disease infestation can be better determined. At present, the survival of some members of a population otherwise decimated by a phytoplasma-induced disorder or by bud rot, for example, gives rise to speculation. Do the survivors possess tolerance or resistance? Are they simply lucky or robust enough due to favorable environmental conditions that impart some kind of vigor-related resistance? The instance of phytoplasma-induced lethal yellowing in the Caribbean, Florida, and Central America illustrates some of these issues quite well. In the Red Malayan Dwarf and Panama Tall, which are widely cultivated in Jamaica and elsewhere, there is known and widely exploited tolerance/resistance to Phytoplasma. This is a quantitative trait with high heritability (Ashburner and Been, 1997). Some of the other genotypes also show tolerance, though usually less than either of the two discussed earlier (Zizumbo et al., 1999). However, there are recent reports that the hybrid has succumbed to the disease in some environments, and it is now known that there are many other phytoplasma-induced lethal diseases worldwide. The organism has been shown, through molecular probing, to have a significantly different form in each of those other locations around the globe. In Southeast Asia, phytoplasma infected palms have been found to be free of symptoms. Red Malayan Dwarf originates here, which raises the important question of whether this genotype acquired resistance to phytoplasma through natural selection during the evolutionary history of the coconut palm. It then appears that in countries where the coconut palm has grown for about 3000 years, India can be considered its heartland because of the palm's adaptation to this major scourge. The adaptation of the organism itself is part of the problem, because, as is common with so many other pathogens, the pathogen has undergone change through mutation and selection. The convergence of adapting host and adapting pathogen may be clarified in the foreseeable future, however, as molecular tools are brought to bear. Such tools would be used to identify markers associated with tolerance on the coconut side and pathogenicity on the side of the pathogen (Cardena et al., 1999). However, in order to achieve this, an understanding of the genetics of resistance and/or tolerance is required.

A Devastating Virus

The coconut cadang–cadang viroid (CCCVd) of northern Philippines and a related form in Guam is a very serious pathogen and is relatively less investigated compared to the phytoplasma scourge. This is due to its limited distribution and slow rate of spread. It has proven to be far more recalcitrant during the stage of detection and proof of its pathogenicity because of (1) the molecular minuteness of the causal organism, as it is the smallest infectious particle known to biological science; (2) the absence of any identified transmitted agent; and (3) the existence of nonpathogenic, near analogues in coconut populations worldwide (Hodgson and Randles, 1999; Rodriguez, 1999). There is but little or no evidence of tolerance to CCCVd. Hence, attention is focused on early detection of infected palms using molecular probes and elimination of these palms in the hope that viroid-free zones might be created in the affected regions.

Adapting the Coconut to Market Needs

Though coconut is a graceful symbol that beautifies the environment, human interest in the nut centers on the market needs. The amenity added to the environment by the coconut palm enhances its value in the tourist market or the real estate market. The nut is so highly valued in some places that a safe form is sought by busy tourist centers, which do not produce large and dangerous fruits that require extensive pruning. As the fruitless coconut is diametrically opposite to the fruitful adaptation sought by all the coconut farmers of the world, its development is not likely to be included soon in the program of coconut industry research institutes.

Yield Potential of the Coconut Palm

The thrust of genetic research in coconut concentrates on high yield and oil content. Related to increased yield expressed in this way is also yield of edible products and coconut water, shell, and fiber. Hence, increased yield covers most of the market interest inbringing about genetic change. Associated characters that have been sought within the broad thrust for increased yield are the following:

a. Earliness of the onset of fruiting (precocity)
b. Large nuts to facilitate processing by hand
c. Uniformity of nut dimensions to facilitate mechanization of processing
d. Higher oil content, which is often associated with enhanced flavor
e. Ease of harvest, as in the case of Dwarfs for production of nuts for drinking water
f. Special aroma or taste characters, for example, edible husk and perfumed water
g. Makapuno, having jellied endosperm for edible purposes, and so on.

The flavor and quantity of toddy or "coconut nectar" tapped from the inflorescence are also potential breeding objectives, not presently identified.

Quality Traits

Within specific markets are often specific quality traits that add value to the product. For example, variations exist in quality and length of fiber for use in the vast array of products derived from coir. There is also interest in possible variants in the mix of fatty acids in coconut oil. There has been an "anti-coconut lobby," especially the saturated oils lobby like that of soybean, which has brought about a "coconut scare," implying the adverse effect of saturated fats (cholesterol) in coconut oil. This lobby created the label "artery-clogging saturated oils" in coconut and specifically targeted coconut oil, which enjoyed a global market. This was based on some flawed research in which a deficiency of essential fatty acids in a saturated fat oil diet supposedly increased the blood cholesterol content. However, recent research underlines three great dietary strengths of coconut oil—it provides a readily digestible source of energy, critical cell wall lipids, and the precursors to monolaurin and monocarpin, which have outstanding antimicrobial and antiviral properties (Enig, 1999, 2000). About 50 percent of the fatty acid component of coconut oil is lauric acid, which has high market value in the manufacture of detergents and other industrial chemicals. Surprisingly, the huge increase during recent decades in the supply of lauric oil from palm kernels and from genetically modified canola seems to have stabilized this high demand, working to the advantage of coconut oil.

The Fatty Acid Mix

Some variability exists in the relative mix of the six main fatty acid components of coconut oil, but the range is fairly narrow for each one. If a particularly good market for one component were to be found, it might be met by increased overall production rather than seeking increased yield of that component by genetic modification.

Coconut as a Food Item

Coconut as a food item has immense possibilities in both domestic markets of the country where it grows and also in the international market. Canned or tetra-packed coconut water of different grades, including tender and intermediate stages up to matured water, and different degrees of dilution and sugar addition has received wide acceptability, including fresh water from the recently harvested fruit. In China, coconut water is labeled the "official drink of the Chinese State banquet" and is widely available in the coastal cities. Demand for coconut milk and cream also will continue to rise as its flavor and nutritional merits become more widely known outside the centers of production. Canned coconut nectar is an attractive, sweet drink increasingly consumed in Asia.

A need exists for concerted argument and tangible efforts against the marketplace opponents of coconut oil products who seek to capture the market share by dubious means, where making negative assertions about its nutritional qualities is a popular route. The attempt of the soybean oil lobby is one such attempt to derail the coconut oil from the world market, as with the "cholesterol scare" discussed earlier in this section. A lot of recent research (Enig, 2000) has shown the outstanding quality of coconut oil, even in suppressing activity of some viruses. Such additional benefits must be made known to the common man, who might otherwise vacillate about the choice of coconut oil as a cooking medium and also of coconut as a healthy food. In this connection, the experience of Kerala State in India needs particular mention. Over centuries, the favorite cooking oil of the people of Kerala was coconut oil, until the early 1980s, when the soybean oil lobby began to drum up the cholesterol scare. Gradually the Kerala people began to move away from the use of coconut oil for culinary purposes. A concerted effort by the Coconut Board of India, situated in Kerela, to dispel the inaccurate notions about the safety of coconut oil is beginning to show dividends. Former skeptics are using coconut oil again. This is a positive sign. Additional determined efforts need to be taken on a global scale to win back the lost confidence. It would be a slow process, but worth pursuing. From the point of view of breeding, all efforts are directed to the overall aim of raising the yield level; any change in quality traits is purely fortuitous.

Research and Development in Coconut Production

Research and development (R&D) in coconut production focuses mainly on farmers' interests. There are many regional institutions, which by themselves cannot advance much on R&D, but together can accomplish much. This is particularly true of the Pacific and the Caribbean. In addition, some institutions exist with a global view of coconut improvement, such as the International Plant Genetic Resources Institute (IPGRI), a United Nations agency, and its specialized offshoot, the Malaysia-based Coconut Genetic Resources Institute (COGENT). The French research agency CIRAD also takes a very broad view of the needs of global coconut research. Funding bodies have been formed in some donor countries from which financial aid is directed to support coconut research. For instance, in Europe, the Bureau for the Development of Research on Tropical Perennial Oil Crops (BUROTROP) is an international nonprofit organization established in 1989 with such an aim. Its mandate is to assist, strengthen, and further develop research on coconut and oil palm. It helps to transfer research results to the production sector, which will benefit the small-scale farmer in the form of improved self-sufficiency and capacity to produce cash crops for local or regional consumption. The BUROTROP Board of Administrators consists of seven representatives from regional organizations and producer countries in Asia, Africa, Latin America, and the Pacific, and seven others from European donor countries and international agencies it operates. The financial support comes from the Directorate General (DG) of Research of the European Commission.

Global Coordination

COGENT has a very specific mandate to facilitate the description and exchange of coconut germplasms and their use in generating populations with improved performance around the world. COGENT has linkages with a high proportion of the world's coconut-growing countries and promotes standardization of morphological descriptors for coconut as well as providing training in germplasm collection, breeding strategy, and technology. It has obtained funding to support regional collections of coconut germplasm in Brazil representing Latin America, Ivory Coast representing the African continent, Kerala State representing the Indian subcontinent, and Papua New Guinea representing the South Pacific.

National Research Centers

There are research centers that have carried out outstanding research in coconut scattered around the world, notably in India since 1916, at what is now the Central Plantation Crops Research Institute (CPCRI) in Kasaragod in Kerala State. The CPCRI has concentrated mainly on genetic improvement, focusing on drought tolerance and also the devastating root wilt, now known as Kerala coconut decline and which has been a scourge for decades, and appropriate soil management.

Research in India and Sri Lanka

In Sri Lanka, the Coconut Research Institute was established in 1929. The institute carried out outstanding work on factors controlling coconut yield and grappled with the challenges of breeding to improve a highly heterozygous population. In India and Sri Lanka, much effort was and still remains directed toward multiple cropping associated with coconut, having achieved significant progress over the decades. Coconut production was beset by many problems in many other countries early in the twentieth century—among these, the coconut decline in the Caribbean and insect attack on almost every part of the palm in diverse populations around the world are significant. But attempts to control these problems were short lived. The most recent example was in Kerala, with regard to a mite attack, which became known as the *Mandari* disease. The symptoms are shriveled nuts with black streaks so that the nuts fetched a poor price. The whole state was a victim of this scourge and no effective control was devised. In fact, there is a comparison between this and the coconut decline. Replacement of the affected palms was the only remedy. In what was then known as the Federated Malay States, higher yield was sought through replacement of Tall varieties with Dwarf ones. The generation of a Dwarf hybrid Fiji helped matters.

Research in the Second Half of Twentieth Century

As competition from other oilseed crops grew from the 1950 onwards, interest in coconut improvement began to spread across the coconut-growing countries. The Philippine Coconut Authority (PCA) was formed to oversee research and has an outstanding track record of improving coconut technology and collaboration with other agencies, especially in the development of high-yielding hybrids. For instance, France, through CIRAD (mentioned earlier in this section, formerly Institute for Research for Oils and Oil Seeds [IRHO]), set up a major facility in Ivory Coast on the African continent early in 1950s, which was later extended to Vanuatu, French Polynesia, Cambodia, and Brazil. The British did likewise in Jamaica, as did Australians in Papua New Guinea. CIRAD continues to play a major role in coconut research with high technology-aided research in Montpellier, France, and the staff are seconded on long-term programs in Ivory Coast and Vanuatu and on short-term programs to other countries where the coconut palm is grown. Unilever, a major private investor, initiated research in 1951 in

Solomon Islands, which was subsequently supported by the colonial government but was later closed down in the 1970s after releasing a very promising hybrid, which was used to replant the island's entire plantation. GTZ of Germany, an outreach funding agency, has been active in supporting research in coconut, especially in Tanzania. Other institutes that are playing significant roles are: Wye College in the United Kingdom, pioneers in tissue culture; University of Adelaide, Australia, which provides outstanding expertise in viroid and virus research; Max-Planck-Institut in Koln, Germany, which contributes significantly to molecular expertise; and the Center for Scientific Investigation at Yucatan (CICY), Mexico, which is a leader in research in phytoplasma-related coconut diseases.

A Look into the Future of Coconut

Coconut, the "tree of life," holds out much promise to humanity. This tree of the people was hijacked by corporate business to meet a desperate shortage in vegetable oil that began in the 1850s and spilled over a century to the 1950s. Other oilseeds were established closer to the foreign markets in mid-twentieth century, and demand for coconut began to decline. Quite ruthlessly, competing oil producers, in what many term an entrenched "oil lobby," took early and, more often than not, flawed research into health effects of different vegetable oils and interpreted them to the detriment of coconut. The resulting cholesterol scare is a real case in point. Without a matching promotional effort by the coconut industry to provide balance in the information reaching the consumers, some of its best global markets simply vanished, both in culinary and food sectors. What happened in Kerala State, India, the home of the coconut in the culinary sector, illustrates this point. It was a great relief for the coconut industry that the superior properties of lauric acid in both soap and detergent manufacture provided some stabilizing influence on global trade in coconut in recent years. However, lauric acid from coconut has found its competitor in the lauric acid from transgenic canola. This simply illustrates that there is an urgent need for an intelligent strategic positioning of coconut oil, not only as a healthy cooking medium, not just for the people of Kerala, where the oil has been in use since millennia, but for the world at large, and also to project the nut as an important source to manufacture a variety of value added products. Only such a strategy will take coconut to it's pre-eminent position.

As a health food, there are many properties related to coconut oil that have been documented by Enig (1900, 2000). But generating adequate resources to gain broad community recognition of the true value of coconut in the diet may require a prolonged campaign. In a comprehensive review titled "Coconut: In support of good health in the twenty first century," Enig (1999) reported some outstanding findings with very important long-term implications. The following briefly touches on this point. Monolaurin (formed in the body from lauric acid in coconut oil) is the antiviral, antibacterial, antiprotozoal monoglyceride used by the human or animal to destroy lipid-coated viruses such as HIV, herpes, cytomegalovirus, influenza, various bacteria including *Listeria monocytogenes* and *Helicobacter pylori*, and protozoa, such as *Giardia lamblia*. Many other investigations cited by Enig (1999, 2000) directly contradict the questionable interpretation of other findings from projects funded by rival vegetable oil industries. These provide the hope that coconut will be restored to its rightful place of pride as a valuable food for all. The challenge that faces the coconut industry now is to adequately educate the consumer, rather than spending funds to further research, because that will help recover the lost ground.

Protection of the Production Base

Once the future of the coconut market is assured, more attention can be diverted to issues concerning production and processing. As earlier described in this chapter, coconut management is well understood, and much progress has been made in adaptation of the crop to different environments, especially those with periodic water deficit. The use of molecular markers now offers the possibility of dealing with many disease entities of coconut with increasing confidence and scientific precision. The phytoplasma group of pathogens has been a serious scourge whose geographical span was not realized well until new molecular technology became available. Global testing is still incomplete; hence, a distribution map for phytoplasma is incomplete. Using molecular techniques, phytoplasma-affected palms can be identified before the onset of symptoms, which will allow early clearing out of infected palms, and molecular markers linked to resistance or tolerance can be used to identify desirable individuals and populations. Much remains to be done to perfect techniques and reduce costs, and the capacity of an entity such as the phytoplasma to evolve into more virulent forms cautions science to sustain its vigilance. Other diseases need to be added to the list for molecular testing, but the outlook is better on this front than it has ever been. *Phytophthora* is another widespread disease, called bud rot, where there are diverse strains that might better be dealt with using the same molecular technology, but little is being done about that at present.

There are several other serious diseases that are more localized than phytoplasma-caused ones or bud rot, where the use of molecular techniques holds out much promise. The list includes CCCVd in Philippines and its variant in Guam, the viral foliar decay of Vanuatu, and the trypanasome heart rot in the Caribbean (Dollet, 1999).

Advances in Processing Technology

There is enormous potential on the processing technology front along the copra pathway where coconut oil extraction is concerned. On a large scale, copra has yielded an oil requiring costly refining, bleaching, and deodorizing in order to be marketable. There is an exceptional opportunity to extract oil from shredded kernel dried only to 12 percent. The avoidance of high temperature and pressure provides a far more attractive aromatic oil for food and cosmetic uses (Etherington and Mahendrarajah, 1998; Etherington et al., 1999). In general, manufacturers accept food standards in the production of coconut milk, cream, and desiccated products. The Asian Pacific Coconut Community (APCC, 1997) has published codes and standards for aqueous coconut products, which will assist in standardizing the range of products and raising consumer confidence. Concerted promotion of such standards by industry leaders is required to achieve widespread adoption.

When quality is maintained, the range of coconut-based products, such as high-value coconut nectar, alcoholic and nonalcoholic drinks (derived from toddy), fiber, shell, mats, wood derivatives, food products such as coconut chips, ornamental products made out of coconut shell, and so on will have a good market domestically in the country of production and also when exported. It would be the responsibility of science to help grow and protect the palm in environments where it currently thrives. This will enable the farmers and the industry to join hands in sustaining the nut in the global market. Coconut, indeed, is a major component of the vegetation of the tropics and semitropics. This is why its future is bright.

Contact Information for Research Centers and Institutes

The following list contains contact information for research centers and institutes across the world that are useful for those engaged in coconut research and development.

Bangladesh

Senior Scientific Officer
Horticultural Research Center

Pomlogy Division
Bangladesh Agricultural Research Institute (BARI)
GPO Box 2235, Joydepur Gazipur—1701
Tel.: 880-2-9800441/9332340
Fax: 880-2-841678
Email: baridg@bttb.net.bd

Benin

Directeur
Institut National Des Researches Agricoles Du Benin (INRAB)
Station de Recherché sur le Cocotier
BP Cotonou, Benin
Tel.: 229-240101
Fax: 229-250266, 225-20 226985/21 248872

Brazil

Head of Coconut Germplasm Bank
EMBRAPA/CPATC
Av. Beira-Mar, 3250
CEP 49025-040
Aracaju-SE Brazil
Tel.: 55-79-2171300
Fax: 55-79-2319145
Email: tupi@cpatc.embrapa.br

China

Director
Wenchang Coconut Research Institute
Chinese Academy of Tropical Agriculture Science
Wenchang Erli
Wenchang City, 571321 Hainan Province
China
Tel.: 86-898-3222416/3224820
Fax: 86-898-3230949/3223918
Email: cricatas@public.hk.hi.cn

Cook Island

Ministry of Agriculture
Government of the Cook Islands
PO Box 96 Rarotonga, Cook Islands

Tel.: 682-28711
Fax: 682-21881
Email: cimoa@oyster.net.ck

Costa Rica

Director, Coconut Research
Ministerio de Agricultura y Ganaderia Siquirres
Frente, Sevicentro Siquirres—Limon, Costa Rica
Tel.: 506-718 6092
Fax: 506-718-7191/768-8410

Cote d'Ivoire

Directeur, Station de Recherché Marc Delorme
Centre National De Recherché Agronomique (CNRA)
Port Bouet, 07 PO Box 13, Abidjan 07, Cote d'Ivoire
Tel.: 225-21 248872/248067

Cuba

Deputy Director
Ministerio de La Agricultura
Instituto de Investigaciones de Citricos y Otros Frutales (IICF)
Ave. 7ma No: 3005 entre, 30 y 32, Miramar, Playa
Havana 10600, Cuba
Tel.: 537-293585/225526/246794
Fax: 537-246794/537-335217
(National Citrus Corporation)
Email: iicit@ceniai.inf.cu

Fiji

Principal Agronomist
Ministry of Agriculture, Fisheries and Forests
Private Mail Bag, Raiwaqa, Suva, Fiji
Tel.: 679-477044 ext.263
Fax: 679-400262
Email: krsinfo@is.com.fj

Ghana

Director, Oil Palm Research Institute
PO Box 74, Kade, Ghana

Tel.: 031-804-710229 (Director)/710226/710228
Fax: 031-233-46357
Email: csur@ghana.com

Guyana

Program Leader (Roots and Tubers)
National Agriculture Research Institute (NARI)
Mon Repos, East Coast Demerara, Guyana
Tel.: 592-202841 /42/43
Fax: 592-204481
Email: nari@guyana.net.g

Haiti

Coordinator, Ministry of Agriculture
Centre de Recherché et de Documentation Agricoles (CRDA)
Damien, Republique d'Haiti
Tel.: 509 (22) 4503
Fax: 509 (45) 4034

India

Director, Central Plantation Crops Research Institute (CPCRI)
Indian Council of Agricultural Research
Kasaragod 671 124, Kerala State, India
Tel.: 91-4994-430333
Fax: 91-4994-430322
Email: cpcri@x400.nicgw.nic.in

Indonesia

Director, Department of Forestry and Estate Crops
Agency for Forestry and Estate Crops Research and Development
Research Institute for Coconut and Palmae
PO Box 1004, Manado 95001, Indonesia
Tel.: 62-431-812430
Fax: 62-431-812587
Email: balitka@mdo.mega.net.id

Jamaica

Director of Research, Coconut Industry Board
18 Waterloo Road, PO Box 204
Kingston 10, Jamaica

Tel.: 1-876-9261770
Fax: 1-876-9681360
Email: conchar@cwjamaica.com/cocindbrd@cwjamaica.com

Kenya

Regional Research Centre (RRC)
Mtwapa Box 16 Mtwapa
Kenya
Tel.: 254-11 485842/39
Fax: 254-11 486207

Kiribati

Chief Agriculture Officer, Division of Agriculture
Ministry of Natural Resources and Development
PO Box 267
Bikenibeu, Tarawa, Kiribati
Tel.: 686-28-139/108
Fax: 686-28-139/21-120

Malaysia

Station Manager
Stesen MARDI Kemaman, Batu 11
Jalan Air Putih PO Box 44, 24007
Kemaman Terengganu, Malaysia
Tel.: 60 09-8646361/148
Fax: 60 09-8646361
Email: abo@mardi.my

Marshall Island

Chief of Agriculture
Ministry of Resources and Development
Agriculture Division
PO Box 1727, Majuro, Marshall Islands
Tel.: 692-625-3206/0740
Fax: 692-625-3005
Email: agridiv@ntamar.com

Mexico

Researcher
Centro de Investigacion Cientifica de Yucatan, A.C. (CICY)

KM7 Antigua Carretera A Progresso, Ex-Hacienda
Xcumpich, Apartado Postal 87, 97310 Cordemex
Merida, Yucatan, Mexico
Tel.: 52-99-813923/813966
Fax: 52-99-813900/813941
Email: cos@cicy.mx

Mozambique

Research Assistant—Entomology
National Agriculture Research Institute (INIA)
Entomologist Crop Protection Section
Box 3658, AV. Das FPLM, INIA-MAPUTO, Mozambique
Tel.: 258(1)460097
Fax: 258(1)460074
Email: sancho@zebra.uem.mz

Myanmar

Acting Director General
Department of Agriculture
Planning Ministry of Agriculture
Thiri Mingalar Lane, Off Kaba Aye Pagoda Road
Yankin PO, Yangon, Myanmar
Tel.: 095-1-665750
Fax: 095-1-663984/651184
Email: dap.moai@mpt.maid.net.mm

Nigeria

Chief Research Officer
Nigerian Institute for Oil Palm Research (NIFOR)
PMB 1030 Benin City, Nigeria
Tel.: 52-440130
Fax: 52-248549

Pakistan

Director Horticulture
Pakistan Agricultural Research Council
Plot No 20, G-5/1, Post Box 1031
45500 Islamabad, Pakistan
Tel.: 92-51-920-7402
Fax: 92-51-920-2968/920-240908
Email: hashmi@reshem.sdnpk.undp.org/mtahir@isb.comsats.net.pk

Papua New Guinea

Acting Director, PNG Cocoa and Coconut Research Institute
PO Box 1846
Rabaul, East New Britain Province
Rabaul, East New Britain, PNG
Tel.: 675 983-9108/983-9131/983-9185
Fax: 675 983-9115
Email: ccri@datec.com.pg

Philippines

Deputy Administrator
Agricultural Research and Development Branch
Philippine Coconut Authority, Don Mariano
Marcos Avenue, Diliman, Quezon City, Philippines
Tel.: 632-920-0415/632-426-1398
Fax: 632-920-0415
Email: cbcarpio@mozcom.com

Samoa

Acting Assistant Director of Research
Ministry of Agriculture, Forests, Fisheries and Meteorology
Research Division
PO Box 1587, Apia, Samoa
Tel.: 685 23416/20605
Fax: 685 23426/20607/23996
Email: apeters@lesamoa.net

Seychelles

Director General
Crop Development and Promotion Division
Ministry of Agriculture and Marine Resources
Grand Anse, PO Box 166, Victoria Mahe, Seychelles
Tel.: 248-378252/378312
Fax: 248-225425
Email: antmoust@seychelles.net

Solomon Islands

Director of Research for Permanent Secretary
Dodo Creek Research Station
Ministry of Agriculture and Fisheries
PO Box G 13, Honiara, Solomon Islands
Tel.: 677-31111/31191/31037

Fax: 677-31039/21955/31037
Email: ibsram@welkam.solomon.com.sb

Sri Lanka

Director, Coconut Research Institute
Bandirippuwa Estate, Lunuwila, Sri Lanka
Tel.: 94-31-57391/253795/55300
Fax: 94-31-57391
Email: rescri@sri.lanka.net/dircri@sri.lanka.net

Tanzania

Director, Mikocheni Agricultural Research Institute (MARI)
Ministry of Agriculture and Co-operatives
PO Box 6226
Dar es Salaam, Tanzania
Tel.: +255 51-700552 or 74606
Fax: +255 51-75549 or 116504
Mobile: +255-812-784031
Email: arim@africaonline.co.tz

Thailand

Director, Horticulture Research Institute
Department of Agriculture
Chatuchak, Bangkok 10900 Thailand
Tel.: +66(2) 579-0583/579-0508/561-4666
Fax: 662-561-4667
Email: hort@doa.go.th

Tonga

Head of Research
Ministry of Agriculture and Forestry
Vainani Research Division, Nuku'alofa
Kingdom of Tonga
Tel.: 676-23038
Fax: 676-32132/24271/23093

Trinidad and Tobago

Ministry of Agriculture
Central Experimentation Centend V 117
Arima PO 52 La Florissante Garden
Dabadie, Trinidad and Tabago

Tel.: 1-809-642-8552/642-0718
Fax: 1-809-622-4246

Tuvalu

Acting Director of Agriculture
Ministry of National Resources and Environment
Department of Agriculture, Private Mail Bag
Vaiaku, Funafuti, Atoll, Tuvalu
Tel.: 688 20-825 or 186
Fax: 688 20-826

Vanuatu

Head of Coconut Division
Vanuatu Agricultural Research Centre
PO Box 231, Espiritu Santo, Vanuatu
Tel.: 678-36320/36130
Fax: 678-36355
Email: labouiss@pop.vanuatu.com.vu/CARFV@vanuatu.com.vu

Vietnam

Coconut Scientist, Oil Plant Institute of Vietnam (OPI)
171-175 Ham Nghi St.
District 1, Ho Chi Minh City, Vietnam
Tel.: 848-8297336/8243526
Fax: 848-8243528
Email: opi.vn@hcm.vnn.vn

4 Cinchona (*Cinchona* sp.)

Origin and Use of Cinchona

The original cure for malaria, quinine and its related alkaloids, is obtained from the bark of cinchona. Quinine is the oldest naturally occurring phytochemical known in medicine. Dr Jaime Jaramillo Arango, a former Rector of the National Faculty of Medicine at Bogota, has established decisively in his book, *The Conquest of Malaria* (1950), that natives in Peru knew of the curative properties of this bark, then called "*quina-quina*," against periodic fevers long before the Spaniards arrived in the Andes. It is believed that around the early 16th century, Father Burtolome Tafur brought some bark with him to a religious conclave in Rome, considered then most susceptible to malaria. It proved effective in the treatment of malarial fever and gained popularity. Another Jesuit priest stationed in the Andes then began to supply the bark regularly to the Holy City and distributed it as "Jesuit bark" or "fever bark" to treat malarial fever. The bark gained much popularity in the next 200 years until two French chemists isolated and purified quinine alkaloid as the most potent compound to fight malaria. This period, therefore, witnessed vast natural populations of cinchona trees continuously uprooted and exploited in the Andes and tons of the bark were shipped to different parts of the world. It brought the high rainfall-receiving mountain ranges of the Andes, mainly in Peru, Bolivia, Colombia, and Ecuador, into the focus of the explorers, herbalists, and botanists, as it was known to hold rich resources, not only of many taxa of cinchona, but also other valuable sources of herbal products and pharmaceuticals. In fact, ipecac root was another source of new of a new pharmaceutical discovered from the region for the treatment of hill diarrhea.

Malaria has been the single-most dreadful scourge for human sufferings, killing a large part of human population annually; thus, the likely scarcity of the source (antimalarial drugs) along with periodic restrictions on its collections and export imposed by local governments brought large investments for raising plantations of cinchona in far-off places during the latter part of the 18th century. This led to the establishment of a lucrative plantation industry for pharmaceutical purposes in Indonesia and India; later, manufacturing capacity increased in Western Europe.

Quinine is the most important alkaloid of the bark of cinchona. In the form of its salts, such as sulfate, bisulfate, hydrochloride, and dihydrochloride, it is used for the prevention and treatment of malaria. In cases of malignant malaria, in which case the disease was later found to be resistant to synthetic antimalarials, quinine

The Agronomy and Economy of Important Tree Crops of the Developing World. DOI: 10.1016/B978-0-12-384677-8.00004-7
Copyright © 2010 by Elsevier Inc. All rights of reproduction in any form reserved.

hydrochloride was given as an intravenous injection slowly for rapid cure. In combination with other compounds, quinine hydrochloride is also used as a sclerosing agent in the treatment of varicose veins and internal hemorrhoids. Quinine sulfate has long been used in preparations made in Europe for the treatment of night cramps, but has largely been given up now due to occasional complications causing thrombocytopenia (Pin, 1998). Quinine water (100 ppm) is used as a gargle to cure sore throats and foul smells.

The bark of cinchona has astringent and bitter tonic properties. Thus, quinine hydrochloride and quinine sulfate have been extensively used as bitter substances in soft drinks, alcoholic bitters, and liquors. In France, ground cinchona bark was used chiefly as a raw material in the manufacture of aperitifs and restorative liquors. Quinine has also been used as an abortifacient (Morton, 1977). Quinine is bacteriostatic and a highly active *in vitro* compound against protozoa. It also inhibits the fermentation of yeast. Quinidine is another chemical from the bark of cinchona, which has been found to be very useful as a cardiac depressant, and its sulfate form is used in the treatment of cardiac arrhythmias. Quinine can be catalytically converted into quinidine, and it was thus used as a major source of quinidine production.

Evolutionary History

The widely circulated legendary history of the discovery of cinchona bark by European settlers is attributed to the first Countess of Cinchon, wife of the Viceroy of Peru, who fell sick in 1638. When the Governor of Loja, Ecuador, heard of it, he sent a packet of "quina-quina" bark that cured her completely. On her return to Spain, the Countess is said to have liberally distributed the cure, and for this reason the bark powder during early times was also called "countessa powder." Haggis (1941), through painstaking investigations, discovered that the first Countess of Cinchon died in Spain 3 years before her husband was appointed as Viceroy of Peru. The second Countess remained quite healthy until she died in Colombia, without ever returning to Spain. Keeble (1997) further discredited this legendary theory. He stated that the first authentic use of cinchona bark against malaria in Europe was made around the year 1630; it arrived in trade in Europe around 1643, and reached England in 1655 (Holmes, 1930). It entered the pharmacopoeia in 1677 and consequently received official recognition. Yet, the name of the Countess of Cinchon lingers on because in 1747 the famous botanist Carolus Linnaeus named the fever bark tree as *Cinchona officinalis*.

There are now 23 distinct and validated taxa that belong to the genus *Cinchona*, ranging from large shrubs to small trees that are distributed from an elevation of 800–3000 m above mean sea level (MSL) all over the slopes of the Andes rain forests. Chemical studies of these taxa, including their innumerable natural hybrids, have shown that only a few species, such as *Cinchona calisaya*, *Cinchona ledgeriana*, *Cinchona officinalis*, and *Cinchona pubescens* (*Cinchona succirubra*), are rich in quinine and cinchonine alkaloids. The range of quinine in bark (of twig, trunk, and root) is 2–8 percent. Prior to the Second World War, almost 90 percent of the world supply originated in Java in Indonesia, which produced and exported cinchona bark and its alkaloids for a value ranging from 16 to 22 million UK sterling pounds annually. India was then the second-most important source with an annual production of 25 tons of quinine salts, though its exports were relatively meager; the average Indian export of quinine salts was worth about Rs 10 million (US$1 = Rs 46, approximated at current exchange rates) annually then. The Belgians, meanwhile, developed huge plantations in Zaire, now known as the Democratic Republic of the Congo, in Africa, which surpassed Indian production. Being richer in quinine, Zaire established its place next to Indonesia in trade and reaped huge profits during the Second World War, when supplies of quinine were short. Bernardino Gomes of Portugal produced a crude crystalline mixture of alkaloids from the bark, and later the French chemists P.J. Pettetier and J.V. Caventon isolated and purified quinine crystals from the bark in 1820. Quinidine was discovered in 1833 and cinchonidine in 1847.

Further Voyage of Cinchona to the East

No sooner had cinchona bark been established as an effective source of quinine alkaloids, useful in malaria control, then the native governments in Peru, Bolivia, Columbia, and Ecuador placed embargos on the export of cinchona seed, seedlings, and other plant materials. However, the Dutch were successful in sending plants to Justus Hasskarl, Superintendent of the Botanical Garden of Java. It is said that a man named Miller, a Dutch tourist (according to his passport), went sightseeing through cinchona territory during 1852, quietly collecting seeds as souvenirs. When he could not freely move into the provinces, where the plant grew abundantly, he approached the governor of the province with certain proposals, which were rejected. Undaunted, he pursued his objective through shady characters in the administration and succeeded in exchanging hundreds of seedlings and a bagful of live seeds for a fabulous sum of money. Returning to Java, Miller (Hasskarl, in all probability) was honored with a knighthood. Interestingly, when a German newspaper uncovered the unscrupulous event, the Dutch government quickly denied it. Burkill (1935) stated that the first individual plant to reach Europe, however, was *Cinchona calisaya* from seed collection during James Weddell's first journey to Bolivia in 1846. It was raised in Paris and handed over to the botanist Hugo De Vries in 1851. It was then planted in Buitenzoirg, just 280 m above MSL, and died there, but not until a cutting was taken from it. This cutting thrived in the mountain garden of Tjibodas, Indonesia, at an MSL of 1300 m, where it was maintained. More plants from Holland followed, raised from part of the seeds deposited from Hasskarl's collection there. By 1860, a million small healthy cinchona trees were reported growing in Java.

An interesting instance concerning an English cinchona bark trader, Charles Ledger, living near Lake Titicaca, was recorded by Saha (1972). Ledger managed to collect the seed of a particular tree (possibly yielding potent bark), belonging to *Cinchona ledgeriana* from the highlands of Coroico, Bolivia. He sent the seed to his brother in London. The British government refused to buy the seed, but two Dutch Consular officers bought them and sent them to Java. In Java, where cinchona cultivation was already established, these new seeds produced 2000 plants. It proved to be a richer source of quinine and laid the foundation of the Dutch monopoly on cinchona trade, which The Netherlands has retained ever since. Burkill (1935) inferred that the best species of cinchona thus reached Europe by accident, for Ledger's interest in the subject was secondary; he did not expect that the seeds that he sold were to those who were very keen to invest in raising cinchona commercial plantations. By 1918, the production of cinchona bark based on the choice taxa of *Cinchona ledgeriana* in Java had reached its zenith. The bark was brought to the Amsterdam factory for extraction and production of quinine salts. This enabled the Dutch to reap huge profits annually. It displaced the quinine market from countries such as Bolivia, Colombia, Ecuador, and Peru. These countries maintained varying genetic material in natural reserves.

Introduction of Cinchona to India and Sri Lanka

In 1856, a British botanist Johan Forbes Royle suggested to the then East India Company that cinchona should be introduced into India without any further delay. The company assigned an expedition to the Andes and chose Clements R. Markham for the job. He organized an expedition on a bigger and more systematic scale (Watts, 1893). He assigned the work to three parties, one led by himself to get yellow bark from the Bolivian forests and Southern Peru. He traveled to the interiors, making tedious journeys and hiring local hands. He collected a few hundred seedlings of the yellow and red bark tree (*Cinchona pubescens*), which were brought to India in October 1859. The second party was led by Spruce and Cross to search for red bark trees and allies from eastern slopes of Ecuador. The third party under the leadership of Woolcock Pritchett was to look for gray bark in central Peru. All were to send the collections to the Royal Botanic Gardens in Kew in the United Kingdom. Markham entrusted his collections to W.G. McIvor, the Superintendent of the Botanical Garden in Ooty (Tamil Nadu, India). Swamy (1953) noted from records that all the plants brought by Markham, when transferred to fields in Dodabetta in the Nilgiri Hills of Tamil Nadu, perished, possibly due to the long sea voyage through the Panama Canal. However, seeds gave encouraging results. Markham's attempt to raise seedlings in the cool climate of Nilgiri Hills succeeded. Spruce and Cross collected plants and seed of *Cinchona pubescens*, which reached India safely. Soon after, McIvor received a fresh consignment of seeds from Pritchett, surprisingly by mail, which arrived safely and in a viable condition. The seedlings established well in the field. The credit for establishing commercial cinchona plantations in India should go to McIvor. He sent a small consignment of seeds to Thomas Anderson, Superintendent of the Royal Botanical Garden in Calcutta, to establish a cinchona plantation in Sikkim. He also sent a small consignment of seeds to the Royal Botanic Garden in Peradeniya in Sri Lanka for trials at Hakgala Botanical Garden. This was the first commercial cinchona plantation in Sri Lanka under the British rule. By 1881, 138 tons of bark were exported to England and the quantity rose to 690 tons by 1885 (Watts, 1893). Anderson raised seedlings from McIvor's seeds near Sikkim in 1861, but, his attempts to establish these in Darjeeling Hills in 1863 failed due to heavy mist over the hills. The trial was repeated next year in Mungpoo, where it succeeded, and so Mungpoo was chosen as the site for commercial cultivation of cinchona. Subsequently, Anderson was sent to Java to procure plants and familiarize with cinchona cultivation, which the Dutch government generously supported. He brought 50 plants of *Cinchona calisaya*, 284 of *Cinchona pahudiana*, and only 4 of *Cinchona lancifolia* and handed them to McIvor and brought 193 plants of *Cinchona pubescens* from his garden to cultivate them in Bengal (Nandi, 1993). Following Anderson's death, the charge was taken over by C.B. Clarke in 1870 and then by George King, both distinguished botanists of their time, who contributed enormously to the establishment and expansion of cinchona plantations in Bengal. By 1871, cinchona plantation's viability was established in the high rainfall area of Bengal. Cinchona needs fertile soils and shade, both of which were abundantly available in the terrain. In 1890 a second plantation was established in Munsung near Sikkim, a third in Rango near Indo Bhutan border in 1938, and a fourth one in Latpanchor in 1943, adjacent to Mungpoo. By 1954, the total area under cinchona was around 3000ha, this being reduced to 2100ha in 1955 due to a slump in the cinchona market. It reduced further to 1820ha by 1970 (Nandi, 1993; Saha, 1972).

In Tamil Nadu, the acreage under cinchona increased from 230ha in 1866 to 860ha by 1931 and by 1970 it was 2316ha (Swamy, 1953). Most (80 percent) of the trees were *Cinchona ledgeriana* (Rao, 1974). In 1867 a cinchona plantation was started at Nunklow in Assam on the eastern slopes of Khasi hills. Cinchona was also tried in northwest Indian hills, but the trees could not stand the frost and low winter temperatures. Due to the dip in the market, attempts to encourage planters to take to cinchona cultivation in South India did not succeed. The cinchona plantations in India were promoted entirely as a state enterprise, as part of a public measure, despite periodic ebbs in demand and price of end products in the world market.

The Other Major Cinchona Producing Countries

In 1927 cinchona was introduced into Philippines, covering an area of 1874ha on Mount Kitanglad, Bukidnon. A census of the plantations during late 1980s revealed a holding of 2.5 million trees, composed of seven species. Burundi on the African continent also had cinchona plantations. Southern China and Formosa (Taiwan) had the cultivation to meet local demand. Belgians raised the plantations in Zaire and the United States in Guatemala. The Japanese occupation of Java disrupted supplies and caused scarcity. This led to the establishment of the plantations in several African countries such as Tanzania, Kenya, Cameroon, and Rwanda. It also captured the interest of South America in the plant. Among these ventures, that of Guatemala was the most successful because of modern technology in the manufacture of quinine and production practices. It was a joint venture between the U.S. government, Guatemalan coffee planters, and a pharmaceutical company (Merck). They mass-produced vegetative propagules from selected clones and used improved cultural practices for high bark yield and superior quinine content to compete in world market. The plantation covered 400ha with around 1.75 million trees by mid-1948 and continued production despite excessive cyclic nature of the market (Popenoe, 1949).

The Characteristics and Chemistry of Cinchona Plantation Species

Evergreen shrubs and small trees growing up to 20 m and having a diameter of 15–30 cm, mainly found in the eastern and northern slopes of the Andes from 19° south to 10° north, extending on the other side of the equator make the Cinchona species. Cinchona species grow in the pre-mountain tract at 800 m and rise to 3000 m above MSL in Bolivia, Colombia, Ecuador, and Peru. *Cinchona pubescens* extends further into northern Venezuela. Taxonomists in the past have placed various numbers of species (30, 40–65) in this genus, possibly due to the existence of interspecific forms in the natural population. However, after a thorough study of the Andean Rubiaceae, Anderson (1998) recognized only 23 valid species. Anderson (1995b) based his inference on cladistic analysis of structural data composed of 48 characters. He transferred a few species to allied genera, created a new genus under the name *Cinchonopsis*, and created a new taxa, *Cinchona amazonica*, making the genus monophyletic in origin. He came to believe that although generic endemism is low in the Andes, cinchona falls among five subendemic genera because of large species endemism (recorded around 59 percent in the tropical Andes). The ancestors probably occupied low forests of the Andes, and later evolved species may have arisen through adaptation to mountain conditions (Anderson, 1995a). These plants bear opposite elliptical to ovate lanceolate leaves with the entire margin. The flowers are small, pink, cream, or brown in color, with small lobed calyx united at the base. The corolla is tubular, made up of five spreading lobes. The flowers emit a delightful fragrance. The fruit is an oblong capsule, containing 40–50 winged seeds. The Andes has given some of the most famous industrially important plants of the tropical rainforests of South America. The basic chromosome number is (x) 17 in all cultivated species, including the four species grown in India (Mathews and Philips, 1979). Several species are polyploids ($2n = 68$). Some hybrids, such as "*Hybrida*" and "*Robusta*," have proved valuable in the plantations because they are hardier and invariably produce higher quinine content than either of the parents. Although a large number of cinchona species have been tested for their yielding ability in the plantation sector, commercial supplies of bark are obtained from the four species listed earlier. Of these, *Cinchona ledgeriana* and *Cinchona officinalis* are species of choice for the plantation industry because of their rich quinine content.

Taxonomy: Diversity Analysis of Source Species

Cinchona calisaya Wedd (Yellow Bark)

This is a large bushy tree with a straight trunk, occurring in the lower reaches of Bolivia and southeastern Peru. It prefers lower elevations (400–1000 m above MSL). It has thick leaves, which are oblong to lanceolate, with smooth surfaces. Inflorescence is a large panicle of numerous pale pink flowers. Capsules are 8–17 mm long and oblong in shape. The bark is thick (2–5 mm), has a grayish outer layer with a broad longitudinal, and a few transverse fissures peeling off in places. The trunk bark contains 3.89–7.24 percent of the total alkaloids, of which quinine content varies from 0.78 to 5.57 percent (Anonymous, 1993). The species show variable quinine content under plantation. Initially it was introduced commercially in Sikkim in northeastern India and Moyar Valley of Nilgiris in Tamil Nadu, but was later replaced by the more consistent quinine yielding *Cinchona ledgeriana.*

Cinchona ledgeriana Moons (Yellow Bark)

It is a weak, highly branched, fast-growing tree, attaining a height of 6–16m at maturity. The species thrive between 1000 and 1900m elevation above MSL in the hills of Darjeeling. Leaves are light green, elliptical, acuminate, with a small curve in the axil. Flowers are small, pale yellow, and appear from May to October. The capsule is oval-lanceolate and 15–19mm long. The bark is similar to *Cinchona calisaya*; it is thick (2–5mm), but cracks are more numerous and less deep. The average total alkaloid content of the root and trunk (10–12 years of age) is 7.21 and 6.01 percent, of which quinine is 5.4 and 1.98 percent, respectively (CSIR, 1992). It is the richest source of quinine, and individual trees containing as much as 14 percent quinine have been located in plantations (Chatterjee, 1982). The species is very exacting in its climatic requirement, and is a parent stock of major plantations in India. Under less favorable conditions, a hybrid, *Cinchona ledgeriana* × *Cinchona pubescens* called "*Hybrida*," is preferred. The hybrid has lower quinine content but produces robust growth (Gupta, 1980). Bark yields of 6–6.25 tons/ha are obtained, which contain 2–2.5 percent quinine (Chatterjee, 1982).

Cinchona officinalis Linn. (Pale or Crown Bark)

A weak, struggling, slender, slow-growing tree, 6–10m tall with dark green leaves, it prefers a cooler climate at 1200–2000m elevation, above MSL in the Nilgiri hills. Leaves are small, smooth, ovate-lanceolate, 4–10cm in dimension with reddish petioles. Flowers are deep pink to rosy in color, 1.4–1.6cm long, borne in small terminal panicles. The capsule is ovoid-oblong, 1.5–2.0cm long. Bark is rough and brown, which is yellow within, around 1.5mm thick; the outer surface has numerous transverse cracks with recurved edges. On average, the trunk bark yields 4–6 percent total alkaloids, up to half of which is quinine; however, the average in India is low, around 1.5–2.0 percent (Chatterjee et al., 1988). Thus, a hybrid between *Cinchona officinalis* × *Cinchona pubescens* called "*Robusta*" is preferred in plantations. This is hardy tree, adapted to a wide range of elevation (1200–3000m above MSL) and temperature regimes (Gupta, 1980). Its average quinine content is 2 percent in the trunk bark (Chatterjee et al., 1988). Another hybrid, *Cinchona officinalis* × *Cinchona ledgeriana,* is also grown in Mungpoo (Directorate of Cinchona and Other Medicinal Plants, 1994).

Cinchona pubescens Vahl. (syn. **Cinchona succirubra** *Pav. ex klotzsch) (Red Bark)*

This is a hardy tree, native to Peru and Ecuador, with a straight trunk, vigorous in growth. Grows to about 18–20m high and does best at elevations between 1200 and 2000m above MSL in the cooler climate of Annamalai's hills in Tamil Nadu.

The species possesses the remarkable ability to withstand both high humidity and drought conditions. Leaves are very large (40–50 cm × 30–40 cm) in dimension, thin, light green in color, and elliptical in shape. Flowers are rosy pink in color and are produced around the year, 1–2 cm long, and the upper surface of the corolla is white with pink stripes. The capsule is oblong, 2–3 cm long. Bark is dull brown, 2–6 mm thick, with well-marked longitudinal wrinkles and a few transverse cracks. It is rich in cinchonine, but poor in quinine content (Chatterjee, 1993; Nandi, 1993). The bark of the root, trunk, and twigs has 7.21, 6.09, and 4.0 percent total alkaloids, respectively, of which the quinine content is 0.76–1.42, 1.1–1.74, and 0.8–1.16 percent, respectively (CSIR, 1992). A new alkaloid, quinamine, is found in the bark. The species is used as a root stock for the grafting of *Cinchona ledgeriana* in Indonesia (Gupta, 1980) because of its hardy and vigorously growing nature at lower elevations in Indonesia (CSIR, 1992).

Chemistry of Cinchona

Cinchona bark contains 6–10 percent of the total quinoline alkaloids. The main constituents of the alkaloids are quinine, quinidine, cinchonine, and cinchonidine, in addition to 30 other minor bases related to quinine. These alkaloids contain quinoline and quinuclidine rings with a vinyl group attached to it. In addition to alkaloids, the bark has coloring matter (up to 10 percent), flavonoids, an essential oil, and polyphenols. The alkaloids exist chiefly as salts of quinine and cinchotanic acids, and their relative concentrations vary in different species. These alkaloids form during the descent of the sap; therefore, their percentage content is lowest in the twigs, high in trunk (bark), and maximum in the root (bark). Of these, the collar portion (30–45 cm in length, near the base) is the richest in quinine content. The alkaloid content in trees increases with age (1–12 years), depending upon species. Among these plantation species, quinine is the major alkaloid of *Cinchona ledgeriana*, *Cinchona calisaya,* and *Cinchona officinalis. Cinchona pubescens* has a larger proportion of cinchonine (3.3–3.4 percent) out of higher (6.0–8.2 percent) total alkaloids. The cinchonidine content is the highest (3.7–4.9 percent) in the bark of *Cinchona kartamanahs* at 12-plus years of age and is also high (2.1–2.2 percent) in the bark of "*Hybrida*" at around 6–8 years of age (Dayrit, 1994). The bark has high (3–4 percent) ash content and its quantitative estimation can identify the source species. Doraiswami and Venkatraman (1982) reported maximum ash content in crown bark grown at 2000–2500 m above MSL, lesser in red bark grown at 1500–2000 m, and least in yellow bark grown at 1000–1500 m elevation in India.

Both quinine (melting point 174.4–175°C) and quinidine (melting point 173.5°C) contain a methoxy group and are stereoisomers. Thus, quinidine is a dextrorotatory isomer of quinine. They show blue fluorescence with oxygenated acids such as sulfuric acid in filtered ultraviolet light. Cinchonine (melting point 264°C) and cinchonidine (melting point 204.5°C) do not possess a methoxy group and therefore do not show fluorescence. Quinine is isolated from the total alkaloids as quinine sulfate. It is a white, crystalline, odorless compound that is highly bitter in taste. It is sparingly soluble in water, but highly soluble in organic solvents. Besides quinine salts, cinchona febrifuge (powder), which contains a mixture of leftover alkaloids after the extraction of quinine, is marketed. Similarly, a standard mixture of cinchonine, cinchonidine, and quinine is sold as "Totaquinine," and is also prescribed for malarial treatment. Quinidine is present in small quantities (0.2 percent) in most of cinchona barks, but is relatively higher in *Cinchona calisaya* and is maximum in *Cinchona tayansis*. It is produced commercially by the chemical conversion of quinine through oxidation, and its percentage of commercial conversion depends upon factors such as the oxidation agent, catalysts, and the reducing agents employed. A rapid, accurate, and inexpensive quantitative, high-performance thin-layer chromatography (HPTLC)/densitometry method is available for the simultaneous determination of four alkaloids in the bark (Dayrit, 1994). Leaves contain up to 1 percent of total alkaloids, with younger ones containing more. Analysis of *Cinchona ledgeriana* leaves revealed five monomeric indole alkaloids (including quinomine, aricine, and 3-epi quinine) and seven quasidimeric indole alkaloids. Analysis of *Cinchona ledgeriana* leaves from trees grown in Kenya and Zaire, of *Cinchona pubescens* grown in Thailand, and that of "*Hybrida*" has shown that the alkaloid content was very low (cf bark) and that no quinoline group of alkaloids was present to merit commercial extraction (Keene et al., 1983).

Toxicology of Cinchona

Both ground cinchona and quinine cause urticaria in rare cases, contact dermatitis, and other hypersensitivity reactions in humans. Ingestion of these alkaloids can result in a chemical disorder known as "cinchonism," characterized by severe headache, abdominal pain, convulsions, visual disturbances, blindness, and auditory disturbances such as ringing in the ears, paralysis, and even collapse (Leung, 1980). Quinidine and related alkaloids are reported to be absorbed from the gastrointestinal tract, and a single 2 8 g oral dose of quinidine may be fatal to an adult (Anonymous, 1993). The use of quinine is discouraged during pregnancy due to fetal and abortifacient effects.

Extraction, Production, and Trade

It was in 1871 that commercial extraction of quinine began in India, after the first quinine factories were established at Naduvattam in Annamalai district, in Tamil Nadu, and Mungpoo in Darjeeling, West Bengal. These factories produced quinine sulfate in 1887 for the first time; later they produced quinine hydrochloride and quinine ethyl carbonate, which is tasteless quinine. Their capacities were enlarged in 1905 to 27 tons (Mungpoo factory) and 20 tons in Naduvattam factory. The latter was closed due to fire, and a new factory was established in 1965 with enlarged capacity. However, these factories never ran at full capacity, owing to a shortage of bark supply.

Extraction

The century-old "Open Type" extraction plants have long been used by Indian units, using mineral oil as solvent. For this purpose, moistened bark is ground to a fine powder (60 mesh) in disintegrators and mixed with slaked lime containing more than 60 percent calcium hydrochloride. Mixing is done thoroughly in mechanical mixers and kept as such for 24 hours. It is charged into extractors, where sodium hydrochloride solution is added and stirred well. This mixture is kept in extractors overnight and extracted with a mineral solvent at 88°C using live steam. The hot extracted oil, containing the alkaloids, is shaken thoroughly with diluted sulfuric acid in lead-lined tanks for transferring the alkaloids present in the oil into the acid. The oil from the mother liquor is separated and recycled, whereas the sulfuric acid layer containing the alkaloids is boiled and filtered. The hot and clear acidic liquor is neutralized with a hot sodium hydroxide solution at 5.5 pH and cooled. On cooling, crude quinine sulfate (gray in color) crystallizes out and is centrifuged. It is refined by boiling with water and activated carbon to obtain pure quinine sulfate (Doraiswami and Venkatraman, 1982). The remaining alkaloids are precipitated by making the mother liquor alkaline. Individual alkaloids, such as quinidine, are crystallized through selective solvent extraction. The production units in India were revamped in the mid-1970s by introducing mechanization and replacing mineral oil with a close circuit extraction technique that uses Toluene. This has improved efficiency and recovery of alkaloids. In European countries, further advancement has occurred and they now use counter-current containers for efficiency and speed.

Manufacturing Capacities

European countries produce the bulk of world's supply of quinine salts and have a global marketing network. Boehringer Mannhein Plant in Germany is controlled by multinational companies and has 125 tons capacity per annum. The Amsterdam Chemi Farma of The Netherlands used to control the Java plantation and has a 100-ton capacity. After nationalization, the company collaborated with the government of Indonesia to produce the end products. It has attempted to regain its preeminent position. The Lake and Cruickshank of England, Societe Chimie Pointel Girard of France, and Bucher and Company of Germany have capacities of 35, 30, and 50 tons, respectively, per annum and depend mainly upon purchases of bark from African and South American countries. India and Guatemala have their own extraction plants. In addition, Bolivia, Ecuador, and Colombia have erected smaller extraction plants with annual capacities of 2, 5, and 1.5 tons, respectively (International Trade Centre, 1982; Saha, 1972).

Global Trade in Cinchona

Globally, production of cinchona bark fluctuates between 5000 and 10,000 tons annually. This estimate is based on a consumption of 300–500 tons of cinchona alkaloids annually. Production of quinine in 1981 by the major manufacturing units was estimated at 265 tons (International Trade Centre, 1982).

Quality Standards

Cinchona bark is not graded, as there is no grading system, although classification is made into industrial bark, which is in small pieces, and "druggist quills," which are up to 30 cm long, 1–8 cm wide, and 2–6 mm thick. Industrial bark is used for alkaloid extraction and represents the bulk of demand for cinchona, whereas "druggist quills" are used in gelatin preparation (International Trade Centre, 1982). In trade, commercial specifications include the country of origin, color of the inner side of the bark, moisture, total alkaloid, and quinine, quinidine, and quinine sulfate contents. Of these, quinine content is given as the percentage content of quinine as anhydrous alkaloid (QAA) and quinine sulfate content as SQ-7 (quinine sulfate with 7 water molecules) or SQ-2 (quinine sulfate with 2 water molecules). In practice, the QAA and SQ are not usually provided by suppliers in developing countries, but are determined by potential buyers from samples submitted by suppliers. Most pharmacopoeias prescribe a minimum total alkaloid content of 5–7 percent in the bark.

Cinchona bark in pieces or chips is packed in jute bags of 50–60 kg capacity, whereas ground bark is packed in polyethylene-lined jute bags or polyethylene sacks of 100 kg capacity. Prior to chipping or grinding, the bark must be dried to a minimum of 10 percent moisture content. Both yellow and red barks have been approved for use in beverages, with a restriction that the total cinchona alkaloids should not exceed 83 ppm (0.0083 percent) in the finished beverage offered for sale. The bark of *Remijia pedunculata* (Rubiaceae) from Colombia also contains quinine and quinidine alkaloids and is used as a source of quinidine. This can be distinguished from cinchona bark by a simple test: the powder of the latter, when heated in a dry test tube, preferably with a little glacial acetic acid, produces reddish fumes, which are condensed in the upper part of the test tube.

The Market Turmoil

The cinchona market in the 19th century was driven by the European colonial powers, which were expanding their empires into the humid tropical regions endemic to malarial infestation. To be commercially viable, a cinchona tree needs to grow for at least 10 years. Raising plantations or expansion of current ones was a risky affair as global markets were also given to commercial turmoil. Market swings were very common, leading to anxiety among planters, manufacturers, and traders. The cyclic nature of demand in cinchona products has led to periodic and violent fluctuations in global price, affecting plantations in the following years. Large production of quinine salts in Indonesia led to the first major slump in 1920. India recorded an increase in production from 12 tons in 1915 to 20 tons in 1920, an annual increase of close to 35 percent. This led to drastic reduction in imports. Regulated release of supplies from Indonesia later led to an increase in demand. In addition, supplies from natural sources had decreased significantly. Soon after, new plantations appeared in the Philippines, Thailand, Taiwan, Zaire, and Guatemala. But the Japanese occupation of Java (the production center) during the Second World War cut off supplies completely, and the prices immediately shot up. This forced many European countries to invest in their African colonies.

During early 1950s, most governments in tropical countries launched public health measures to eradicate mosquito breeding grounds through liberal use of DDT, as mosquitoes were identified as vectors of malarial parasite Plasmodium. These measures led to a slump in the demand for antimalarial drugs. By then, though synthetic drugs appeared in the market, they did not affect demand immediately. However, a slump in demand occurred again between 1952 and 1962, and the international price of quinine was reduced to half of the market price in India, which had by then lost its export market. This led Sri Lanka, Burundi, Rwanda, and other African countries to give up cinchona plantations and even cut down immature plants. Indian government lowered its plantation targets. A pharmaceutical inquiry committee, set up by the Indian government, recommended replacement of existing trees with selected clones of high quinine content. Research on plantation management to lower costs was also initiated. Thus, systematic research on cinchona was initiated only after this period. In 1963 political turmoil erupted in Indonesia. Reports from several countries indicating that a malignant malarial strain, resistant to synthetics but responsive to natural quinine, surfaced. The alkaloid quinidine was also found beneficial in cardiac ailments. These developments led to market revival and demand.

The war in Vietnam increased further demand for natural quinine. Indian quinine was once again sought after by trade, and its export rose significantly from 1966–67 to 1971–72. Quinine price increased from Rs 110 to Rs 525 per kg by the end of 1965. This led to the rehabilitation of old plantations. Expansions were planned both in West Bengal and Tamil Nadu in the following years.

The demand for quinine sulfate in soft drinks grew rapidly in Western countries. In general, the price of quinine rose to Rs 1088 per kg in 1986 and reached an all-time high of Rs 4067 per kg 3 years later. However, synthetic antimalarials had a much greater impact on the market during this period and lowered the demand of quinine sulfate as drug. In fact, synthetics such as chloroquine and primaquine showed the advantage of killing both gametes and the diploid stage of plasmodium, whereas quinine kills only the asexual form of the parasite and the consequent effectiveness of reducing the chances of recurrence of malaria in the patient. Bruncton (1995) estimated that late in the 1980s, nearly half of cinchona bark deliveries were diverted to the food industry for the production of tonic bitters and additives, and between 30 and 50 percent were converted into quinidine; therefore, only a small proportion was channelized into making antimalarial drugs. A new drug, lidocaine (procainamide hydrochloride), came to market to compete with quinidine for its antiarrhythmic property. The liquor industry also obtained a new source of bitter in quassia bark (stem wood of *Picrasma excelsa* from Jamaica, West Indies), which largely replaced quinine from the soft drink industry. The quassia bark contains quassin and allied quassinoids, which are 50–60 times more bitter than quinine sulfate. The decline of the market began around 1982, and it gradually reached a level of collapse by the early 1990s.

During this period, the Chinese bought huge quantities for use as abortifacients in their population control program, sustaining the market for a few years, but the program failed and was given up by 1989. Soon after, the Amsterdam-based factory collapsed and liquidated its assets. In India, the Nilgiri and Annamalai plantations were given back to the state forest department, whereas the plantation program in Darjeeling was cut to a minimal output. The plantation land in Indonesia also lost ground to other crops owing to continuing losses. However, quinidine still continues to find a market in North America.

Research and Development Efforts in Cinchona

A Development of Improved Cultivation Practices

It was in 1839 that the British botanist Sir William Jackson Hooker wrote a dissertation on cinchona. He maintained that the cinchona tree stands to coppicing (from base) well. The tree generated new growth that could be harvested again after 6–10 years. It was also discovered that cut and regrown cinchona has higher levels of effective alkaloids in its bark. This method of harvesting was adopted by plantations and has remained in practice ever since (Brockway, 1979). Moens, a Dutch biochemist working in Java, identified chemical traces in the progeny of *Cinchona ledgeriana.* These were named cv. *chinidonifera* and cv. *cinchonidifera* based on the presence of cinchonine and cinchonidine alkaloids in their barks, respectively. This was a landmark development in recognizing intraspecific chemical taxa in medicinal plants research (Teteny, 1970). The Dutch in Indonesia devised a program on selection and clonal propagation of vigorously growing high assay trees that resulted in uniformly high alkaloid extraction. The work was done in the corporate sector and was never published. It was known that cinchona growing in moist substratum generated substantial hydrostatic pressure in the root system, and this depended on a continuing supply of magnesium and orthophosphates from the soil. These roots could not absorb water that is bound to soil colloids. The young leaves and buds required daily exposure to rain or mist to soften or remove the natural varnish-like coating that inhibits or even deforms new growth, if allowed to dry (Dawson, 1991). In Java, the richest trees (the parent stocks) were never allowed to seed but were grafted to other seedlings of *Cinchona pubescens.* The high-yielding trees were the conserved and nurtured on plantation scale. However, when these trees of high-yielding stock were planted in India, the quinine yield was lower (half or even less) than in Java. It was suggested that volcanic soils may influence productivity. Although *Cinchona ledgeriana* has been the preferred choice in India and Indonesia, it was found to suffer from low winter hardiness and poor rooting ability.

Breeding and Selection

There is no report on systematic breeding. However, both "*Hybrida*" and "*Robusta*" have outperformed their parents. The hybrids based on *Cinchona ledgeriana* have outperformed others in yield. Individual trees of *Cinchona ledgeriana* have shown 14 percent or more quinine in those found in Annamalai and Darjeeling plantations. These selected trees, as in Java, were used as parental material in new plantations to improve their yielding ability (Rao, 1974). Clonal selection provided an easy method of raising the output remarkably in Indonesia, Zaire, and Guatemala. Work in Balai Penelitiam in Indonesia on correlations between morphological characters and production capacity of cinchona clones has shown that the quinine percentage in the bark and total alkaloid content per ring of 6-year-old trees were closely correlated with the yield of quinine. Sukarja (1997) suggested that this be used as the main criterion to select cinchona mother trees. Among selected clones planted in a field at a site 1450 m above MSL, "clone cib-5" produced maximum yield based on the growth rate (girth) and yield of trees (Sukarja and Munir Supralo, 1977). Polyploids have been evolved in three commercial species, *Cinchona ledgeriana*, *Cinchona calisaya*, and *Cinchona pubescens*, through colchicine treatment in Guatemala. The polyploids showed little visible differences and gave only marginally higher quinine content over the control and so were given up from further breeding efforts.

Soil and Climate

Cinchona needs well-drained, deep, fertile organic soils over open gravelly substratum, rich in humus, and of pH ranging from 5 to 6.5. The plant thrives over gentle slopes and can grow also on steep slopes, but protection against soil erosion is needed. The Darjeeling hills have clay loam acidic soils rich in organic matter, organic carbon ranging from 1.5 to 3.34 percent, but deficient in P and K. The available levels of N, P, and K of the land chosen for planting were 315–414, 1.01–21.0, and 0.8–13.7 mg/kg soil, respectively (Datta et al., 1990). Cinchona prefers a cool climate with well-distributed high rainfall of 190–500 cm per annum. Vigorous growth occurs when showers alternate with sunshine frequently. Chatterjee (1993) reported that high quinine-yielding *Cinchona ledgeriana* gave relatively low content over the hills of South India in comparison to Darjeeling in Eastern Himalayas. This may possibly be because of occasional high temperature and drought in the southern hills where soil is semialluvial red loam containing medium-high organic carbon. A temperature of 21–24°C is ideal for luxuriant growth. Temperature should not go below 8°C and should be higher than 30°C in summer. The range of variation in day and night temperature should be minimal during the growing period. All species are susceptible to frost as well as sub soil waterlogging.

Raising a Cinchona Plantation

Raising a Nursery

The cinchona plant produces a large quantity of fruits over clustered panicles. The seeds are collected, cleaned, and stored in a dry place. One gram of seed contains 300–400 seeds. The seeds begin to lose viability after 6–8 weeks and lose it completely in 1 year. To raise seedlings, first raised beds of 3.6 m × 1.2 m are made. Leaf compost and manure are thoroughly mixed in the soil. About 25–50 g seeds are thickly broadcast during February–March in Darjeeling area on the surface of the bed and covered with a thin layer (0.2–0.5 cm) of soil under partial shade provided by thatched roofing. Water is sprinkled over the beds through a fine nozzle. Seeds germinate in 20–40 days, germination rate varying from 50 to 85 percent. It is estimated that 400 g seeds can produce more than 1 lakh seedlings. In 3 months the seedlings grow a pair of leaves when they are transferred to another nursery in rows at 5 cm × 5 cm spacing. In 5 months, the seedlings attain 10 cm height, bearing 3–4 pairs of leaves. Then they are transferred to another nursery at 10–12 cm spacing either way; this is to control mortality due to closer planting. Partial shade (roofing) is removed after 3 months, and the seedlings are allowed to harden in the open for the next 4 months. It thus takes 16 months to attain 20–25 cm height at which point they are ready for field planting during the wet season (Nandi, 1993). Although several reports suggested allelopathic inhibition to seed germination where cinchona is planted, Aerts et al. (1991) rejected this notion after detailed field study. They found that the actual concentration of alkaloids in plantation soil is very low, and this level plays no role in inhibiting seed germination. However, propagation through seeds is often unsatisfactory as the viability period is short and progenies of alkaloid-rich clones often produce very low alkaloid content, suggesting a high percentage of heterogeneity in the seed.

To raise pure clonal blocks, vegetative propagation of high-yielding trees is resorted during the wet season when root initiation occurs in 40–60 days; however, this propagation method adds to the cost considerably. Further, rooting is poor (around 10 percent) in high-yielding trees. Top-worked cuttings have responded to rooting better, and the rate of success is 70–80 percent. Patch budding is practiced on a plantation scale in Java. Similarly, cleft grafting of *Cinchona ledgeriana* is done over stock of *Cinchona pubescens* or even over *Cinchona* × *Robusta* and has given up to 90 percent success. *Gootee* planting is easier and is practiced in south India (Rao and Veeraraghavan, 1954). It is done on 2- to 3-year-old branches using 2 percent indole butyric acid (IBA) in lanolin paste for 24 hours before dressing with balls of soil (Thakurta and Dutt, 1944). It produces roots in 4 months with 88 percent success. Nandi (1993) reported investigations on top-worked primary lateral stem cuttings of *Cinchona ledgeriana* × *Cinchona hybrida*, which were planted during April–May in an experimental nursery. The cut ends were treated with growth hormones, namely IBA, indole acetic acid (IAA), and kitinase (KN) at different concentrations (100, 200, 300, 500, and 1000 ppm). The results indicated that *Cinchona* × *Hybrida* is more responsive to treatment compared to *Cinchona ledgeriana.* Wargadipura and Sutandjona (1977) found mixing equal proportions of top soil and composted cow manure to give the best result for survival and growth of plants in a nursery in Java. Thus, patch and slit budding, cincturing, air layering, veneer grafting, and mount layering were studied and recommended for raising new plantations (Directorate of Cinchona and Other Medicinal Plants, 1994; Mukerjee and Chatterjee, 1960). Mass propagation through tissue culture now offers a better alternative. Koblitz et al. (1983) developed a protocol of micropropagation for *Cinchona ledgeriana* × *Cinchona pubescens*, but they found that the transfer of rooted shoots into soil was complicated because of inadequate development of the root system. More research is needed to exploit this avenue.

Planting and Management

In Darjeeling, seedlings with 4–6 pairs of leaves which are half defoliated to reduce transpirative losses are planted early in the monsoon season (June), which helps survival rate. Pits of 1.2m × 1.2m × 2.0m × 2.0m dimension are dug based on the slope. The seedlings with soil sticking to the roots are planted and staked to obtain clear erect boles. In general, 3000 plants/ha are stocked, saving room to plant shade trees. Trees come to flowering after the fourth year. Debudding of trees increases laminar growth and alkaloid content (Nandi, 1993). In the first year, two–three weedings are done until November. Light to deep forking of land is done from the second year onward yearly for soil aeration. Usually 100kg/ha each of N and P as inorganic fertilizers is applied during the second and third years. Application of 10mg/l of Mn and Mg is reported to augment growth and alkaloid content (Nandi and Chatterjee, 1983). Use of micronutrients, such as Mn, Fe, Mo, B, and Zn, has been found to enhance quinine and alkaloid contents (Chatterjee et al., 1988). Planting shade trees is recommended. *Crotalaria anagyroides* and *Tephrosia candida* are sown at 3m × 3m × 4m × 4m spacing in the first year itself and these fast-growing leguminous plants have the added advantage of fixing atmospheric N as well and act as a cover against soil erosion. Temporary shade trees are thinned out after the third year to 12m × 12m spacing and finally removed when the cinchona plant attains a good height. Permanent shade trees, such as *Alnus nepalensis*, *Aleurites montana*, *Aleurites fordii*, and *Mallotus philippinensis*, are planted at 12m × 12m spacing and maintained in the plantation. A plantation should have 80 percent mature trees of cinchona and 20 percent of shade trees (Nandi and Chatterjee, 1991).

After thinning operations in the third or fourth year, when half of the growing trees are removed by uprooting and their bark is collected, the first crop is obtained. A second harvest is taken after the seventh or eighth year when only one strong shoot is allowed to grow and others are cut at base. The final harvest is done at 12 years of age, when the remaining trees are uprooted. It may be noted that the bark of the trunk and branches has maximum alkaloid content starting from the fourth up to the seventh year, but declines thereafter. The root at final harvest has a much higher alkaloid percentage. The bark is removed after ringing the stem at 45–60cm intervals from the base. The ringed material is kept for 1–2 days in partial shade to loosen the bark. The bark is easily stripped by knife, leaving the woody central portion intact. Extraction of bark is done during dry weather to facilitate drying of the produce. Initial drying is done in subdued sunlight for 3–4 days, when 70–76 percent weight loss occurs, and then it is dried in the open for the next 3–4 days. The dry produce for storage should contain not more than 10 percent of moisture. In Java, the bark is shaved off as near the cambium layer as possible without injuring it. This way the bark is quickly removed. A modified method involves stripping and mossing (covering by moss) which is said to facilitate early growth. It is found that the renewed bark is never as thick as the original, but has high alkaloid content.

In *Cinchona ledgeriana,* nearly 33–40 percent of the total produce is made up by root bark. It is superior in quinine content. Bark yield from a well-managed plantation in Darjeeling at 1000–2000m MSL elevation ranges from 3900 to 5000kg. In *Cinchona* × *Robusta* grown at 1300–1500m MSL elevation, it ranges from 2900 to 3250kg/ha (Chatterjee et al., 1988). Growth of Cinchona × *Hybrida* was better than the parent species (*Cinchona ledgeriana*) and this is reflected in its high bark yield and quinine content (Nandi, 1993). In Guatemala, plantations yield 9–16 tons of bark/ha (Popenoe, 1949).

Pests and Diseases

Damping off is common and the most noticeable diseases in nursery beds particularly at lower elevations are *Rhizoctonia solani*, *Macrophomina phaseoli*, and *Phytophthora* spp.; *Fomes lamoensis* and *Pythium vexans* are responsible for this condition. These fungi penetrate the seedlings through their roots and cause symptoms of sudden wilting and rotting of young succulent seedlings. Good drainage beds, spraying with Bordeaux mixture, and fumigation with chloropicrin are recommended for control (Sarma and Chatterjee, 1987). Soil application of thiram or pentachloronitrobenzine (PCNB) at 20kg/ha and spraying seedlings with 0.2 percent Zineb or other copper fungicides (0.2 percent) reduces infection. Blight caused by *Phytophthora palmivora* is occasionally reported in India. Necrosis of terminal leaves and branches in the seedlings leads to plant mortality. This fungus also causes girdle canker in older trees. Dusting with Bordeaux powder at regular intervals is recommended. Stem bark disease due to *Phytophthora cinnamomi* is a major disease in Central African countries (Rwanda and the Democratic Republic of Congo). Studies of infected material showed that anthraquinones were present in infected bark, but not in healthy ones. This suggests that these compounds in cinchona act as phytoalexins. The alkaloid content of the infected bark is lower (2 percent) than that of healthy ones (Wijnsma, 1986). Stripe canker (*Sclerotium rolfsii*), stem blight (*Sporotrichum* spp.), top blight and girdle canker (*Phytophthora parasitica*), and collar rot (*Fomes noxious*) have been reported from Annamalai hills. Of these, *Phytophthora parasitica* is the most devastating because the organism spreads from infected plants and causes wilt on older trees as well. Application of copper-based fungicide over the soil is effective. It is recommended that diseased plants be uprooted and burnt. The damage due to pests is less because young *Cinchona ledgeriana* plants contain cinhophyllene-type indole alkaloids with small amounts of 5-methoxy-tryptamine. These compounds provide a chemical defense to the growing plants against herbivorous insects (Aerts, 1992). However, caterpillars of *Helopeltis* spp. and *Catalpa sphinx* cause damage to the nursery plants. Spraying of dimethyl phosphoric ester controls the infestation. The leaf bug *Disphinctus humeralis* occasionally attacks tender foliage, but can be controlled by spraying 0.2 percent malathion.

Role of Biotechnology in Cinchona Improvement

Using stress, precursors, elicitors, auxins, enzymes, and use of a hairy root system, many investigators have attempted to produce cinchona alkaloids in high concentration in cell-suspension cultures. A large body of research in this area has been carried out in European countries during the past decade, though commercial success is still a distant goal. Chung and Staba (1987) established leaf shoot and organ cultures of *Cinchona ledgeriana* on MS media containing benzaldenine and studied the effect of age and growth regulators on alkaloid production. They found an increase in the content of alkaloids with an increase in the age of the cultured leaf shoot. The 32-week-old tissue cultures contained the same amount of alkaloids as 1-year-old plants. Quinine production was favored by the presence of benzaldenine (5 mg/l) and gibberellic acid (5 mg/l). Quinidine production was higher in the presence of indole 3 acetic acid (5 mg/l). High concentrations of abscisic acid and mefluidide inhibited growth and alkaloid production. Feeding various precursors to 8-week-old leaf shoot cultures increased the total alkaloid content by approximately 66 percent with tryptophan, 42 percent with secologanin, and 5 percent with strictosidine type. A decrease of 10 percent was registered with methoxy-strictosidine type alkaloid intermediates. Earlier, Hay et al. (1986) reported a fivefold increase in quinine and quinidine yield on feeding unlabeled 1-tryptophan (500 mg/l) to root culture suspensions in this species. Harke (1985) found that higher auxin levels improved production of indole alkaloids (cinchonamine) in the callus, although many times growth may not occur in light. Importance of light and dark periods was highlighted by Payne et al. (1987) in cultures transformed with *Agrobacterium tumefaciens* in a medium free of exogenous phytohormones. Growing transformed cultures in the dark resulted in 50 times more alkaloid accumulation than in cultures grown in light. This dark-dependent accumulation was not confined to a particular time in the growth cycle, although the extent of stimulatory effect increased when cultures were kept longer in darkness. Blue light was detrimental to alkaloid accumulation, but in red and green light, the level of accumulation was equivalent to that in dark. Alternating cultures between light and dark conditions for up to a 28-day period resulted in alternate periods of low and high alkaloids production. Isaac et al. (1987) isolated two nicotinamide adenine dinucleotide phosphate [NADP(H)]-dependent enzymes from suspension cultures of *Cinchona ledgeriana* in an interesting investigation that has bearing on biosynthetic pathways of indole alkaloids in plants. These enzymes catalyze the reversible reduction of cinchonidine to form both cinchonine and cinchonidine, and one of them shows reversible activity with its 6-methoxy derivative quinidine to form quinine and quinidine. Geerlings (1999) successfully integrated TDC (tryptophan decarboxylase) and STR (strictosidine synthase) enzymes from periwinkle (*Catharanthus roseus*) in terpenoid indole and quinoline alkaloid biosynthesis. The products of TDC and STR (tryptamine and strictosidine) were found in high amounts (1200 and 1950 mg/g dry weight, respectively). Quinine and quinidine levels rose to 500 and 1000 mg/g dry weight, respectively.

A Look into Cinchona's Future

Cinchona's natural alkaloid quinine has played a great role both in the prevention and treatment of malaria during the past four-plus centuries. Prior to the discovery of quinine, malaria was treated by the release of humors, that is, bleeding, purging, and the use of emetics. Robert Talbor, who used quinine first in England, was regarded as a quack by English physicians. He won recognition only after successfully treating Charles II of France from a severe malarial attack (Keeble, 1997). Even the synthetics that have replaced the original quinine are formulated on the novel model of quinine molecule, which is a great tribute to its unique nature. The natural drug is produced by a tree, which takes years in a plantation to produce the target alkaloid, while the production of a synthetic drug can be regulated in a laboratory taking much shorter a time period. Yet laboratories in several European countries are working to economically produce quinine and cinchonine through cell cultures, which, hopefully, might stand a fair chance of succeeding in the long run. Cinchonine has been used in population control programs because of its abortifacient nature but with uncertain results. It has not been given up. In fact, safer and cheaper birth control drugs are in great demand in developing countries, and if cinchona alkaloids or their natural analogues succeed in clinical trials, it will open the way for their comeback in pharmacy. Quinidine, on the other hand, has only limited demand and is a drug of choice in North America. Cinchonidine has a weaker action than quinine and has been found useful in rheumatism, neuralgia, and sciatica. It is also used as an antispasmodic in whooping cough. The demand for natural products is on the increase in Western countries. Cinchona bark is prescribed as a gargle for sore throat. In homeopathic medicine, it is given for nervous ergotamints, anemia, and convalescence. In European herbal medicine, cinchona bark is a top antiprotozoal medicine, given in all kinds of fevers. It provides relief in enlarged spleen, liver, and gall bladder disorders. It is used as a stimulant in hair-growth products. The bark has made a comeback in "bark tea" for the management of malaria. This remedy calls for a cup of boiling water to be poured over approximately 1 g of ground natural bark and allowed to steep for 10 minutes. One cupful of infusion is recommended to be taken half an hour before meals to stimulate appetite or after meals to treat indigestion (Witch, 1994). The popularity of natural products is gaining ground, and it is believed that cinchona bark will remain in the medical chest among herbal medicines for a very long time to come.

5 Cocoa (*Theobroma cacao* L.)

Next to coffee and tea, cocoa is a very important beverage crop. Its importance is in the confectionery industries and is the main ingredient of chocolate. Cocoa of commerce is the cured beans. Its composition is given in Table 5.1.

The stimulating effect of cocoa is due to the presence of the alkaloid theobromine. It is a tropical crop cultivated mostly in underdeveloped countries, especially on the African continent. Ironically, the majority of its value-added products are consumed by the population of the developed west. There is a steady increase in cocoa-based products of about 12 percent every year, and there is immense potential in this crop's cultivation.

Origin and Distribution

Cocoa is a native species of tropical humid forests on the lower eastern equatorial slopes of the Andes in South America (Cheesman, 1944). Domesticated by the natives of Central America, it was considered divine in origin. The generic name *Theobroma* literally means "Food of the Gods." Cocoa was domesticated and consumed for the first time by the Maya and Aztecs. The first Europeans to drink cocoa were the Spanish who invaded and conquered the Aztec empire in Mexico in the 16th century. The Spanish learned from the Aztecs the technique of making *xocoatl*, a drink made from cocoa beans after roasting and grinding. The word chocolate is supposed to have originated from *xocoatl*. The word *cacao* also was used by the Spanish and it probably originated from *cacahuatl*, a word that Aztecs used for cocoa beans. Even before the Spanish conquest, cocoa was taken to different regions by the Maya, Aztec, and Pipil–Nicarao people (Coe and Coe, 1996; Young, 1994).

From the center of origin, the species spread out creating two main groups: the *criollos*, which resulted in the dissemination through the Andes toward lowlands of Venezuela, Colombia, and Ecuador and toward the north to Central America and Mexico, and the *forasteros*, which resulted from dissemination toward the Amazon Valley in Northern Brazil and the Guaynas (Alvim, 1987). *Criollos* spread to Central America and to a large number of Caribbean Islands, including Trinidad in 1525 and thereafter to Jamaica. The dissemination to Venezuela and Costa Rica was made by the Spanish (Pittier, 1993). Introduction to Martinique and Haiti was by the French. Planting in Belem and Bahia in 1750 was attempted by the Portuguese.

The Agronomy and Economy of Important Tree Crops of the Developing World. DOI: 10.1016/B978-0-12-384677-8.00005-9
Copyright © 2010 by Elsevier Inc. All rights of reproduction in any form reserved.

Table 5.1 Composition of Cocoa

Constituents	Percentage
Fat	57.0
Protein	7.0
Carbohydrate	7.0
Theobromine	1.7
Moisture	6.0
Total ash	2.7
Minerals	1.1
Pectin	4.1
Fiber	2.1
Cellulose	1.9
Pentosans	1.2
Mucilage and gums	1.6
Tannins	6.2

Natural hybridization between *criollo* and *forastero* led to the origin of *Trinitario*. It has been reported that the *criollo* population from Venezuela and the Amelonado-type *forastero* from Guayana could have been involved in hybridization leading to the production of *Trinitario*. In 1822, cocoa seeds were taken from the Portuguese colonies of South America to the islands of Sao Tome off the West Coast of West Africa. It also spread to the neighboring island, Principe. Cocoa cultivation was started in Fernando Po in 1840. The most successful introduction into African mainland was made by the Ghanian Tetteh Quarshie in 1879. He brought a pod from Fernando Po, and the early population of cocoa in Ghana is considered to have originated from this pod. From Ghana, it spread to other African countries, the most important of which are Ivory Coast, Cameroon, and Nigeria. In these countries, there was immediate increase in area and they eventually turned out to be the largest producers of cocoa in the world. As it stands today, about 68 percent of the total world production of cocoa beans come from these African countries where the crop was introduced relatively very late. A characteristic of African cocoa used to be, especially up to the 1950s, the homogeneity of cocoa populations with pods resembling melon (*amelonados*).

The Venezuelan *criollo* was introduced in Celebes by the Dutch in 1560. They also introduced the crop to Java. The Spanish took the *criollo* types from Mexico to the Philippines in 1614. It was introduced into Sri Lanka from Trinidad in about 1798. From Sri Lanka, cocoa was taken to Singapore and Fiji in 1880, Samoa in 1883, Queensland in 1886, and Bombay and Zanzibar in 1887. Cocoa was introduced into Malaysia in 1778, and to Hawaii in 1831 and to India in the early part of 20th century, but its cultivation began on a large scale fairly recently in 1960.

Table 5.2 Country-Wise Cocoa Production

Country	Production ('000 tons)	Percentage of Total
African continent		
Ivory Coast	1175	41.92
Ghana	398	14.20
Nigeria	202	7.21
Cameroon	121	4.32
Other African countries	40	1.43
Total	1936	69.08
Central and South America		
Brazil	130	4.64
Other Central and South American countries	177	6.31
Total	307	10.95
West Indies	33	1.17
Asia and Oceania		
Indonesia	393	14.02
Malaysia	79	2.82
Papua New Guinea	35	1.25
Other Asian countries and Oceania	20	0.71
Total (Asia and Oceania)	527	18.80
World total	2803	100.00

Source: Cocoa Market Report, ED & F Man (April 2002), No. 370.

Area and Production

Table 5.2 summarizes world-wide cocoa production. The world production of dry cocoa beans was around 1500 million tons per annum in the 1970s and varied from 1.5 million tons to 2.5 million tons in the 1980s. It remained at around 2.5 million tons during the 1990s up through 1994. Based on the production statistics of 2000–01, the major producing countries were Ivory Coast, Ghana, Indonesia, Brazil, and Nigeria, their contribution being 82 percent of the world's production (Table 5.2). The African countries produced 68 percent and the Central and South American countries produced 14 percent of the total. The Asian countries produced the remaining 18 percent. India's contribution is negligible (0.21 percent).

The Botany of Cocoa

Cocoa is one among the 22 species assigned to the genus *Theobroma*, a member of the family Sterculiaceae. *Theobroma cacao* is the only species of economic importance. *Theobroma bicolor* Humb. & Bonpl. is cultivated for the edible pulp around the beans, and the beans are used like those of cocoa. The beans of *Theobroma angustifolium* Moc. & Sesse. are mixed with cocoa in Mexico and Costa Rica, and the sweet pulp around the beans of *Theobroma grandiflorum* (Wild. ex Spreng.) Schumann is used for making a drink in parts of Brazil and is also eaten.

The Root

Cocoa has a tap root that grows predominantly downward with only few branches. When the soil is deep and the growing conditions are favorable, the tap root can grow to a depth of about 150 cm. The primary function of these roots is anchorage. The main feeding roots are those which arise from the tap root and grow laterally. Most of these roots are concentrated just below the soil surface up to a depth of 15–20 cm (Wahid et al., 1989). The lateral spread of such roots will be up to a depth of about 120–150 cm around an adult cocoa plant. As the bulk of feeding roots is concentrated near the soil surface, any form of digging around the cocoa plant can harm the plant.

The Stem

The cocoa plant grows in tiers. The shoot of a seedling that grows upward is called "chupon." After growing to a height of 1–1.5 m, the growth of the chupon ceases, and three to five lateral branches arise. These lateral branches are called "fans" or "fan branches." The point at which fans arise is called "jorquette" and the process of the formation of fans from jorquette is called "jorquetting." A layer of fans may be called a "tier." If allowed to grow unchecked, new chupon buds arise on the main stem below the first jorquette and grow up to jorquette again. Chupons may be distinguished from fans by their nature of growth and leaf arrangement. Fan growth will be predominantly to the sides, whereas chupons grow vertically up. Leaves of the fans are arranged in one plane and are alternate. Leaf arrangement of the chupons will be spiral with a phyllotaxy of 3/8. Chupon leaves will have longer petioles than those of fans and will also have a more pronounced pulvinus. Normally, buds arising on a chupon give rise to chupons, except at the jorquette when jorquetting occurs. Similarly, fans produce only fan branches. Occasionally they produce chupons.

Inflorescence

The thickened leaf axils on stems called the "cushions" bear the flowers. The number of flowers per cushion is up to 50. The cocoa flower is a compressed cyme and has five each of sepals and petals, ten stamens in two whorls, and a superior ovary of five united carpels. Among the ten stamens, five of the outer whorl are sterile, and five of the inner whorl, which occur opposite the petals, are fertile. These fertile stamens occur concealed in the pouched portions of petals. The ovary is simple and five-lobed. The number of ovules per flower ranges from 40 to 60. The style has five stigmatic lobes. Cocoa flowers are produced in large numbers, but only a few of them develop into fruits. Those that are not fertilized fall off within 24 hours. The flowers are ill-adapted for pollination by natural methods as well as for self-pollination as the fertile stamen is surrounded by a ring of staminodes. The flowers are devoid of scent or nectar, and the pollen grains are sticky. Natural pollination occurs only with the help of small crawling insects. The most important of the pollinating insects are the ceratopogonid midges of the genus *Forcipomyia*. The insects are small and barely visible to the naked eye. The midgets are attracted by the pigmented tissues of the staminodes and the guidelines of the petals. The midges moving on the guidelines near the anther pick up the pollen grains, and when they crawl to the staminodes, some of the pollen get transferred from their body to the stigma. Though midges are the most important pollinating agents, other insects, such as ants, are also implicated as probable pollinating agents. There is a probability, though slight, of wind pollination as well. Flowers start to open late in the afternoon and are fully open by noon the next day. As such, most pollination occurs in the early hours of the day.

Self-Incompatibility in Cocoa

A unique feature of cocoa plant is its self-incompatibility in some types, first reported by Harland in 1925. Upper Amazon and Ecuador types introduced in Trinidad were self-incompatible. Most of the self-incompatible plants were also crossincompatible. Most of the homozygous types, such as West African Amelonados, are self-incompatible. Though the self-incompatible types may have the advantage of better fruit set under varied situations, self-incompatibility is important in commercial hybrid seed production. Incompatibility in cocoa is unique in that the site of incompatibility is the embryo sac (Cope, 1962). After incompatible pollination, the pollen tube grows faster and delivers the gametes into the embryo sac in a normal fashion. The embryo sac is in no way abnormal, and the rejection is due to the failure of male nuclei to unite with the egg. This incompatibility is referred to as "prefertilization inhibition in the ovule," and it is genetically controlled. Fusion or nonfusion is controlled by a series of alleles operating at a single focus (S), showing dominance or independent relationships (Purseglove, 1968). In incompatible matings, the flowers drop 2–4 days after pollination. If a population of cocoa is examined for self-incompatibility reactions, it could be observed that the majority of the plants belong to the self-incompatible group. Crossincompatible types frequently occur between two individuals with different genotypes, and it occurs only in diploid gametophytic systems when individuals share the same S genotype (Richards, 1986). Mallika et al. (2002c) studied compatibility relations among 16 selected parents. Out of 128 crosses attempted, 23 were crossincompatible. Crossincompatibility is an indirect measure of the degree of closeness between genotypes. When the parents used in crossing happen to be genetically similar, the incompatibility mechanism operates.

Fruits

The cocoa fruit is botanically a drupe, often called a "pod." The mature fruit consists of a thick husk containing 30–50 seeds. The seeds are covered with a sugary, mucilaginous coating called "pulp." The seeds are held in position with the help of a placenta. The pods can be of green or reddish color when immature. Green pods change to yellow when mature and reddish pods to orange or yellow in color. The pericarp (husk) is fleshy and thick.

Seeds

Seeds, which are called "beans," constitute the economic part of this crop. The pulp covering the seed contains about 10–15 percent sugar. The size of the beans is of practical significance, and a minimum average bean size of 1 g or a bean count of not more than 100 is usually taken as the standard. The number of beans per pod ranges from 30 to 60, each seed contains two convoluted cotyledons, a small embryo and a thin membrane, the remains of the endosperm, and a leathery test (shell).

Classification

The cultivated and wild cocoas were classified as follows:

1. *Criollo*: When ripe, the pods are red or yellow, deeply furrowed, markedly warty, conspicuously pointed, pod wall too thin, seeds large, plump, and almost round, cotyledons white or pale violet, which are less astringent. The beans ferment quickly, but yield is poor. It produces the highest-quality cocoa. It is susceptible to stress and not adaptable to all situations. It can be sub divided into Central American *criollo* and Venezuelan *criollo*.
2. *Forastero*: Unripe pods are green, turn yellow on ripening, inconspicuously ridged and furrowed, surface smooth, ends rounded or bluntly pointed, pod wall thick, seeds flattened, fresh cotyledons deeply pigmented and dark violet, giving an astringent product. The trees give high yields and are hardy. Quality is not comparable with *criollo*. The beans take 5–6 days for fermentation.
3. *Trinitario*: These originated in Trinidad from a genetic mixing of *criollos* and *forasteros*. These are heterogenous and exhibit a wide range of morphological and physiological characters. It is difficult to specify the characters of *trinitarios* as they may have pod and bean characters ranging from those typical *criollos* to those of *forasteros*.

In addition, several other subgroups fall under *forasteros*. The best known *forasteros* are the "*Amelonados*" of the African region, which were the predominant types traditionally cultivated in West African countries since the 19th century. *Amelonados* are typically self-compatible and the pods have a melon shape with nearly smooth pod surface. The Amazonians showed wide genetic variability and are highly useful for breeding work in the major cocoa-producing countries.

Germplasm Collection

International agencies are responsible for the distribution of cocoa. The International Cocoa Genebank, Trinidad, and the collection at Centro de Ensenanza Investigacion de I ICA, Turrialba, Costa Rica (CATIE), are designated as "universal collection depositories." The core of Trinidad collection is Pound's Ecquadorian and Peruvian collection, which forms 70 percent of it, the 1952 Anglo Colombian collection, representatives of Chalmers' and Allen's material, and selections from cultivated cocoa in Trinidad and other Caribbean islands. The core of the Turrialba collection is selections from cultivated cocoa, especially the United Fruit Company clones and their derivatives from Costa Rica, similar material from other American countries and *criollo*. Large collections of primary material are also maintained in Colombia, Ecquador, French Guiana, Venezuela, and Brazil. Field collections are maintained in Puerto Rico, Ivory Coast, Jamaica, Malaysia, Grenada, Nigeria, Papua New Guinea, Ghana, and India. In Kerala Agricultural University, India, 544 diverse types of cocoa are being maintained. The germplasm has been distributed from Trinidad and Costa Rica. Large quantities of seed were distributed from Trinidad to Ghana in 1944 and to Nigeria and Papua New Guinea in the 1960s. Long-distance distribution is done using intermediate quarantine facilities at the Royal Botanic Gardens, Kew (University of Reading) and the US Department of Agriculture in Miami, Florida. These transfers are carried out by authorized organizations such as the International Board for Plant Genetic Resources (IBPGR).

Crop Improvement and Management

Priorities in Breeding

Yield improvement is the prime objective of all cocoa-growing countries. With the onset and spread of deadly diseases like Witch's Broom (WB), cocoa swollen shoot virus (CSSV), vascular streak dieback, black pod, and so on, which are difficult to manage with chemicals, more emphasis is being placed on evolving disease-tolerant types of cocoa. Emphasis is also being placed on retention of traditional flavor, adaptation to local environment, early and sustained bearing, tree shape, pod size, and bean characters.

Breeding Methods

A major portion of cocoa production is derived from countries situated away from its center of origin, and thus the introduction of diverse types plays an important role in cocoa improvement. The genetic base of cocoa is very narrow; as such, crop improvement programs are not yielding spectacular progress. The introduced types are evaluated in the field for yield, pod and bean characters, incompatibility reaction to pests and diseases, and adaptation to the environment. The superior types emerging from this can be utilized for commercial planting or may be included in the breeding program.

Since cocoa is crosspollinated, it is not advisable to import germplasm as pods or seeds for breeding purposes. Vegetative materials such as bud wood can be imported, which is then budded to stock plants to get true-to-type plants. The introduced materials are maintained in isolation in quarantine houses until they are certified as free from pests and diseases. After proper evaluation, the introduced material may be directly released as improved clones. Some of the introductions can be used as sources of desirable genes for disease, pest, and drought resistance, quality, or other valuable characteristics, which may then be incorporated into adapted varieties through hybridization procedures.

Selection

There is ample scope for selection in cocoa because of the highly heterozygous nature of the crop. Immense variability exists in the seedling populations. The variability is so high that, in a seedling population, about 75 percent yield is obtained from 25 percent of the trees. The remainder of the trees will be of low productivity. The yield is influenced by changes in environmental conditions. Yield per tree varies with spacing, shade, soil conditions, nutrient supply, and so on. Longworth and Freeman (1963) suggested to consider tree yield along with trunk diameter for better efficiency in selection. In addition to pod production, the weight of cured beans per pod also may be considered for selection. Number of beans per pod is a trait which is not influenced by environment but the bean weight is influenced by the environment to a considerable extent (Pound, 1931). A good approach to yield improvement in cocoa is to select plants superior in yield and their subsequent development into clones. For selecting individuals from populations, certain criteria have been fixed; plants yielding not less than 100 pods/tree/year, each pod weighing 350–400 g or more with a pod value of not more than 10, and with 35–40 beans having a fermented dry weight of 1 g are selected as parents. In general, cocoa is well adapted to vegetative propagation by grafting, budding, or cuttings.

A number of superior clones have been selected throughout the world, and these are getting very high acceptability among the growers. Kerala Agricultural University has initially selected 70 clones, out of which 7 were released for cultivation (Cadbury Cocoa Research Project CCRP 1–7).

Hybridization

In cocoa, hybrid vigor between parents showing good combining ability can be exploited. A large number of crosses have been made in countries such as Trinidad, and the potentials of the parents have been assessed. Posnette (1951) demonstrated interpopulation heterosis in cocoa. The initial crosses involving Pound's seedling collection showed extraordinary vigor, precocity, and high yield in Ghana. These observations and similar ones in Trinidad were attributed to hybrid vigor (Bell and Rogers, 1956; Montserrin et al., 1957). In trials with diverse crosses in Brazil, Costs Rica, Ghana, Ivory Coast, Nigeria, and Papua New Guinea, there were significant additive components for yield. A number of hybrids with high yield and other desirable characters have been evolved in different countries. Hybridization work in Kerala Agricultural University in Thrissur, India, led to the release of three hybrids (CCRP 8–10) with high yield and tolerance to vascular streak dieback disease. Heritability estimates ranging from moderate to high and large additive components of variation indicate easy progress toward high yields, at least in the early years of the program. However, after an initial boom in the first phase of hybridization, yield improvement in hybrids during the second phase of breeding was not as spectacular as expected. This is duc to the poor genetic base. Most of the hybrids and clones are derived from a relatively lower number of types, and lack of yield improvement in intrapopulation crosses is due to inbreeding.

Hand Pollination

The method is described by Mallika et al. (2000). In artificial pollination, a flower bud that will open the following day, recognized by its whitish color and swollen appearance, is selected. The bud is covered with a hood of plastic tube or hose pipe piece 5 cm × 1.5–2.0 cm, which is sealed to the bark using materials such as plasticine or glaze putty. The tube is covered with muslin cloth at the top, kept in place with a rubber band. This ensures circulation of air and exclusion of insects. Opened flowers are collected from the desired male parent and stamens are carefully taken out by pushing the corresponding petal. One entire anther with a part of the filament is deposited on the stigma. One or two staminodes may be pinched off to give access to the stigma. Emasculation is not necessary due to the presence of self-incompatibility. For selfing, hand pollination is done using stamens from the same flower. The pollinated flowers are labeled using tin foil pieces fixed in the cushion using ball pins. The hoods are removed 24 hours after pollination, and in 3–5 days fertilization is confirmed by the visual swelling of the ovary. In order to prevent undue shedding and wilting of fruits from hand pollination, it is usual to remove all developing fruits on the tree produced by open pollination. Developing pods are covered with wire mesh after 6–8 weeks to protect them from mammalian pests. The pods are collected at maturity, and beans are extracted and sown in the nursery.

Pre-Selection Method

In cocoa, interrelationship between vegetative characters and yield is positive (Enriquez, 1981; Glendinning, 1966; Ngatchou and Lotode, 1971; Paulin et al., 1993). However, Francis (1998), Sridevi (1999), Verghese (1999), and Amma et al. (2002) recorded contradictory results, concluding that after bearing fruits, vegetative growth slows down, and the correlation between growth reduction and yield became positive. This points to the need to evolve a viable pre-selection method in cocoa.

Selection of Superior Hybrids

The seedlings selected based on vigor or disease tolerance are field planted. On attainment of a steady yield, the hybrids are evaluated for their performance. The highest-yielding hybrids with other desirable attributes are multiplied and released as new clones. The parents selected for hybridization programs are tested for their general combining ability (GCA) and specific combining ability (SCA). To test GCA, all selected clones are crossed with a standard variety and the progenies are evaluated both in the nursery and in the field. A few best combiners are then selected and crossed in all possible combinations to assess SCA. Parents of promising hybrids are identified as best combiners. The best combiners are multiplied and used as parents in seed gardens to produce quality hybrid seeds.

Clonal Seed Gardens

The parents used in seed gardens are selected based on the results of progeny trials. The search for the best combiners involves the screening and selection of a large number of crosses, both at the seedling and adult stages. Having selected the parents, they are propagated vegetatively. The female parent should be self-incompatible. The desired crosses can be ensured either by hand pollination or by the proper design of the seed garden where natural pollination is relied upon. With two self-incompatible parents, all the pods resulting from crosspollination can be used for seed. Where one parent is self-incompatible, seed collected from the self-incompatible parent only (and in such cases, the pollen parent) is planted in a ratio of 1:5 female parent trees. The seed garden must be isolated to some extent from other cocoa; a distance of 200 m is considered sufficient to prevent unwanted crosspollination.

Problems in Hybridization

Cocoa is a perennial crop with an outbreeding nature. Most cocoa types are self-incompatible. Selection of self-incompatible parents in breeding programs makes hand pollination easy, as emasculation is not necessary. The existence of crossincompatibility between some of the parents often poses problems to the breeder. Hand pollination often leads to no pod sets due to crossincompatibility, and certain proportions of the developed pods wilt due to delayed incompatibility. Each successful pollination gives rise to a pod, which contains about 35–60 beans. When a large number of crosses are made, the number of hybrid seedlings produced will be too high to be planted in the field on account of limitations in space and cost. This necessitates the development of selection criteria based on early growth parameters of the hybrid seedlings in the nursery itself, which bears positive correlation with final yield. This must, however, ensure that a valuable hybrid produced out of crossing is not lost while screening the seedlings.

Inbreeding

Inbreeding forms a part of the breeding activities not only to breed parents with some degree of homozygosity for the production of hybrids but also breed materials homozygous for such desirable traits as disease resistance. Self-incompatibility makes inbreeding difficult or impossible. In cocoa, certain incompatible trees are encountered in a population, and in these plants selfing is possible. The selfing needs to be continued up to six to seven generations to attain homozygosity, and thereafter these plants can be utilized for crossing to exploit hybrid vigor. An inbreeding program has been in progress in the Kerala Agricultural University (Mallika et al., 2002a) since 1987.

Inbreeding for Disease Resistance

Five major diseases, namely Witch's Broom (WB), black pod (*Phytophthora* pod rot [PP]), moniliasis pod rot (MO), cocoa swollen shoot virus (CSSV), and vascular streak die back (VSD), affect the crop, causing about 40 percent yield loss per year. Selection for disease resistance under field conditions is time consuming, and environmental factors plus genotype × environment interaction may affect the genotypic variation in host resistance. Screening tests on seeds (CSSV, Ghana), young seedlings (WB, Brazil), and seedlings (VSD, India) are of practical use in selection programs. Selection for host resistance requires standardization of the environment and inoculation methods to reveal maximum genotypic expression of major components of host resistance. A close correlation of the results of the pre-selection test with mature plant resistance should exist. The important diseases and their extent of damage and progress in breeding are outlined in the following sections.

Black Pod (BP) (Phytophthora palmivora *and* Phytophthora megakarya)

The most devastating disease, worldwide, of cocoa is the black pod and pod rot. *Phytophthora palmivora* occurs in the center of origin of cocoa and causes 44 percent global crop loss. *Phytophthora megakarya* is restricted to Cameroon, Nigeria, Togo, and Ghana, causing about 10 percent crop loss. *Phytophthora capsici* parasitizes cocoa in Central and South America. It is the predominant cause of pod rot in Brazil where it is less aggressive than *Phytophthora palmivora. Phytophthora citrophthora* also attacks cocoa in Brazil. Much progress has been made recently in the study of the nature of variation in host resistance as evidenced by field scores and artificial inoculation tests on *Phytophthora palmivora* in Trinidad (Iwaro et al., 1997), Costa Rica (Phillips-Mora, 1996), Papua New Guinea (Tan and Tan, 1990), and Ivory Coast (N'Goran et al., 1996). Similar studies have been conducted on *Phytophthora megakarya* in Cameroon (Nyasse et al., 1996).

Witch's Broom (WB, Crinipellis perniciosa)

This is endemic to wild cocoa and is restricted to the Western Hemisphere, causing about 21 percent crop loss. It is prevalent in the center of diversification of cocoa in the Amazon and Orinoco River basins, Ecuador, Bolivia, Peru, Venezuela, Guyana, Surinam, Brazil, Trinidad, Tobago, and Grenada. The disease caused up to 30 percent crop loss in these countries (Rudgard and Lass, 1985). The disease occurs in all species of *Theobroma* and the closely related genus *Herrania*. The fungus spreads through seeds as well, and hence quarantine measures should be strictly enforced.

Search for disease resistance started with F.J. Pound's expedition to the Amazon basin and was continued in Trinidad. Investigations are in progress in Brazil (Gramacho et al., 1996), Trinidad (Laker et al., 1987), Ecuador (Aragundi et al., 1987), and in the United Kingdom (Wheeler and Mepstead, 1988). No genotypes are completely immune to WB infection, and the resistance is generally of a quantitative and incomplete nature. Some indications of the role of a few major genes are also available. The level of resistance of SCA 6 to WB has remained the same for more than 60 years. However, under Ecuadorian conditions, SCA 6 shows a lower level of resistance.

Cocoa Swollen Shoot Virus (CSSV)

There are several viruses that infect the cocoa plant, such as the cocoa necrosis virus found in Ghana and the cocoa swollen shoot virus (CSSV) in Sierra Leone. CSSV is the most serious disease in Ghana, Nigeria, and Togo. It occurs also in Ivory Coast and Sri Lanka. In Ghana, CSSV caused severe economic problems that spilled over as a political turmoil because cocoa is the mainstay of the country. In other West African countries, the onslaught of the virus has been far less severe and the countries managed to live with it. Mild strains of CSSV have been found in Trinidad, Sri Lanka, and some other *Trinitario* clones in Indonesia. Work on CSSV has been reported from Tafo (Adu-Ampomah et al., 1996; Sackey, 2000), Nigeria (Williams and Akinwale, 1994), and Togo (Djekpor et al., 1994). Results of these investigations indicate that there is no immunity or high level of field resistance, but, in some Upper Amazon genotypes with *Iquitos* mixed *calabacillo* (IMC), Parinary (PA), or Nanay (NA) parentage, the rate of spread was only one-quarter of that of *Amelonados. Amelonado* cocoa is generally more susceptible to African CSSV than Upper Amazon and *trinitario* types. Some resistance sources have been reported in Upper Amazon types. So far, only a limited number of CSSV-resistant genotypes have been utilized. It is, therefore, necessary to increase the efforts to detect other resistant progenitors.

Vascular Streak Dieback (VSD) (Oncobasidium theobromae)

VSD is the most important disease in Indonesia, Malaysia, and Papua New Guinea, causing about 9 percent crop loss. More recently, it has spread to all other South Asian countries. Kerala State in India and also Philippines have reported its serious occurrence. Inadequate plant quarantine measures appear to have played an important role in the rapid spread of the disease. The threat of the disease to the cocoa industry in Southeast Asia is very much reduced with the detection of partial levels of resistance in several Upper Amazon and *trinitario* genotypes. Results of investigations in Papua New Guinea (Blaha, 1996), Malaysia (Lamin et al., 1996), and India (Mallika et al., 2000, 2002a) indicate that the nature of resistance is of a quantitative and incomplete nature. A high level of resistance exists in SCA 6 and 12, NA 33 and KA2-106. The resistance is inherited in an additive manner, and heritability is high. Resistance has been reported in *trinitario* cocoa grown in Papua New Guinea and Upper Amazon selections. The resistance has been stable over 25 years and its use controlled the spread of VSD in Papua New Guinea. Thousands of hybrids tolerant to VSD have been planted in the farms of Kerala Agricultural University in India, and some of these are precocious with very high yield potential.

Moniliasis or Frosty Pod Rot (MO) (Moniliophthora roreri)

The fungus is endemic in wild *Theobroma* and *Herrania* species. This disease is becoming increasingly serious in Ecuador, Colombia, and Central America, causing about 5 percent crop loss. It is also prevalent in Peru, Venezuela, Panama, and Costa Rica. The number of genotypes with resistance under natural field conditions is low in Ecquador and Costa Rica. It appears that some relationship exists between WB and MO resistance. Clone EET 233 was consistently resistant in Ecuador. Other sources or resistance to this disease are Scavina 6, PA 169, UF 296, IMC 67, SPA 9, and EET 59.

Ceratocystis Wilt (Ceratocystis fimbriata)

This disease has been most devastating in the center of diversity of cocoa. It has spread to Ecuador, Venezuela, Colombia, Trinidad, and Southeast Asian countries. Upper Amazon selections IMC 67 and Pound 12 are resistant (Lass and Wood, 1985). *Criollo* types are highly susceptible.

Climatic Requirements for Cocoa

Temperature and rainfall are the climatic parameters that mostly affect cocoa plant. Other factors such as altitude and latitude influence the growth of cocoa mainly through their effect on temperature and rainfall.

Effect of Temperature

The range in the mean monthly temperature of the majority of cocoa-growing regions is found to be from 15°C to 32°C, and this range is considered to be the optimum for the growth of cocoa. The absolute minimum for any reasonable period is taken to be 10°C, below which frost injury is likely. These temperature limits set the latitude limits for cocoa plant's best growth to within 8° north and south of the equator. Temperatures of a location are modified by altitude. Therefore, when deciding the elevation limits up to which cocoa can be grown, the temperature regime in the plains can be used as a guide. The extent of decrease in temperature with increasing elevation over land will be around 4.5–5°C/km rise in elevation.

Effect of Rainfall

The two parameters that are related to rainfall are the total amount and its distribution. In most of the cocoa-growing areas of the world, the total annual rainfall is in the range of 1500–3000 mm. Values lower than 1500 mm may mean that the supply of water through rain may be insufficient to meet the evapo-transpiration demand, and regular supplementary irrigation may be necessary to support unrestricted growth of the plant. Values beyond 3000 mm may result in excessive and continuous rain during reasonably long periods of the year, which will favor incidence of diseases such as black pod (*Phytophthora palmivora*) and vascular streak dieback (VSD). Heavy rainfall is the prime cause for the severity of VSD in Papua New Guinea and its rapid spread in India; no other country in the world has shown such severity of the disease. In India, concentration of the rainfall to 2–3 months aggravates the spread of the disease. To ensure good growth, rainfall distribution is more important than the total amount received. In other South American, African, and Southeast Asian cocoa-producing countries, rainfall distribution is more or less even. It is so well distributed that about 10 cm of rainfall is received every month, in particular, in Brazil, Ghana, and Malaysia.

Soil Type

Cocoa can adapt well to a wide range of soils, especially in Africa and Asia. In regions of high rainfall, soils that are coarse and acidic are obtained. Very coarse sandy soils are not usually put to cultivation of cocoa in countries where other soils of choice are present. Also, virgin, freshly cleared forest soils are used for cultivation of cocoa. These make soils of most of the cocoa-growing regions of the world rich in organic matter and nitrogen, well drained, and acidic to neutral in pH. Since cocoa roots penetrate to a depth of about 1.5 m, soil depth of this magnitude is required.

Propagation

Cocoa can be propagated both by seed and by vegetative means. Both are discussed here.

Seed Propagation

This is the easier method between the two. But when this method is followed, the seedlings obtained will be genetically highly variable. The chances for the recovery of better progeny are high from elite parents with the following selection criteria.

1. Trees of *forastero* type having medium or large pods of not less than 350 g weight or 400 ml volume; yield to be not less than 100 pods/tree/year
2. Husk thickness of pods to be not more than 1 cm
3. Pod value (number of pods required to give 1 kg of wet beans) to be not more than 12
4. Number of beans per pod to be not less than 35
5. Bean weight (average weight of fermented and dried beans) to be not less than 1 g

A more desirable procedure for ensuring quality will be to collect seeds from biclonal or polyclonal seed gardens. Seeds of cocoa are nondormant and lose viability quickly within 7 days of extraction. If the seeds are to be stored for more than 7 days, they should be kept in moist charcoal and packed in polyethylene bags. The beans may be extracted, testa with pulp removed, and the beans stored in polyethylene bags. This saves space and weight of material while transporting. The best method is to store as pods. The potting medium of farmyard manure, sand, and soil in equal proportion is good enough to raise cocoa seedlings. Though seeds will germinate any time of the year, nurseries may most conveniently be sown by December–January so that 4- to 6-month-old seedlings will become available by the onset of monsoons for field planting. Keshavachandran (1979) recorded 94.5 percent germination when fresh seeds were sown in February–March. Seeds are to be sown with the hilum end facing downward or sown flat. The depth of placing is to be such as to just cover the seeds with soil. Removal of pulp by abrasion with suitable materials has been found advantageous in enhancing the germination percentage. The extent of advantage from this practice is only marginal. However, peeling was suggested to have the advantage of rejecting non-*forastero* types with white or pale purple cotyledons. The seeds germinate in about 6 weeks' time and germination will continue for another week. Germination will be around 90 percent.

Four- to six-month-old seedlings are suitable for planting. Experience has generally shown that when field conditions are suitable for seedling growth, there will be little difference in subsequent field growth of cocoa, when seedlings of varying nursery are used. When seedlings are raised under dense shade, and if shade intensity is likely to be substantially lower in the field, hardening the seedlings by exposing them to higher illumination levels will be necessary. The period of hardening may be about 10 days.

Vegetative Propagation

Prior to hybrid seed production programs, vegetative seed propagation has been practiced in almost all cocoa-producing countries. This ensures uniformity. As most of the plants in a population are self-incompatible, the use of vegetatively propagated material from a single plant cannot produce any yield. Hence, when budded plants or grafts or rooted cuttings are used, it is to be ensured that multiclonal blends are planted. The different methods, namely budding, rooting of cuttings, and grafting, are successful.

Budding

Budding on rootstocks of 6–12 months growth is most often resorted to, though green budding on seedlings of 2–4 months' age is also possible. The procedure for budding on older rootstocks has been described by Nair et al. (1995). Precuring the bud wood by cutting off the laminae of all the leaves of the selected branch to a distance of about 30 cm from the tip 10 days before budding increases bud take. The bud wood, just hardened showing brown bark and just hardened green leaves, is to be selected. When bud sticks are transported long distances before budding, they are dipped in benzyl chloride and washed in water. The cut ends are sealed using molten wax and wrapped in wet cotton wool, wet tissue paper, or blotting paper. The bundle is then packed in a box using wet packing material. The packet is then covered with polyethylene sheets. This helps to extend viability about 10 days from the date of collection.

Bud wood from fan branches or chupons can be used for budding. However, when bud wood is taken from fan shoots, the budded plant will have a bushy appearance. Plants growing from chupon buds will grow like seedlings. Even with this advantage of chupon buds, fan branches are often selected for budding just because the availability of bud wood from chupons will be less than that from fans. Any of the common methods of budding are suitable for cocoa. In Ghana, T-budding is considered to give the highest percentage of success, whereas the experimental results at Kerala Agricultural University favor the adoption of the patch method. These results are probably arising out of the differences in the skill of the workers. About 3 weeks after budding, the grafting tape is cut off, and if bud union has occurred, a vertical cut is made halfway through the stem above the bud and the stem is snapped back. Continued connection of root stock with the terminal leaves through the intact side is essential for bud sprouting and further growth. The stock portion is cut back after the bud has grown to a shoot, and at least two leaves have hardened. It is then allowed to grow for a further period of 3–6 months, after which it may be transplanted. Under normal situations, 70–90 percent success is assured.

Rooting of Cuttings

Cuttings are more frequently taken from fan branches, though chupons also can be used. Fairly young branches of 10–15 cm length, green at the top and brown below, are selected. The number of leaves on the stem should be at least four, the stem cutting should be 20–30 cm long. The apical half of the laminar portions of the upper leaves are cut off, and all the leaves of the lower portion are removed, leaving four to five leaves at the tip. The extra length of 2 cm of the stem is then cut, and the freshly cut end is dipped in a mixed solution containing 5 g of 1-naphthalene acetic acid and 5 g of indole butyric acid dissolved in 240 ml each of water and ethyl alcohol. The treated cuttings can be planted in polyethylene bags filled with potting mixture. A small hole is made in the center of the pot and the hole is filled with sawdust. Cuttings are then planted on the sawdust portion. It takes about a month for the cuttings to produce roots. Until this time, the cuttings are to be kept in a completely humid atmosphere and the soil is to be kept moist. This can be achieved by keeping them in a chamber that has provisions for spraying water at frequent intervals, or, preferably, they may be kept covered with a polythene sheet and watered once every 3 days. The sides of the polythene sheet are to be kept pressed on the ground by weighing them down on the sides (McKelvie, 1957). When this method is followed, the cuttings are to be kept in a heavily shaded area allowing only about 10 percent light to avoid excessive heat development. Cuttings strike roots in about a month's time. The rooted cuttings are to be hardened by exposure to sunlight up to about 10 a.m. in the first week, up to about 11 a.m. in the second week, and up to noon in the third week. From the fourth week onwards, they are kept exposed throughout the day under shade. The cuttings will be ready for planting in about 6 months.

Grafting

Four-month-old seedlings are used as root stock. Scion may consist of a shoot of comparable thickness that has just turned brown and has at least two hardened leaves. These leaves are cut back to stumps of about 5 cm length just prior to grafting. The stock stem is cut at a convenient height (about 5 cm from ground level) and a longitudinal slit of about 2 cm length is made. The stem portion of the scion shoot is also given a slanting cut on two sides to a similar length and it is made into a wedge.

The scion portion is then inserted into the stock portion and tied round with budding tape to keep it in position. The stock is then kept covered with a polythene bag and tied round the stem to keep it in position. This bag is to be removed after about a week. It takes about 3 weeks for the graft union to be completed before the tape can be removed. There will be 90 percent approximate success. The advantage with this method is that the skill required for operation is less than that required for budding. The fact that it needs a longer scion for each plant to be established and that the unsuccessful stock cannot be reused are the main disadvantages.

Shaping Clonal Cocoa Plants from Fan Shoots

Plants derived from vegetative propagation using fan branches have diffused branching systems and are asymmetrical in growth habit. If a better shape of the plant is desired, which may be beneficial in the long run, appropriate formation pruning may be necessary. This involves the identification of a chupon arising on a fan shoot, allowing it to grow, and removing the original, lower fan-like shoots in stages. This, however, has to be done slowly, as an early drastic pruning will inhibit growth.

Field Maintenance

Planting

Cocoa seeds can be sown directly or seedlings planted at any time of the year if soil moisture conditions are suitable. Under Indian conditions, the best time for planting would be by the onset of pre-monsoon showers (May–June) in Southern India. Although early planting would necessitate watering before the onset of monsoon, this will be beneficial for better establishment as in the case of other plants. Each country has its own method of planting, as in the case of Sao Tome where big pits of up to 200cm are dug, filled with soil and manure, and seedlings are planted. On the other extreme, in Ghana, seeds are pushed into the soil or seedlings are planted in pits that are just big enough to contain the ball of earth of the polybag seedling. In general, experimental results indicate a lack of any significant advantage out of making planting pits if soils are naturally deep enough and are fertile. However, if soils are gravelly or if hard pans occur within the depth of penetration of roots, such a practice may be advantageous. Again, if soils are naturally low in fertility, especially on the surface, there may be an advantage arising out of the incorporation of manures that usually accompany the filling of planting pits. When soils are of low fertility, and where gravelly laterite zones occur at varying depths, it is better to dig pits of 50cm depth, length, and width, fill them with a mixture of surface soil and organic manures, and plant the seedlings at the surface level. A point to be noted is that cocoa seedlings are to be planted on the soil surface as the feeding roots of cocoa get concentrated on the surface irrespective of the zone at which seedlings are initially planted. Except for India, Malaysia, and the Philippines, cocoa is planted as a monocrop under natural or planted shade trees.

Spacing

The spacing adopted for cocoa in the major cocoa-producing countries is highly variable. It was generally very low in the *in situ* system of planting in Africa. Experimental work at the Cocoa Research Institute of Ghana has indicated that for the *Amelonado* cocoa, a close spacing in the range of 1.7 m × 1.7 m to 2.7 m × 2.7 m was found to be the optimum. Within this optimum range, closer spacing was advantageous in the early years, especially for the unshaded cocoa. For the Amazonian types, a wider spacing in the range of 2.7 m × 2.7 m to 3.3 m × 3.3 m is recommended in Ghana. On the other extreme, a relatively wide spacing 5 m × 5 m is adopted in Sri Lanka. The advantage of closer planting in earlier years is generally noted in the spacing experiments. From a consideration of experimental results on spacing and the practices followed in cocoa-producing countries, Wood and Lass (1985) concluded that spacing in the range of 2.3 m × 2.3 m to 3 m × 3 m would suit the crop. The ultimate spacing and population level depends upon the extent of canopy development, the variety used, and the type of management. For the African situation, where the less vigorous *Amelonado* is predominantly cultivated, and where practically no costly input is used, a closer planting may be beneficial. This will also mean a better crop in the early bearing period. Among the other cocoa-producing countries, such as the Philippines, Papua New Guinea, and Malaysia, where cocoa is cultivated along with coconut, the spacing followed in Malaysia is rather a close one, there being two rows of cocoa in between the rows of coconut at a plant-to-plant distance of around 2 m. The coconuts are spaced 8–10 m, and the cocoa population in a hectare would be about 1000. Cocoa is also planted at a spacing of 2.7 m × 2.7 m and the usually adopted spacing of cocoa also is the same. The general experience is that such a spacing results in crowding of cocoa canopy. As a tentative recommendation for the Indian situation, a row of cocoa may be planted in between two rows of arecanut. For the space-planted coconut population, a row of cocoa at a plant-to-plant distance of 3–4.5 m is recommended for Indian cocoa.

Shade for Cocoa

Cocoa is a plant that originally rose in shade and did well under shade. The shade levels at which this crop was cultivated had been highly variable. The results of a number of shade trials taken up since 1950s in cocoa-producing countries have, however, shown that the shade requirement of this crop varies widely depending on the stage of growth, with it requiring as much as 75 percent shade in the early stages. This is gradually brought down to about 25 percent when cocoa comes to production. Such variations in shade levels are provided in the sole crop situations in Costa Rica, Ghana, Colombia, and Brazil by providing different types of shade plants and by their selective thinning. The investigations by the Kerala Agricultural University (Nair et al.,1996) on the response of cocoa to shade indicated that the girth of stem and yield increased with increases in illumination levels. The results suggested that it is possible to cultivate cocoa without shade under the prevalent weather conditions in the State of Kerala, and that the productivity will be the highest under shade-free situations. However, shading may be necessary in the early years after planting using temporary shade plants.

When it is a monocrop of cocoa, permanent shade trees are planted or left without removal at a wide spacing of about 13–15 m, and temporary shade plants are planted at the same spacing as the cocoa plants, alternating with it. The common temporary shade plants in African countries are banana, tree cassava or cera rubber (*Manihot glaziovii*), and cocoyams (*Colocasiaesculenta*). These shade plants are gradually removed as cocoa grows and the canopy develops. The permanent shade tree commonly used for planting in Ghana is *Terminalia ivorensis*. The other shade trees that may be used are *Gliricidia maculata* planted at a spacing of 3 m × 5 m, dadap (*Erythrina lithosperma*), and *Leucaena leucocephala*. With both permanent and temporary shade plants, the level of shade will be high, resulting in the best vegetative growth of the young cocoa plant. When temporary shade plants are removed as cocoa comes to bear, illumination will increase, stimulating production.

Manures and Fertilizers Used in Cocoa Cultivation

Most of the major cocoa-producing countries do not use manures or chemical fertilizers on a regular basis. The primary reason is the inherent fertility of the soils of the major cocoa-producing countries, because they are cleared forest soils rich in humus. The presence of shade trees, the dense canopy development of cocoa, and the large turnover to the soil of the cocoa litter prevent any substantial soil erosion or loss of nutrients. In an investigation carried out by the Central Plantation Crops Research Institute (CPCRI) in Kasaragod, Kerala State, India, the amount of organic material returned to the soil as cocoa litter was estimated at 818 and 1985 kg/ha/year, respectively, on a dry weight basis in the single and double-hedge systems of planting cocoa. The quantities of fertilizer nutrients contained in organic material from double-hedge planting were estimated as 50 kg N, 11 kg P_2O_5, and 35 kg K_2O per hectare. The quantities of N, P, and K removed by cocoa pods per kg of dry beans will work out to 43.80, 8.04, and 64.29 g, respectively. For a crop yielding about 2 kg dry beans per plant (about 60 pods) per year, the average crop removal by pods would be around 85, 37, and 154 g each of N, P_2O_5, and K_2O, respectively. The fertilizer recommendation for cocoa under average management is 100:40:140 g N, P_2O_5, and K_2O, respectively, per plant per year, which tallies with the crop removal figures. For cocoa under better management where the average annual yield is more than 60 pods, double this dose is tentatively recommended.

Method and Time of Fertilizer Application

Within a depth of 15 cm and to a radius of 125–150 cm the feeding roots are concentrated. Hence, it is best to apply fertilizers in a shallow basin of 120–150 cm radius and raked in without serious damage to the roots. The general recommendation in most cocoa-producing countries is to broadcast fertilizers in the entire field without any soil tillage. Although this may be suitable when the soil surface is wet, there is the possibility of immediate solubilization, which leads to both physical loss due to leaching and volatile loss, which is mostly with the case of N. An immediate mixing of fertilizers with soil will reduce the chances of a volatile loss of N, especially when easily soluble urea is used as the nitrogenous source. To time fertilizer application, the stages of crop activity and the seasons of moisture availability may have to be considered. When the crop is unirrigated, flushing occurs mostly in the rainy season starting from May–June and continues up to November–December. Though essentially the same trend persists in irrigated cocoa, there will be some flushing in summer as well. For rainfed cocoa, availability of soil moisture will impose another restriction. Taking these into account, fertilizer application for rainfed cocoa may be done in two splits, first coinciding with pre-monsoon (southwest monsoon) period during May–June and the second by the close of the monsoon period in September–October. For irrigated crop, fertilizers may be applied in four equal splits during May–June, September–October, December, and February. Such four-split applications have been found beneficial for coconut also and this practice can go well in coconut–cocoa mixed cropping.

For young cocoa in the field, the fertilizer dose may be one-third the annual dose for adult plants during the first year and two-thirds in the second year. As cocoa under good management will start giving reasonable yield from the third year onwards, it may be logical to apply full dose of fertilizers from this point on. Application of organic manures for adult cocoa may not be essential, as there is a large return of organic debris to the soil by cocoa plants and a consequent substantial enhancement of soil organic matter content. However, for young cocoa, organic manures will be useful. These may be applied in the planting pits when seedlings are field planted and to the shallow basins afterwards.

Pruning

To restrict growth to a convenient height, pruning is done to have the first tier developed to the desired height and to remove excessive and inconvenient development of branches. Nair et al. (1994b) observed significantly higher yield in unpruned control plants during the fourth and fifth years, which ceased to be statistically significant during the subsequent years though the trend of superiority of control continued with decreasing magnitude with advancing age. It is, hence, only for convenience that cocoa is pruned. Among all the major cocoa-producing countries, the only country in which cocoa is regularly pruned is Brazil. Pruning is not a part of cocoa management in any of the African countries.

The advantages of pruning are the convenience of harvest, easier application of plant protection measures, and general cultivation. If unrestricted growth of the tree is allowed, harvesting would require workers to use knives attached to poles. Climbing the cocoa plant to harvest would not be feasible lest damage occurs to the flower cushions. It will be more convenient to undertake spraying if plant growth is restricted. Having the first tier developed at heights lower than 1–1.5 m will make cultivation operations difficult. With these considerations in mind, the following pruning operations may be suggested:

1. The first tier should develop to a height not less than 1.5 m. If plants jorquette at lower heights, the stem with the developing fans may be nipped off just below the jorquette. New chupons will arise on the main stem. One healthy chupon shoot may be allowed to grow up and jorquette. All the other chupon shoots may be removed. This process of nipping

chupon shoots may be continued until the desired height is reached. The height at which jorquettes form is decided by the nature of the plant, but environmental factors also play a role. In general, jorquetting is higher for plants under heavy shade. Nonavailability of mineral nutrients and water tends to lower jorquette height. Some plants tend to grow very tall before jorquetting. No known method exists by which jorquetting height can be lowered in such cases. However, such plants are rare in population.

2. There should be only one main chupon stem. At times, additional chupons arise from the main chupon, which are to be removed periodically.
3. Vertical growth is to be limited to a single tier. However, a second tier may be allowed to develop if the first tier is damaged. Arresting further vertical growth would require continuous removal of chupons that develop from below the jorquette. This will have to be a continuous process as the normal tendency is for the plants to put out new chupons. Normally, chupons arise from chupon stems only, and fan laterals arise only from fans. Rarely, chupons arise from fans, and should be removed at the early stages.
4. Drooping fan branches may be cut off at a suitable distance from the jorquette once a year, preferably when the crop load is low. December–January and July–August may be convenient under the Indian conditions, as cocoa will be nearly pod free during this period. Removal of part of the foliage may also help reduce transpiration in the summer.
5. "Centering" as part of pruning was recommended previously. This involves removal of all lateral fan branches that arise on the fans to a certain distance around the jorquette. The distance from the jorquette that is often suggested is in the range of 30–50cm. This allows sunlight to fall on the main stem, which is beneficial to enhance flowering. This speculation, however, lacks experimental support and, further, even if it enhances flowering, it may not mean enhanced production as flower production is rarely a limiting factor in cocoa production. The general experience is that flower production is profuse even in plants with dense canopies shading the main stem. A more probable advantage from centering may accrue from better aeration of the fruit-bearing main stem. This may reduce the incidence of black pod in the rainy season.
6. Although the previous recommendations may apply in the case of young cocoa, mature plants may be pruned gradually without much shock. The suitable period for this operation for Indian cocoa may be in December–January. Similarly, if the second tier that is already developed is to be removed, it may be done in phases, removing the fans one by one.

Control of Weeds in Cocoa Plantations

The usual weed control operations common in major cocoa-producing countries, where the crop is grown sole, slashing is done twice a year at a height of 5–15cm. Experiments on near-complete removal of weeds in comparison to slashing done in Ghana have indicated no additional advantage to complete removal. In the long run, it may also have the disadvantage of enhanced erosion of bare soil. Slashing will have the advantage over clean cultivation of being less labor intensive. Weeds compete with the main cocoa crop for water, nutrients, and light. Under conditions in which cocoa is grown in the major producing countries where rain-free period is short and fertility super optimal, weeds may not seriously compete with the crop.

Irrigation Requirements

As much as irrigation is concerned it is not a usual practice in cocoa plantations. Only container grown seedlings are irrigated. Experimental results also generally indicate the lack of response to irrigation. Obviously, the reason for such a lack of response must be attributable to the well-distributed rainfall in these cocoa-growing regions, the water stored in the root zone being sufficient to make the plant function normally. In countries where the rain-free period extends from 4 to 6 months, irrigation in summer months would be beneficial.

Top Working

Top working is useful to rejuvenate old and unproductive cocoa plants and also to convert genetically poor yielders to high yielders. This technique was standardized by the Cadbury-Kerala Agricultural University Cocoa Research Project, in Vellanikkara, Thrissur District, Kerala State (Nair, 1994). This technique consists of snapping back the desired trees below the jorquette after cutting halfway. The snapped canopy continues to have contact with the trunk. A number of chupons arise below the point of snapping and this is triggered by the breakage of apical dominance. Patch budding, as practiced in the nursery, is done on three to five vigorous and healthy shoots using scions from high-yielding, disease-resistant clones, and the remaining chupons are removed. The polyethylene tape is removed 3 weeks after budding and the stock portion above the bud union is snapped back. The snapped portion is removed after the development of at least two hardened leaves from the bud. When sufficient shoots are hardened, the canopy of the mother tree can be removed completely. Because of the presence of an established root system and trunk with reserve food, top-worked trees grow much faster and give prolific yield 1 year after operation. Though top working can be done in all seasons, it is preferable to do it during the rain-free period in irrigated gardens. In the case of rainfed crop, it may preferably be done after the receipt of pre-monsoon showers.

The top-worked trees start yielding heavily from the second year onward. About 50 percent improvement in yield is obtained in the second year (Nair et al., 1994a). About 100 percent increase in yield is obtained in the third year (Jose et al., 1998). Loss of crop for 1 year during the operation is thus compensated by bumper crops in the following years. The main stem will continue to belong to the original plant, and the fruits borne in this area belong to the poor yielder. Better yields are, however, obtained from the fan branches of the high-yielding clone used for top working. Analysis of performance of these top-worked trees over a period of 8 years indicated that these trees continued to yield heavily (Nair et al., 2002).

Flowering and Fruit Set

There is no uniformity in flowering in cocoa throughout the year, and there are peaks during some months of the year in flowering. These periods of peak flowering are often different for different regions indicating strong association with climatic factors. For example, in Ghana, normal flowering reaches its peak in May–June, starting from March and extending through April. In addition to this normal flowering is the "crazy" flowering that may occur during any period of the year. Several factors are considered responsible for the seasonal trend.

Genetic Factors

Cocoa types differ both in flower production and seasonal pattern. The African Amelonado is generally the least floriferous and is normally out of flower or nearly so from August to December in the Ghanian climate. During the early part of the year, Amelonado may remain without flowers generally; but stray flowering is noted in some years. The more floriferous cocoa types, in addition to producing large number of flowers, often show some flowering every month with peaks from March to May.

Environmental Factors

1. *Moisture Stress*: Cocoa flowers profusely by the onset of the wet season following a dry spell. Flowering is more profuse when the dry period is extended. In Ghana, the wet season starts normally from February following a dry period extending over December and January. On receipt of rains, a spurt of flushing is followed by the flowering period. In some cases, decreased flowering appears to result from decreased soil moisture following a period of excess moisture.
2. *Temperature*: Mean monthly temperatures below 23°C are considered to suppress flowering.
3. *Radiation*: Solar intensity is an important parameter in cocoa production. Limitation in illumination intensity of the cocoa canopy is not reported as a factor affecting flowering. In fact, in locations and seasons with continuous monsoon rains and cloudy weather, flower production is significantly inhibited.

Internal Factors That Influence Cocoa Production

Information available so far suggests a strong association between carbohydrate, nutritional, and hormonal reserves in the plant and seasonal patterns of flowering. Induction of flowering, following fertilizer application, shows that mineral nutrient status influences flowering. Flowering is reported to follow heavy flushing by the onset of the wet season in Ghana and the consequent rise in leaf area of plants. Increased leaf area index (LAI) would support higher photosynthetic activity and carbohydrate status. Hence, carbohydrate supply is considered a factor responsible for inducing flowering in such situations. The fact that fruit load affects flowering intensity, there being a decrease in flowering during intense fruit development periods, is taken as another indication of carbohydrate status as a factor.

Putting all the previously detailed factors together, the seasonal flowering pattern of cocoa can be explained in most of the situations. Thus, in Ghana, cocoa is considered to enter into a period of activity from February to May with the beginning of the wet season following the dry period extending over December and January. The relief from moisture stress induces flushing, leading to an increase in LAI. After about a month of flushing, when leaves harden and start photosynthetic activity, an induction of flowering occurs during this period. Competition from developing pods acts as the inhibitory factor up to November, which is the month of peak harvest in Ghana. The dry spell ensuing from December prevents flowering up to February.

Cherelle Wilt

Among the large number of flowers that are successfully fertilized, only a small percentage is carried to maturity. Until it attains a length of about 10cm, the young cocoa fruit is called a "cherelle," and more than 80 percent of the cherelles formed on a mature tree usually wilt. This phenomenon is called "cherelle wilt" and several reasons are attributed to its occurrence. Several fungal organisms and insect pests are found associated with wilting of cherelles, especially fungi. Attempts to correct cherelle wilt using fungicides and insecticides, although they could almost eliminate these organisms, may reduce it only slightly. The major cause for this is considered to be physiological, involving competition mainly for carbohydrates and, to a lesser extent, for mineral nutrients. On account of physiological reasons, cherelle occurs only up to a certain stage of growth of the developing pods, there being two peaks, 50 and 70 days after fertilization. No wilting occurs after 100 days. The first wilt coincides with cell division in the endosperm and the second one with the rapid growth of an embryo.

Indications of the involvement of mineral nutrient supply as a factor affecting cherelle wilt are given by the observations of a decrease in percentage of wilt following fertilizer application. However, fertilizer supply appears to have a much larger effect on setting of fruits than on cherelle wilt. The major factor considered responsible for incidence of nonpathological cherelle wilt is the competition for carbohydrates. It has been reported that the highest degree of wilting occurred during or just after periods of leaf flush. Another observation indicating competition as the factor is the decrease in wilting following removal of pods. Probably more important than the association with the vegetative sinks is the competition offered by older pods for the carbohydrate supply.

Indications of competition from pods as sink are evident in the rate of growth of young pods. The general trend is that the pods that set early grow fast whereas the late sets are slower in growing. The period required for wilting will be longer in early set pods than those of late set ones. The pattern of competition discussed previously for carbohydrates would mean that if a tree has massive flowering, a large number of cherelles will set compared to when fruiting is scattered. When hand pollination is used in hybrid seed production, this fact is put into practice. The total annual yield, however, may remain the same as the trees that are heavily loaded with fruits tend to show less profuse flowering and more cherelle wilt in the period that follows. In the large-scale hybrid seed production in Ghana, hand-pollinated mother trees are allowed to rest in alternate years as a safeguard against excessive carbohydrate drain and the probable consequent damage to the trees. This would also facilitate having a higher fruit load when hand pollination is used.

Plant Protection

More than 1500 insect pests have been found to attack the cocoa plant in different cocoa-growing regions of the world. However, a small number is only of economic importance. Among the major pests infesting cocoa, the significant ones are the red borer, tea mosquito bug, mealy bug, gray weevil, cockchafer beetle, rat, striped squirrel, and a host of others.

Insect Pests

Red Borer (Zeuzera coffeae)

This pest infests mainly the young cocoa plant. Larvae bore into thick shoots and into the main stem below the first jorquette along the center and cut a traverse tunnel before pupation. Small lateral galleries with openings are seen at intervals of about 25 cm on the stem. Infestation of the main stem causes the plant to dry up. The affected fan shoot shows complete withering of leaves, and these subsequently break off. Plants can be protected from the red borer by spraying 0.1 percent Carbaryl on the stems. Pruning and burning of the affected parts are to be done as soon as the symptoms are noticed.

Tea Mosquito (Helopeltis antonii)

These bugs mainly attack the pods. Circular water-soaked spots develop on the infested pods around the feeding punctures. These punctures later turn pitch black. Multiple feeding injuries cause deformation of the fruits. This pest can be controlled by spraying 0.05 percent Endosulfan.

Mealy Bugs (Planococcus lilacinus)

These bugs occur in groups on the tender shoots, flower stalks, foliage, and developing pods. Cherelles are often severely attacked. Maturing pods infested by the bug develop irregular sunken patches leading to the formation of scabs. This pest occurs throughout the year, but attains peak levels during July–October. It can be controlled effectively by the application of 0.05 percent Quinalphos. The insecticide may be applied only after collecting mature pods.

Gray Weevil (Myllocerus *spp.*)

A number of *Myllocerus* weevils infest cocoa in the different cocoa-growing regions of the world and cause considerable damage. The peak periods of infestation occur between July and September. Young plants are most susceptible. Adults occur in groups on the underside of the leaves and feed on the green matter, leaving the veins intact. The flaccid young leaves are generally not affected. The entire foliage will be badly skeletonized, causing growth retardation. Attack on young plants up to 2 years can be controlled by prophylactic spraying of 0.1 percent Carbaryl. Spraying may be concentrated on the under surface of the leaves. Application of the insecticide may be done twice a year, once during May and again in September.

Cockchafer Beetle (Leucopholis *spp.*)

Grubs of this insect feed on the surface of the roots and tap roots of young cocoa plants causing yellowing and drooping of foliage. Adult plants are also sometimes seen infested. Infestation by this pest is more frequent in coconut–cocoa intercropping, as the pest attacks coconut as well. Seedlings can be protected by application of 10 percent Carbaryl dust in pits around the root zone. Adult plants may be drenched with 2.5 ml/l of Chlorpyrifos.

Citrus Aphid (Toxoptera aurantii)

These aphids occur mainly during June–October. They infest flower stalks, tender flaccid leaves, buds, and tender chupons. On the foliage, the aphids are confined to the lower side. Chemical control of these aphids is not essential. Severely affected plant parts are to be collected and destroyed.

Red Branded Thrips

Adult and nymph thrips appear in colonies on the under surface of the leaves and also on the pods. The thrips feed on the fluid exuding from scraped tissues. Infested leaves turn pale green to pale brown and subsequently dry up. The thrips can be controlled by the application of 0.05 percent suspension of Quinalphos, Phosalone, or Fenthion.

Storage Pests

Cocoa beans stored for more than 2 months are found to be damaged by many species of insect pests. The most significant among them is the rice meal moth (*Corcyra cephalonica*). The larvae of this moth feed on the internal contents of the beans and construct silken galleries using frass and broken-down particles of the beans. Direct application of insecticides to cocoa beans or to the containers is not recommended as it would lead to food poisoning. The cocoa beans, when meant for long-term storage, can be mixed with neem leaves, 2 percent by weight, which helps to protect the beans up to 6 months in storage.

Noninsect Pests

Striped Squirrels (Funambulus tristriatus)

These rodents cut irregular holes on the walls of maturing pods and completely extract the contents. They feed on the mucilaginous pulp around the beans. Continuous trapping using attractants and poison baiting will be effective to check the population. Since the squirrels damage only ripe fruits, damage can be reduced by harvesting mature pods at regular intervals. Mechanical protection of the pods by covering with punched polybags of 150 G smeared with bitumen–kerosene mixture is also partially effective.

Rats

Rats cause damage similar to that caused by squirrels. The holes made on the pods are surrounded by areas of endocarp exposed by scraping the fleshy portions of the pods. Rats also prefer pods in the post-bronzing stage for feeding. Harvesting pods at the right stage when the furrows start bronzing will reduce damage considerably. Baiting with rodenticides in the garden is recommended. Rain proof preparations are preferred. Fumarin bars (rain proof) tied to the base of the inner frond of coconut and setting up bamboo noose traps in coconut crowns can be quite effective.

Civet Cat (Paradoxurus hermaphroditus)

Unlike rodents, civet cats gnaw holes on pods and bite and break the husk. Pieces of broken chunks are 2–3 cm in diameter. Infested pods show two distinctly spaced (about 1 cm apart) markings caused by the canine teeth and a row of small dots representing the markings of the incisors. The civet cats swallow the beans, and as such no trace of beans will be seen under the cocoa plant. Instead, piles of defecated beans are seen scattered around the plantation. Civet cats can be controlled by poison baiting with Carbofuran granules. Ripe bananas are split longitudinally into two halves, about 0.5 g of Carbofuran added, and the halves are closed properly. Two such bananas may be kept on the trunk at five to six cocoa trees per hectare.

Diseases

Cocoa is affected by many diseases. Loss due to diseases in cocoa has been estimated at 21 percent (Hale, 1953). Diseases may debilitate or kill the plant depending upon the type of pathogen involved. The significant diseases are described here.

Phytophthora *Pod Rot* (Phytophthora palmivora, Phytophthora capsici, Phytophthora megakarya, *and* Phytophthora citrophthora)

This is a very serious disease that occurs in the rainy season. Infection appears as minute, translucent, water-soaked spots on the pod surface, which turn chocolate brown, darken, and increase in size. Ultimately, the whole pod is invaded by the fungus, and the pods turn completely black. The beans in a ripe pod may escape partially or wholly from infection as the beans get separated from the pod husk on ripening. Periodic removal and destruction of the infected pods will help reduce spread of the disease. Cultural practices, such as proper pruning and regulating the overhead shade to reduce humidity and improve aeration, have been recommended for the control of the disease. Copper fungicides, such as Bordeaux mixture, copper oxychloride, cuprous oxide, or copper hydroxide, are generally used in the control of the disease. Spraying of 1 percent Bordeaux mixture at 15-day intervals starting from the onset of monsoon along with periodic removal of infected pods is effective in controlling the disease in severely affected gardens. Extracts of *Allium sativum*, *Cinnamomum zeylanicum*, *Lawsonia inermis*, and *Adenocalymma allicea* have been found to be effective in inhibiting lesion development on detached cocoa pods. Antagonistic effects of *Pseudomonas fluorescens* against *Phytophthora palmivora* have also been reported.

Genetic resistance offers better prospects for the control of this disease. Selections of cocoa, such as SCA 6, SACA 12, Pound 7, Catongo, and K82, have shown some resistance to this disease. Investigations conducted in Java have indicated that the cocoa types DRC 16, SCA 6, SCA 12, and ICS 6 were resistant to *Phytophthora* pod rot.

Colletotrichum *Pod Rot* (Colletotrichum gloeosporioides *(Penz.) Sacc;* Colletotrichum theobromae, Colletotrichum lurcificum, Colletotrichum eradwickii, Colletotrichum incarnatum, Colletotrichum fructitheobromae, *and* Colletotrichum thobromicolum)

Infection starts from the surface of the pod, usually from the end of the stalk or from the tip of the pod. The lesions develop as dark brown areas with diffused yellow halo. Infection spreads to the stalk and advances to the cushion. The internal tissues of the pod become discolored. In certain cases, infection may initiate from other parts of the pod also. Dark brown sunken lesions later coalesce to form larger ones. In severe cases of infection, the whole pod surface is infected. The pod shrinks and remains on the plant in a mummified form. Carbendazim and Mancozeb are reported as fungicides with promise to control the disease.

Botryodiplodia Pod Rot/Charcoal Pod Rot (Lasiodiplodia theobromae *(Pat) Griffon and Maubl.* Botryodiplodia theobromae *Pat*)

This pod rot occurs more frequently during the dry season. Symptoms initially appear as pale yellow spots on the pods, which enlarge into chocolate-brown larger lesions. In general, infection originates at the stalk end or at the tip of the pod. At times, the lesion develops from other parts of the pod. In most cases, the entire pod becomes black and exhibits a sooty covering all over consisting of the spores of the fungus. Infected pods become mummified and remain attached to the plant. The disease is also found to affect young twigs causing dieback.

Since the disease mainly affects wounded pods and pods of the plants under stress, better management practices will reduce the incidence of the disease. Use of Rovral (Iprodione) 2000 ppm at monthly intervals for 6 months during the dry season is suggested as control measure. Since the fungus is a wound pathogen, spraying 1 percent Bordeaux mixture along with an insecticide will also be useful.

Witch's Broom (Crinipellis perniciosa) *Stahel Singer*

The disease was first reported from Suriname in 1895 and was largely responsible for the destruction of cocoa both in Suriname and Guyana. This disease has spread to Bolivia, Brazil, Colombia, Ecuador, Guyana, Peru, Suriname, Venezuela, Grenada, Tobago, and Trinidad. The characteristic symptom is the development of brooms or shoots due to the hypertrophic growth of the infected bud. Infection of axillary bud or terminal bud leads to the production of vegetative broom. Such infected buds develop into dense, curved growth with excessive lateral shoots and short internodes. Leaves remain small with swollen stalk and pulvinus. Stipules are generally larger than normal and persistent. Some shoots grow vigorously from the infected region, as a result of which a "grown-through" broom develops. Brooms develop on the fan branches as well as on the chupons. If the bud dies due to infection, the broom remains inconspicuous. Abnormality occurs also in the internal tissues. Infected flower cushions sometimes produce leafy broom with less proliferation of axillary buds or abnormal flowers known as "star broom." Pods are infected by hyphae ramifying through the stalk from the cushion or directly by the penetration of the germ tube. As a result, pods exhibit abnormal shape, such as carrot shape, or one-sided distortion, which varies with the age of the pod. Sometimes infected cherelles do not ripen normally. Larger pods become hard when dry. The source of inoculums can be reduced by removal of the brooms. It is essential to start phytosanitation before the disease becomes severe. Thus, early removal of the brooms helps in minimizing the cost of disease control. All diseased tissues should be removed and burned, since brooms on the ground still produce basidiospores. Pruning of brooms should be done twice a year during dry periods. This disease has not been noticed in India.

Monilia Pod Rot (Moniliophthora roreri *Evans*)

Monilia pod rot or moniliasis is prevalent in South American countries. The fungus infects only young pods. The initial symptoms appear as small water-soaked lesions. Such lesions coalesce to form dark brown necrotic spots with an irregular margin, which later spread and cover the entire pod surface. Later, the lesions become covered with whitish mycelium consisting of abundant conidia. In certain cases, the diseased pods may look healthy; but at harvest, the internal portions will be found rotten. The disease can be controlled effectively by weekly removal and destruction of the diseased pods. The intensity of the disease can be reduced by improved drainage, regulation of shade, frequent and light pruning, and timely weeding of the cocoa plantation. Control of the disease through planting of resistant/tolerant cultivars seems promising.

Mealy Pod Rot (Trachysphaera fructigena *Tabor & Bunting*)

Only in Central Africa is this disease prevalent. Infected pods develop brown lesions similar to those of *Phytophthora* pod rot. However, this pod rot can be distinguished from *Phytophthora* pod rot based on the nature of sporulation. The white encrusted mass of spores produced on the infected area turns pink very rapidly. Vegetative parts of the plant are not affected. It has been found that fungicidal application for the control of *Phytophthora* pod rot can give good control of mealy pod rot also.

Seedling Blight [Phytophthora palmivora *(Butler) Butler*]

Symptoms of the disease develop on the leaves and stems of seedlings. On young leaves, initial symptoms appear as small water-soaked lesions on the under surface of the lamina. These lesions are either scattered all over the leaves or seen at the distal end and margins of the leaves. Subsequently, defoliation occurs. On mature leaves the water-soaked lesions appear along and near the veins. These later turn to dark brown, resulting in leaf blight and defoliation. On immature stems, initial symptoms develop as water-soaked linear brown lesions, which later turn black. Infection starts from the tip of the seedlings, spreads downward, and results in defoliation and dieback. Infection at the cotyledonary region spreads both upward and downward, resulting in wilting.

The disease can be controlled by improving drainage facilities in the nursery, by adjusting the shade, and by drenching or spraying the seedlings with Bordeaux mixture or copper oxychloride just before the onset of the monsoon and thereafter at frequent intervals. Severely infected seedlings should be removed and destroyed. Potassium phosphonate is also found effective in reducing the severity of the disease.

Vascular Streak Dieback (Oncobasidium theobromae *Talbot and Keane*)

Vascular streak dieback (VSD) disease is a destructive one in Papua New Guinea and many Southeast Asian countries, such as Malaysia, the Philippines, Southern Thailand, China's Hainan Island, several provinces of Indonesia, and India. This disease may occur on the main stem of a seedling or on a branch of an older tree. The first symptom is yellowing of the second or third leaf behind the growing tip with the development of green spots or islets scattered over a yellow background. The infected leaves fall off within a few days and, subsequently, the leaves above and below turn yellow and are shed, resulting in a distinctive situation where the leaves on the middle portion of the shoot fall. Lenticel enlargement, axillary bud proliferation, and interveinal chlorosis are the other symptoms. If the bark is peeled off from the infected region, the cambium turns rusty brown very quickly. The xylem vessels show several brown streaks when the affected stem is split open. Hyphae of the fungus are seen in xylem vessels of the infected stem and leaves. Genetic resistance offers good prospects of controlling VSD disease in the long run. Cultivars of Upper Amazon and of *trinitario* origin are, in general, less susceptible than *Amelonado* or its hybrids.

Phytophthora *Canker* [Phytophthora palmivora *(Butler) Butler*]

This disease is reported from Sri Lanka and India. Stem canker appears in different parts of the tree, including jorquette and fan branches. The external symptom appears as grayish-brown, water-soaked lesions with broad dark brown to black margin on the bark. A reddish brown liquid oozing out from such lesions dries up and forms a rusty deposit. The tissues beneath always show a characteristic reddish-brown discoloration. Lesions in the tissues coalesce leading to extensive rotting. The infection spreads from the cortical tissues to the vascular tissues and reaches the wood. Wood infection appears as grayish-brown to black discoloration with black streaks. When canker girdles the stem, dieback occurs. Leaves wilt, turn yellow, and fall off. Pods also show wilting. Finally, the whole tree dies. Spread of infection in the internal bark is faster than the spread on the surface of the bark.

The disease can be managed effectively if detected early. Removal of affected tissues and application of Bordeaux mixture is an effective control measure including Difolatan or any other copper-based fungicide. Care should be taken to remove and destroy the infected tissues completely. Cut and remove the infected small branches. Proper measures of control of *Phytophthora* pod rot will also help in reducing the canker disease.

Pink Disease (Corticium salmonicolor *Berk & Br*)

Fan branches and small twigs are generally infected. The first indication of the disease is the death of the branch. On the bark of the infected branch, characteristic pinkish, powdery encrustations of the fruiting bodies of the fungus can be seen. After a long time, the pinkish color turns to grayish-white. Prior to the appearance of visible external symptoms, many fine, white, silky mycelia would have already spread over the surface and into the cortex of the bark. This leads to the defoliation and death of the distal part of the branch.

Removal and destruction of the infected branches and wound sealing with Bordeaux paste will check spread of the disease. Avoidance of excess shade and provision of adequate drainage are important for better control of the disease. Because the pathogen has a wide range of hosts, care should be taken not to use susceptible species as shade or as a cover crop. In areas where severe incidence is recorded, the disease can be prevented by regular spraying of 1 percent Bordeaux mixture, especially during rainy season.

Thread Blight

Three types of thread blight, namely white thread blight, horsehair blight, and *Koleroga* (normally encountered in the case of Cardamom, mainly in the south Indian state of Karnataka), have been reported from different cocoa-growing countries.

White Thread Blight [*Marasmiellus scandens* (Massee) Dennis & Reid]

The young branches of infected plants contain white mycelial threads of the fungus, which spread longitudinally and irregularly along the surface of the stem. Under highly humid conditions, the fungus grows very rapidly on the stem and enters the leaf at the nodes through the petiole. On the leaf lamina, the fungus spreads in the form of fine threads, as a result of which the affected portions turn dark brown to black, ultimately leading to the death of the leaves. Such dead leaves in a branch eventually get detached from the stem, but are found suspended by the mycelia threads in rows. Extensive death of young branches and hanging leaves in mycelia threads in rows are the common field symptoms. Severity of the disease can be reduced by the removal of the dead tissues, pruning of the affected parts, and proper shade regulation. In severe cases of infection, spraying of copper fungicides will be helpful in controlling the disease.

Horsehair Blight (*Marasmius equicrinis* Muell.)

Young branches of the affected plants are found to be ramified with tangles of the black hairlike growth of the fungal mycelium. The black branched mycelial network of the fungus spreads over the leaves, petioles, and twigs. The mycelia remain hanging loosely. As a result of the infection, the midribs of the leaves show a brown discoloration. From the midrib, the discoloration spreads to the veins and veinlets. Later, the affected leaves dry up. The infected twigs also show drying. Such dried-up leaves and twigs get detached from the branches and remain suspended by mycelia strands. Since horsehair blight disease is observed in neglected gardens, incidence of the disease can be reduced by proper management practices. Removal and destruction of the affected parts are quite useful.

Koleroga Thread blight

The symptoms of *Koleroga* thread blight caused by *Pellicularia koleroga* Donk are similar to those of white thread blight except that the fungal threads are brown.

Colletotrichum Leaf blight [*Colletotrichum gloeosporioides* (Penz.) Sacc]

Colletotrichum infection on cocoa leaves was reported as one of the serious problems of cocoa in Colombia and Ghana. In India, *Colletotrichum* causes three types of foliar symptoms, namely leaf blight, shot hole, and irregular leaf spot. Of these, leaf blight and shot hole are widespread and occur on plants of all age groups. Shade regulation is found to be an effective method of control of the disease in Ghana. The disease can be effectively controlled by proper pruning, and in case of severe incidence, by spraying Carbendazim or Mancozeb.

Chupon Blight and Twig Dieback [Phytophthora palmivora *(Butler) Butler*]

Infection usually initiates anywhere on the young leaf lamina, in the axils of the leaves, on the petiole, or at the tip of the twigs or chupons. The characteristic symptom is the appearance of water-soaked lesions, which later turn brown-black. In severe cases of infection, the lesions coalesce to form large lesions. Severe infection on the leaves leads to defoliation. When the lesions girdle the stem, the portion above the point of infection dies, causing twig dieback or chupon blight. Shade regulation by proper pruning is found to be the best method of reducing the disease incidence. Pruning should be done prior to the rainy season. When incidence of the disease is severe, the disease can be controlled by spraying Bordeaux mixture or any copper oxychloride formulation.

Cocoa Swollen Shoot Virus

Any strains of cocoa swollen shoot virus exist, which differ in the symptom expression, the vectors that transmit them, and the range of their other hosts. The virulent strains observed in Ghana produce various types of leaf necrosis, root and stem swellings, and dieback in Amelonado trees. Such infected trees are usually killed within 2–3 years. Avirulent strains rarely produce leaf symptoms and are not lethal. However, root and stem swellings are often produced.

Swollen shoot virus strain IA (new Juaben strain) produces swelling on the fan branches, chupons, and roots. This strain on young flush leaves of Amelonado causes red vein banding and vein clearing. Later, a fern leaf pattern is produced, and mature trees at this stage have a yellowish appearance. Pods on the infected tree become mottled and are smoother than normal, containing only half the normal weight of beans. Different strains of cocoa swollen shoot virus produce different symptoms on leaves. The symptom variation on leaves produced by the different strains helps in strain identification. The Ghanaian strains are more virulent and damaging than the Nigerian strains. This disease is more damaging to cocoa trees under stress.

More than a dozen species of mealy bugs are reported to transmit the virus. The important mealy bug species transmitting the disease are *Planococcus citri*, *Planococcoides njalensis*, *Planococcus hargreavesi*, *Ferrisia virgata*, and *Pseudococcus concavocerarii*. Spread of the disease can be arrested by removal and destruction of infected plants. However, this method does not prevent new outbreaks. The use of resistant lines is probably the ultimate solution to this malady. Apart from cocoa swollen shoot disease, other virus diseases, such as cocoa necrosis virus, cocoa mottle leaf virus, and cocoa yellow mosaic virus, have also been reported.

Ceratocystis Wilt (Ceratocystis fimbriata *Ellis & Halst*)

This disease is usually associated with damage by beetle borers or pruning wounds. The disease occurs in almost all cocoa-growing countries. Wilting of the whole or part of the tree followed by rapid death are the visible symptoms of the disease. In the initial stages of the disease, mature leaves change from their normal horizontal position and become pendulous. The wilted leaves then dry and remain attached to the dead branch for several weeks. The disease is always associated with borer holes on the stem made by *Xyleborus* beetles. The holes are about 1 mm in diameter, with a small amount of wood dust around it. The internal wood tissue surrounding the wound will be discolored to brownish-red or purplish. Neither chemical control of the beetle or fungus, nor the destruction of the infected plants has proved a useful method of control of the disease. The most practical method of preventing the disease incidence is to minimize wounding during pruning and harvest. It has been reported that *criollo* cultivars are more susceptible than *forastero*.

Verticillium Wilt (Verticillium dahliae *Kleb.*)

This disease is widespread in Brazil and Uganda. The first symptom is the drooping of the leaves without any loss of turgor, so that the leaves hang down without any flaccidity. Subsequently, the leaves dry and roll inward and later fall off. The infected small branches break off gradually. In the early stages of disease development, a marked reduction occurs in the root system and wilting of young pods. Necrosis of tap and main lateral roots occurs only after defoliation of the shoot. Discoloration of the xylem vessels of the petiole, pedicel, stem, and roots is also observed. Severe incidence of the disease, especially following moisture stress because of drought or waterlogging due to excess moisture due to rainfall or lack of drainage, can cause death of the cocoa plant within a week.

Cushion Galls

Cushion galls are important in some countries of South and Central America. Cushion gall is a collective term for a number of forms of flower cushion hypertrophy.

Five types of cushion galls (green point gall, flowery gall, knob gall, disk gall, and fan gall) have been identified and described. Of these, only fan gall and knob gall have been reported from India. Among the different cushion galls, green point and flowery galls are significant and are reported to be caused by *Fusarium rigidiuscula*. Causal agents of the other cushion galls have not yet been identified.

Root Diseases

A number of root diseases have been reported in cocoa. Rarely do these diseases cause any substantial damage to the crop. Different pathogens are known to cause the root diseases. Their primary source of inoculum comes from forest trees cleared prior to planting or infected shade trees. The above-ground symptoms of all root diseases caused by different pathogens are sudden wilting of the leaves of the tree followed by death of the plant. However, root diseases can be identified by the presence of fruiting bodies present on the roots or on the collar region of the affected plant. Root diseases have so far not been reported from India. The major root diseases are:

1. Brown root disease: *Phellinus noxius (Fomes noxius)*
2. White root disease: *Rigidoporus lignosus* (*Fomes lignosus*)
3. Black root disease: *Rosellinia pepo* Pat.
4. Collar crack: *Armillaria mellea* Vahl.

The spread of the diseases can be checked by the removal of infected plants, including roots. Digging trenches around the infected tree is also recommended. The infected plants with their roots should be removed and burned. There is no recommendation for chemical control of the disease, except that the cut surfaces of all stumps of cocoa and shade trees should be painted with an aboricide, such as sodium arsenite or 2,4,5-T, at the time of felling to kill the stumps. If 2,4,5-T is used, a 3 percent solution of copper fungicide should be added.

Harvesting

The pods mature in about 5–6 months' time from the date of pollination. This period varies depending on environmental conditions. A highly significant negative correlation between the number of days from pollination to harvest and the mean temperature during the period of fruit development has been established. According to Alvim et al. (1972), the days to maturity may be calculated from the equation $N = 2500/T–9$, where N = number of days from pollination to harvest and T = the mean temperature in Celsius.

The stage of maturity of pods is best judged by change of color of the pods. Pods that are green when immature turn yellow when mature, and the reddish pods turn yellow or orange. The change in color starts from the grooves on the pods and then spreads to the entire surface. Although pods can be harvested as color changes, the pods may remain on the tree without damage up to a maximum of about 1 month. The intervals between harvests can therefore be extended up to 1 month. However, it is safer to harvest at fortnightly intervals. In areas prone to damage by mammalian pests, harvesting intervals may preferably be shorter. When black pod incidence is serious, shorter harvesting intervals are preferred to ensure field sanitation.

As fruits are borne on the cushions and as damage to flower cushions is to be avoided, harvesting is done using a knife. When trees are tall, harvesting may be done by using knives attached to poles.

Post-Harvest Storage and Breaking Pods

The harvested pods can be stored for 2–5 days. This enhances prefermentation activity inside the pods and helps to facilitate rapid rise in temperature during fermentation, reduces acidity, and imparts stronger chocolate flavor (Arikiah et al., 1994). The harvested pods are broken by hitting against a hard surface, and beans are extracted without placenta and kept for fermentation immediately. Only mature, well-developed pods contain good beans. Pods showing symptoms of damage from black pod on the surface need not be discarded if the beans inside are unaffected. The color of the pulp will be a good indication of suitability as damaged pods show discoloration.

Primary Processing

Primary processing denotes production of dry cured beans for the market. This involves fermentation and drying.

Fermentation

Raw cocoa beans are covered with sugary mucilaginous pulp, and the beans with the pulp around them are called "wet beans." The kernel, which is also called the "nib," is the economically important part. Fresh nib is bitter and is not suitable for manufacture of different products. When raw, it does not have any flavor, aroma, or taste of any of the cocoa products. Chocolate flavor is developed during fermentation, which is done by the manufacturer. All of the standard methods of fermentation essentially involve keeping together a mass of reasonable quantity of wet beans for periods ranging from 4 to 6 days. In most of these standard methods, mixing of the mass of beans occurs usually on alternate days. One of the consequences of fermentation is the loss of most of the pulp around the beans; however, more important is the series of biochemical reactions occurring in the beans, which are necessary for inducing the characteristics of the cocoa products.

Biochemical Changes Occurring during Fermentation

The pulp contains about 84.5 percent water, 10 percent glucose and fructose, 2.7 percent pentosan, 0.7 percent sucrose, 0.6 percent protein, 0.7 percent acids, and 0.8 percent inorganic acids (Hardy, 1960). The pulp is sterile initially, but the presence of sugars and high acidity (pH 3.5) provide excellent conditions for the development of microbial populations. A wide range of microorganisms infect the mass of beans through the activity of fruit flies and contamination from the fermentary. Initially, yeasts proliferate and convert sugars to alcohol. The cells of the pulp start to break down soon after the fermentation process begins, either through an enzyme change or by simple mechanical pressure, and the watery contents of the pulp, which are called "sweating," drain out. This continues for 24–36 hours. The sweating constitutes 12–15 percent of the weight of wet beans. The activity of yeasts leads to the production of CO_2; at this stage relatively aerobic conditions prevail and allow the development of lactic acid bacteria, which assist in sugar breakdown.

The activity of bacteria leads to the production of organic acids. When the sweating has run off, the conditions become more aerobic and the acidity is reduced by citric acid removal. The presence of oxygen allows acetic acid bacteria to take over from the yeasts and convert alcohol to acetic acid. These reactions cause a rise in temperature in the bean mass. A positive correlation exists between sizes of the relevant microbial populations and the amount of acids produced during fermentation (Samah et al., 1993). Temperature rises to peak after first mixing to about 48–50°C and falls slowly until the next mixing. With the next mixing also temperature rises again, but often to a lower peak around 46–48°C, which falls again slowly towards the completion of fermentation. Variations are likely depending on the method of fermentation, location of the beans in the ferment, and environmental conditions. The rise in temperature should be taken as indicative of the necessary biochemical reactions.

The pH of the beans and pulp also varies conspicuously. The fresh cocoa bean pulp is acidic with a pH of around 3.5. The pH of the cotyledons is very much higher, around 6.5. After the beans have been pulped, components of pulp diffuse through the testa into the beans, and the acids that are synthesized from pulp move into the beans to lower the pH of the nibs still further. The pH of the nib on third day would be around 4.8. With further progression in fermentation, pH tends to gradually increase to values around 5.0 by the end of the fermentation period. Although a decrease in pH occurs in the cotyledons, the pH of the pulp increases from the initial level to a final value equal to the nib.

The acetic acid diffusing through the tests causes a breakdown of the polyphenol and lipid membranes of the vacuoles of the cell, and the cell contents get mixed. Various enzymatic reactions take place, and the polyphenols get oxidized. This reaction is partially responsible for the removal of the bitter taste from the beans. The production of volatile compounds arising from the reaction of amino acids with sugars leads to induction of aroma. The exact nature of compounds responsible for this reaction is unknown, although more than 300 compounds are considered to have their influence. The most important change that occurs in the cotyledons during fermentation is the appearance of the chocolate flavor precursors. The proteins in the cotyledons undergo hydrolysis, giving rise to amino acids and conversion to insoluble forms by reaction with polyphenols. Voigt et al. (1994) found that free amino acids and oligopeptides are essential aroma precursors. The combined action of two enzymes, aspartic endoprotease and carboxy peptidase, on cocoa bean protein appeared to be required for the generation of cocoa-specific aroma precursors.

Factors Affecting Fermentation

Ripeness of Pods

Harvesting at intervals of 1–2 weeks ensures quality. Only healthy, ripe pods should be harvested. Use of overripe pods must be avoided as these may contain germinated beans, which may allow the entry of molds and insects. The underripe pods do not ferment properly, and the temperature of the fermenting mass continues to remain at 35°C after an initial rise to 45°C (Knapp, 1926). Alamsyah (1991) observed a weak chocolate flavor and low pH of cured cocoa beans from unripe pods.

Pod Diseases

Most pod diseases can lead to complete loss of beans. Even when the beans are not destroyed, it is undesirable to use the beans for fermentation.

Type of Cocoa

Criollo gets fermented in a relatively shorter period of 2–3 days, whereas *forastero* takes 5–7 days. Hence, mixing of the beans of these two types should be avoided.

Quantity of Cocoa

The heat generated during fermentation is retained by insulation, but this becomes more difficult to achieve with small quantities. About 50 kg beans are required for satisfactory fermentation.

Duration

The duration varies depending on the genetic structure of cocoa mass, climate, volume, and the method adopted. The duration of fermentation ranges from 1.5 to 10 days.

Turning

Turning ensures uniform fermentation. Dias and Avila (1993) recorded faster fermentation when turning was done every 24 hours. Frequent mixing, at 6–12 hours intervals, produced a higher number of well-fermented beans than other treatments (Senanayake et al., 1997).

Seasonal Effects

Temperature rises more slowly in wet weather in June–July. Dias and Avila (1993) recorded higher volatile acid contents in May than in June. Fermentation during the dry season was better than fermentation during the wet season.

Cocoa Bean Acidity

Cocoa products processed from some samples of cured cocoa beans are found to have a detestable acid taste. This is often designated as cocoa bean acidity. Cocoa bean acidity has been reported often from Malaysian cocoa. It has also been found that the beans giving acid taste to the products generally have low pH. The low pH of cocoa beans is also strongly related to titrable acidity.

It has been established that the organic acids responsible for cocoa bean acidity are mainly acetic and lactic acids. These are produced from sugars present in the pulp during the fermentation process. Acetic acid produced during fermentation is an essential component of the fermentation process as the acids contribute to the death of the bean, prevent colonization by putrefactive microorganisms, and create an environment conducive to the formation of flavor and aroma precursors within the cotyledons of the bean. However, excessive quantities of acetic and lactic acids produce an acid taste in the cocoa products, as these are not adequately dispelled in the roasting and conching processes.

The problem of cocoa bean acidity is reported mostly from Malaysian cocoa. The best chocolates are produced from Ghanaian cocoa, in which the acid taste is almost never observed. Bean pH ranges from 5.3 to 5.5 for the Ghanaian cocoa in contrast to a range of pH from 4.4 to 4.7 in the case of Malaysian cocoa.

Bean Maturation

Bean maturation is described as the process involving the removal of acid from cocoa beans by keeping the fermented beans warm, moist, and with good air supply. By maintaining this at desirable levels, it had been possible to raise the pH of the beans to acceptable levels in the range from 5.0 to 5.5. Two methods are suggested to reduce the acidity and to improve flavor of box-fermented Malaysian cocoa.

Box Maturation

The beans set for fermentation in boxes are to be mixed as usual on the third and fifth days. Five extra turns may occur on the sixth and seventh days, and the beans may be taken out on the eighth day.

Drier Maturation

Beans may be kept to thickness of 25 cm and dried at 50°C. Stacking to depths lower or higher than 25 cm results in poorer quality cocoa, presumably because of too rapid drying of beans in the former and lack of adequate aeration in the latter.

Methods of Fermentation

The method of fermentation and its duration will depend largely on the variety of cocoa and the season. *Criollo* cocoas in general need a shorter period of fermentation as compared to *forasteros*. Season influences the duration of fermentation, mainly through its effects on temperature and humidity. At lower temperatures and high humidity, fermentation period will be usually longer. Among the various methods adopted for fermentation in the different cocoa-producing countries, the heap, tray, and box methods are considered as the standard, widely adopted methods.

Traditional Standard Methods

Heap Method

This method is widely practiced in West African countries. The heap method essentially involves heaping a mass of 50–500 kg of cocoa wet beans over a layer of banana leaves. The banana leaves are spread over a few sticks to keep them slightly raised over the ground level to facilitate the flow of sweating. The leaves are folded and kept over the beans, and a few wooden pieces are placed on top to keep the leaves in position. The purpose of keeping the beans covered with the leaves is to conserve the heat produced during the fermenting process. The heaps are dismantled and the beans are mixed on the third and fifth days. It needs about 6 days for the completion of fermentation, and the beans can be taken out for drying on the seventh day. As soon as the beans are heaped, flow of sweating starts and continues for the first 2 days. Color of the pulp changes on the surface of the mass of beans to a depth of about 10 cm by the third day, with the bulk of beans inside retaining the whitish color. This change of color indicates the beginning of acetic acid production, which is limited to the surface layers where aeration is adequate. With the mixing of beans on the third and fifth days, beans on the surface whose pulp has almost drained away get mixed with the other pulpy beans. Diffusion of air is thus enhanced and acid fermentation occurs deep in the bean mass also.

Tray Method

This method was developed based on the early observation that when the beans are heaped to ferment, a change in color of the beans occurs up to a depth of about 10 cm when beans are not mixed. This was taken as an indication that there will be adequate aeration of the bean mass up to this depth without mixing, and that if the beans are kept only up to this height, mixing can and probably should be avoided. Based on this, beans were filled in trays of 10 cm height, holding reasonable quantities of beans, and trial fermented. It was found that when such trays are stacked one over the other, adequate development and conservation of heat occurs and that fermentation would be over in a shorter period.

The usual size of wooden trays is 90 cm × 60 cm × 13 cm. Battens or reapers are fixed at the bottom of the trays with gaps in between that prevent beans from falling through, and allow for free flow of sweatings. Allowing for the space required for the reapers, net depth of the beans inside should be about 10 cm. The length and width of the trays could be increased to any extent theoretically, but the standard dimensions given earlier will make the size suitable for handling. Each tray can contain about 45 kg of wet beans. Thus filled, the trays are stacked one over the other. The minimum number of trays required for a stack is about six. An empty tray is kept at the bottom to allow for drainage of the sweatings. After stacking, the beans of the topmost tray are covered with banana leaves. After 24 hours, the stack of trays is covered with gunny sacks to conserve the heat that develops. There is no need to mix the beans. Fermentation will be normally completed in 4 days. On the fifth day, the beans are taken out for drying.

The minimum number of trays required to be stacked is about six, but as many as 12 trays can be used simultaneously. It is a 6-day stack, the total quantity of wet beans required for effective fermentation will be about 270 kg. When a 12-day stack is used, the minimum quantity will be about 540 kg.

Box Method

This method is suitable for handling large quantities of beans. It is common in Malaysia where cocoa is grown as large plantations. The boxes are made of wood with a standard dimension of 1.2 m × 0.95 m × 0.75 m. Boxes of this size can hold about 1 ton of wet beans. Holes are provided at the bottom and on the sides of the box to allow flow of the sweating and to facilitate aeration. The beans are to be mixed on alternate days, transferring the beans from one box to another at the time of mixing. This would necessitate having a minimum of three boxes. These may be arranged in a row, in which case the beans are to be transferred from boxes after lifting them. To make transfer of beans convenient, the boxes are sometimes arranged in tiers, and shutters are provided on one side of the boxes so that the beans falling from the box at the top will run to the lower box on removing the shutters. The beans are mixed thus on alternate days, on the third and fifth days, and are taken out to dry on the seventh day after 6 days of fermentation.

Although the box method of fermentation will be convenient for handling large lots of beans, the quality of box-fermented beans is often rated as inferior to those obtained from the heap and tray methods. The factors responsible for lowering the quality are often related to inadequate aeration of the fermenting beans, which results in induction of acidity in the beans.

Other Traditional Methods

Basket Method

Beans are fermented in woven baskets after lining them with a layer of banana leaves. The beans are mixed on alternate days. The fermentation period lasts about 6 days. Although variable quantities of beans can be used in this method, good quality beans result only when the quantity is reasonable.

Curing on Drying Platform

This method has been adopted in Ecuador. It consists of spreading the beans on drying trays during the day and heaping them at night. The quality of beans is reported to be reasonable in the main cocoa season, but not in the off season.

Judging the Endpoint of Fermentation

1. *Color of the beans*: *Forastero* beans turn brown with a pale brown center that has a brownish ring around the outside.
2. *External shell color*: The pulp, which is whitish initially, turns to pinkish white after sweatings have run off. At the end of fermentation, the shell surface will attain a reddish brown color.
3. *Smell of fermenting mass*: The fresh pulp has a faint sweet smell, which changes to a characteristic acid smell as fermentation proceeds. This color persists until the end of fermentation. Overfermented, putrifying beans will produce the smell of ammonia.
4. *Development of heat*: After setting for fermentation, the temperature of the mass increases steadily to reach a peak at 47–49°C by the third day. Temperature falls slowly until the end of fermentation to a range of 45–46°C. However, in the methods of fermentation involving mixing of beans, there will be a rise in temperature following mixing, which again will tend to drop steadily.

5. *Plump nature of the beans and color of the exudate*: Well-fermented beans will appear plump and full. On squeezing, a reddish-brown exudate flows out.

All beans in a fermenting lot will not be at the same stage of fermentation, and hence, all the beans in a sample drawn will not show the indices detailed earlier. When 50 percent of the beans show such signs, it can be presumed that fermentation is complete.

Small-Scale Methods of Fermentation

All the standard methods of fermentation require a minimum quantity of wet beans, which, invariably, must be a fairly big lot. Even using the heap method, the smallest batch size is 50kg or the produce from about 500 pods. In areas where cocoa is grown in small holdings, a more convenient alternative would be to adopt a method of fermentation involving small quantities of beans. Development of a small-scale method is not easily accomplished, as very small quantity of beans will make it difficult to develop adequate temperature for the fermenting mass. In the standard methods of fermentation, the conditions in the bulk of the fermenting mass remain anaerobic in the early part of the fermentation period. This also is difficult to be simulated with very small quantities of beans.

Small-scale methods of fermentation were developed using bean lots substantially smaller than those required for the standard methods. Some of these methods have been found to be successful, as judged from the development of temperature of the ferment, pH of the beans, and cut test (Kumaran et al., 1981).

Drying

The fermented beans will have a moisture content of about 55 percent. Such a high moisture content is unsuitable for storage of the beans as putrefaction may set in. The moisture content has to be brought down to about 6 percent for safe storage and transportation. Drying should commence immediately on cessation of fermentation. Unless the beans are skin dry, within 24 hours after fermentation, molds set in and damage the beans.

Drying of fermented cocoa beans is, however, something more than just driving out the moisture, as biochemical oxidation of acetic acid from the beans continues during drying. Thus, a very quick drying or excessive heating of the beans will not be suitable. A very slow drying also will not suit as the beans get moldy if they continue to remain moist for too long. The methods used to dry cocoa can be divided into two main types—first, sun drying and second, artificial drying.

Sun Drying

Sun drying is the simplest and the most popular method in most of the cocoa-producing countries. Depending upon climatic conditions, the beans are exposed to the sun for about 12–20 days. This method generally gives good-quality beans in traditional areas of cocoa production where the weather is sufficiently sunny. In West Africa, the beans are simply laid out in the sun, spread in a thin layer on mats raised off the ground or on concrete floors. After 2 days, the beans are stirred and dried again. In West Indies and South America, drying is done on wooden floors with moveable roofs, referred to as *boucans* in Trinidad and *barcacas* in Brazil. Drying can also be done on moveable trays that can be pushed under a fixed roof. Attempts to improve the efficiency of sun drying have been made using solar cabinets. A rocking dryer designed by small farmers of Ivory Coast consists of a bamboo platform with wooden edges covered with PVC sheet that can be removed to facilitate mixing. The platform pivots about its center so that it can be directed toward the sun.

Artificial Drying

With the introduction of cocoa into newer areas where the climate remains unsuitable for drying in the peak season, artificial drying methods became necessary. The relatively large requirement of space also made sun drying difficult when cocoa cultivation was extended on a plantation scale. Several types of artificial dryers came to be used, and some work on the best drying conditions has been done in different cocoa-growing countries. The results reveal that the major conditions recognized are temperature, rate of air flow, depth of the beans, and extent of drying. However, these result in high acidity. Suitable drying conditions are thus a balance between the economy in drying and bean quality. The maximum permissible temperature for drying is generally taken as about 60°C. It is better to use deeper loads of beans, the depth being convenient enough for stirring. Stirring of the beans also has been found necessary both for uniformity of drying and its efficiency. A convenient thickness could be about 12–15 cm when mixing is done manually.

Recovery of Cured Beans

Recovery varies from 30 to 46 percent depending upon the season and variety of cocoa. During the dry season recovery is high. *Amelonado* records as high as 44 percent recovery, whereas *Amazon* records only 38 percent recovery. Bean size and recovery are inversely correlated. Ripe pods give high recovery of dry beans when compared to unripe pods. To a certain extent, the grower can manipulate the rate of recovery of dry beans by harvesting fully ripe pods and proper post-harvest technology.

Cleaning and Bagging

The dried beans are bagged in jute bags of 62.5 kg capacity, after removing flat and broken beans.

Storage

Dry cocoa beans can be stored for long under suitable conditions. However, the period of safe storage will depend mainly on the relative humidity and temperature of the atmosphere in which the beans are stored. In the temperate climate, where relative humidity is low, the storage life is considered almost infinite. In tropical regions of high humidity, it would be difficult to store the beans for a considerable length of time. As in the case of other plant products, cocoa beans also attain equilibrium moisture content in a given atmosphere either by gaining or losing moisture.

It has been found that the bean moisture content will exceed 8 percent when the relative humidity of the atmosphere reaches 85 percent. This moisture level of 8 percent in the beans is critical, as growth of mold sets in when it is above this level. This means that it would be difficult to store cocoa beans without damage in atmospheres whose relative humidity exceeds 85 percent for a considerable period of the year unless special precautions are taken to prevent contact of the dry beans with air. An increase in pH of the beans during a 28-week storage was recorded by Premalatha and Mohanakumaran (1989). The damage due to molds and insects increased after 36 weeks in storage (Bopaiah, 1992). Storage of cocoa beans beyond 36 weeks requires redrying and packing to prevent deterioration.

The following storage precautions are stipulated by international standards:

1. The ambient humidity must not exceed 70 percent.
2. The bags must be stored at least 7 cm higher than the ground level, preferably on a duck board to allow free circulation of air.
3. There must be a passage at least 60 cm wide between the walls and the bags and between bags of different types of cocoa.
4. Protection against storage pests and rodents must be ensured.
5. Steps must be taken to avoid contamination by odors, off flavors, and dust.
6. The moisture content should be checked at regular intervals.

Secondary Processing

Secondary processing denotes the steps involved in conversion of cured bans into different finished products, the main product being chocolate. Secondary processing of cocoa beans is done in specialized factories. The principles of chocolate manufacture in large factories have been described by Wood (1975), Wood and Lass (1985), and Mossu (1992). The essence of cocoa and chocolate manufacture lies in the development of flavor by roasting the beans followed by extraction of cocoa butter from the nib to produce cocoa powder and addition of cocoa butter to the nib and sugar to produce chocolate. The major steps involved in the various processes are the following.

Cleaning and Sorting

When the beans arrive in the factory, they are cleaned to remove any foreign matter and sorted to separate the small or broken beans by passing them over a continuously vibrating screen. This is well aerated and is filled with powerful magnets. The metallic foreign matter, dust, and broken beans are removed.

Alkalization

When beans are used to manufacture cocoa powder, the cocoa liquor is generally treated with alkali to improve color and develop flavor. Alkalized cocoa is known commercially as "soluble cocoa." The amount of alkali used for the preparation of soluble cocoa is adjusted to bring about partial rather than complete neutralization. Saturated solutions of sodium or potassium carbonate or bicarbonate are most commonly used, whereas ammonia, ammonium carbonate, magnesium oxide, magnesium carbonate or bicarbonate, or mixtures of these chemicals are favored by some manufacturers. Alkali may be introduced prior to roasting or at the nib or chocolate liquor stages. However, it is more economical to mix it with chocolate liquor.

Roasting

Roasting of cocoa beans, more correctly termed as treatment of cocoa beans in hot air, is one of the most important operations in the processing of cocoa, the degree of treatment required being adjusted to the degree of ripeness of the beans concerned and any other pretreatment that they may have undergone (Riedel, 1977). The true purpose of roasting is not only restricted to the loosening of the shells, but also to develop positive flavor as well as the removal of excess moisture and other undesirable volatile matter. Roasting enables moisture content to be reduced to 1.5–2.0 percent. Different methods of roasting can be employed, which lead to the production of different end effects, some of which are more applicable in the case of some specific bean varieties than in the case of others. The manner of roasting must ensure equal treatment to all beans in a specific batch. Cocoa roasting for chocolate manufacture can be done optimally between 120°C and 125°C (Riedel, 1974). Optimal temperature also depends on the time allowed for roasting. Both temperature and time factors greatly influence color and flavor of the beans. Between 120°C and 135°C, chemical changes take place in the nib. Roasting can be done using direct or indirect heating. Gas, steam pipes, or hot air can be used for direct heating.

Kibbling and Winnowing

The shell is separated from the cotyledon by a process known as kibbling. The purpose of winnowing is to separate the shell and germ and to split the cocoa into its natural segments (cocoa nibs). Roasted cocoa beans can contain between 10 and 15 percent shell, depending on the source, and about 1 percent germ. The separation of shell and germ can be carried out separately or together, depending on the choice of commercial plant. Cocoa beans are first cracked by passing through rollers or rotating cones. An air current is then used to blow away the lighter shell. The velocity of this air stream is critical; it should be sufficient to remove the undesirable shell, but not too high to blow off the costly nib, and must be varied to suit the changing size of cocoa bean from differing sources.

Blending and Grinding

When it is necessary to have a blend of beans produced in different regions, blending is done before grinding. The composition of the beans after blending is a trade secret of individual chocolate manufacturers. The cotyledons or "nibs" are ground to get "mass" or "liquor." Cocoa mass contains about 5–58 percent fat, which is known as "cocoa butter." This butter has the characteristic of melting at body temperature. The nibs are ground at a relatively high temperature. Normally cocoa is subjected to pre-grinding followed by fine grinding (Bauermeister, 1978). The particle size of the finished product has a pronounced effect on its suitability as an ingredient of different food products (Minifie, 1968). During grinding, heat is generated by friction, which melts the cocoa butter. Cylinder rollers of three or four stages are used for normal grinding or a ball mill is also used. The ball mill gives better overall performance in terms of fineness of grinding, which is simple to maintain (Bauermeister, 1978).

The cocoa mass can be kept as a fluid under hot conditions or molded and cooled before storage. It is the raw material for conversion into commercial cocoa. Often this is made in the cocoa-producing countries for export, which is then used to produce cocoa butter, cocoa powder, and chocolate.

Extraction of the Butter from the Cocoa Mass

Cocoa butter is extracted from mass or liquor with the help of a hydraulic press. Screw presses have been employed on nibs but not too successfully. Another method of fat removal is solvent extraction. The powder and butter that are obtained by solvent extraction contain solvents, which may cause undesirable flavor changes as in the case of screw pressing. A cocoa butter extractor for small-scale use was devised by Ganesan (1982), which utilizes the pressure developed by a hydraulic jack for extraction of the butter. The equipment can extract 44.8 percent of the butter by applying a pressure of 248.72 kg/cm^2 at 70°C. Broadbent et al. (1997) used a Brazilian made, small-scale, portable expeller to extract cocoa butter.

The cocoa butter obtained by employing any of the above methods is filtered, if necessary, neutralized and refined, deodorized, and tempered. It is then molded and cooled. At this stage it is hard in consistency, waxy, and slightly shiny, pale yellow in color, and oily to touch. It melts at a temperature close to 35°C, giving a clear liquid.

Making Cocoa Powder

The cake left behind at the bottom of the presses after extraction of butter contains a further 20 percent butter. This cake is milled and sieved. Cocoa powder is of two types: high-fat powders containing 20–25 percent fat and low-fat powders containing 10–13 percent fat. High-fat powder is used in drinks, whereas low-fat powder is used in cakes, biscuits, ice creams, and other chocolate-flavored products. In Thailand, high-fat powder is used for the manufacture of cigarettes.

Production of Plain Chocolate

A large number of unidentified compounds are considered to be involved in inducing the characteristic chocolate taste and aroma of cocoa products. The relative abundance of these is expected to vary, depending on each step in the process of manufacture. Precise standardization of conditions is therefore required to make cocoa products of standard and reproducible quality. These compounds are also considered to be responsible for the large brand-related variation in the taste of cocoa products. In simple terms, chocolate is produced by mixing sugar with nib or mass to which cocoa butter is added to enable the chocolate to be molded. The proportion of mass, sugar, and cocoa butter varies with each manufacturer, and it remains a trade secret. The mixture of mass and sugar is ground at elevated temperatures to such an extent that the chocolate is very smooth. The mixture is then refined. This gives an absolutely homogenous mixture and a very fine grain size. It is carried out in cylindrical grinders, which are placed on top of one another and adjusted to operate at increasingly closer spacings rotating at different speeds of around 200 revolutions/minute. The mass then becomes dry and flaky. It is kneaded again in a blender, and at this stage cocoa butter is added along with flavoring agents, if necessary. This mixture is then subjected to a mixing process called "*conchin*." It is carried out in large vats—the conchs. The original conch was a shell-shaped tank, hence the name. In this conch, a roller is pushed to and fro on a granite bed for several hours or even days at temperatures ranging from 60°C to 80°C. The time spent in conchs determines the texture of the chocolate. Most of the cocoa butter and lecithin needed is added at the final stage of conching. Conching removes volatile acids contained in the beans and makes chocolate homogenous.

Tempering

This stage consists of reducing the temperature of the chocolate to 28–30°C in automatic tempering vats.

Dressing

This includes molding, where the tempered chocolate passes into a weighing hopper that distributes it into molds; tapping, which causes the molds to be continuously shaken in order to distribute the mass evenly without air bubbles; refrigeration at 7°C; and finally removing the chocolate from the molds. This is done by turning out the molds onto a felt conveyor belt, which receives the chocolate.

Packaging

The chocolates are wrapped in attractive packages. These operations are fully automated.

Milk Chocolate

The method of preparation remains the same as described earlier for plain chocolate. The only difference is that milk or milk powder is added at the first stage of mixing cocoa mass with sugar. The milk can be condensed with sugar; mass is then added, and the mixture is dried under vacuum. This product is called "crumb," which is ground and conched with additional cocoa butter as described earlier. A typical crumb contains 13.5 percent liquor, 53.5 percent sugar, and 32 percent milk solids.

White Chocolate

This is made of milk, cocoa butter, and sugar.

Other Cocoa-Based Products

A number of products are now available in the market—drinking chocolate, enrobing chocolate, chocolate milk, and so on.

Nutritional Value of Chocolates

Chocolate-based products have high energy value in relation to their volume. They contain a proportion of carbohydrates and protein together with B complex vitamins. Milk chocolate also contains milk protein, calcium, and other minerals. Plain chocolate contains 64.8 percent of carbohydrates, 29.2 percent of fat, 4.7 percent of protein, Na, P, K, Ca, Mg, and P in the ratio of 11, 300, 38, 100, and 140 mg per 100 g, respectively. A 100 g bar has energy equivalent of 500 calories. It contains theobromine and caffeine, which are responsible for stimulatory effects. Chocolate has a restorative, energy producing, and tonic effect on the body. Some studies indicate that plain chocolate has a cholesterol content of 1 mg/100 g and therefore plays only a very negligible role in cholesterol intake. However, chocolate is to be avoided by diabetics.

By-Products of Chocolate

Processing of cocoa both at the primary and secondary levels results in a large quantity of waste materials. The disposal of these is one of the problems in major cocoa-growing countries. Research on utilization of these waste materials indicates that several useful by-products can be made from cocoa waste. The important waste materials are pod husk, sweating, germ, and shell.

Pod Husk

About 70–75 percent of the pod is constituted by pod husk. This generally is discarded after collection of the beans. Nambuthiri and Shivshankar (1987) reported that the pod husk contains crude protein (5.69–9.69 percent), fatty substances (0.03–0.15 percent), glucose (1.16–3.92 percent), sucrose (0.02–0.16 percent), pectin (5.30–7.06 percent), nitrogen-free extract (44.2–151.27 percent), crude fiber (33.19–39.45 percent), theobromine (0.20–0.21 percent), and ash (8.83–10.18 percent). The pod husk contains less theobromine than the cocoa shell, which makes it less dangerous as a feed stuff. Incorporation of a 20 percent pod husk in cattle feed has shown beneficial effect (Sampath et al., 1990). The nitrogen and phosphorus contents of the pod husk suggest its use in paper manufacture, but its low fiber length of 0.3–0.5 mm rules out this possibility. Pod husk as a source for production of furfural (9 percent) is not comparable in yield with materials such as oat hull, corncob, and cotton seed. Hence, production of furfural from pod husk is not commercially viable. The dry pod husk contains 5.3–7.08 percent pectin. This is high when compared to established raw materials such as orange pulp, lemon pulp, and apple pumace. The quality of endocarp pectin is superior to that of pectin from sweatings.

Sweatings

There is about 2–3 percent of alcohol and about 2.5 percent acetic acid in the sweatings. The sweatings contain 79.2–84.2 percent water, 15.2–20.8 percent dry substances, 0.77–1.52 percent citric acid, 11.60–15.32 percent glucose, 0.11–0.92 percent sucrose, 0.90–1.19 percent pectin, 0.56–0.69 percent proteins, and about 0.41–0.54 percent each of K, Na, Mg, and Ca. The pH ranges from 3.2 to 3.5. Sweatings can be used to make jelly or jam. The pectin from sweatings shows slow-setting characteristics.

Shell

The availability of bean shell is of the order of 11–12 percent of dry beans. It contains 2.8 percent starch, 6 percent pectin, 18.6 percent fiber, 1.3 percent theobromine, 0.1 percent caffeine, 2.8 percent total nitrogen, 3.4 percent fat, 8.1 percent total ash, 3.3 percent tannins, and 300 IU vitamin D. The yield of furfural is about 5–6 percent. Although it is possible to extract protein, tannin, and red color from shells, it is not economically viable. The scope for use as animal feed is limited due to high theobromine content. As fertilizer, shells act as a humus-forming base. They do not decompose readily. This can be overcome by heaping for one season. Theobromine is extracted commercially and methylated to form caffeine, which has greater demand than theobromine. As fuel the calorific value of the shell is about 7400–8600 BTU, which is slightly higher than that of wood.

Germ

The cocoa germ has 3.5 percent fat, 6.5 percent ash, 24.4 percent protein, 2.9 percent crude fiber, and 3 percent theobromine. The composition of the germ varies considerably depending upon the country of production.

Research and Development Organizations

As per current estimations, more than 70 percent of world's cocoa is grown by small farmers using traditional labor-intensive systems of husbandry. Improvements in productivity are therefore largely dependent on the work of research institutes in the producing countries and of the few plantation companies that have expertise and facilities. The volume of research carried out by these bodies is insufficient to provide the assurance needs of the future. It is therefore essential to contribute through sponsorship of research and arranging for the results to be disseminated widely. The important research institutes on cocoa are the following:

The Cocoa Research Unit, The University of West Indies, St. Augustine, Trinidad, Tobago
Cocoa Research Institute of Ghana (CRIG), PO Box 8, New Tafo Akim, Ghana
Centro de Pesquisas do Cacao (CEPLAC), Cacao Research Centre (CEPEC), Ilheus, Bahia, Brazil
Fondo Nacional de Investigaciones Agropecaurias (FONAIAP)—Centro Nacional de Investigaciones Agropecaurias (CENIAP), Itabuna, CEP 45650-000, Venezuela
Cocoa Research Institute of Nigeria (CRIN), PMB 5244, Ibadan, Nigeria
Papua New Guinea Cocoa and Coconut Research Institute, PO Box 1846, Rabaul, Papua New Guinea
Instituto Colombiano Agropecuario (ICA), Programa de Cacao, AA 1017, Bucaramanga, Colombia
Cocoa Research Station, Quoin Hill, Tawau, Department of Agriculture, Sabah, Malaysia
Centre de Co-operation Internationale en Recherches Agronomiques pour le Developpement (CIRAD)-IRCC, BP 5035, 34032, Montpellier, Cedex, France
CIRAD-IRCC, BP 1827, Abidjan, Cote d' Ivoire
CIRAD-IRCC, BP 701, 97387, Kourou, French Guiana
Institut de Forests, Department Cafe Cacao (IDEFOR-DCC), BP 1827, Abidjan 01, Cote d'Ivoire
Central Plantation Crops Research Institute, Regional Station, Vittal, 574 243, Karnataka State, India
Kerala Agricultural University (KAU), PO 680656, Thrissur District, Kerala State, India
Coffee and Cocoa Research Institute, J.P.B. Sudirman, Jember 68118, Indonesia

Manufacturers of Cocoa Products

Cocoa products are manufactured by many multinational companies, namely Cadbury, Nestle, Amul, Lotus, and Sathe.

A Look into Cocoa's Future

With a fast-changing economic scenario, cocoa can be expected to make a real contribution to global economy if some of the following points are taken into consideration.

1. Since the early 1930s, a lot of work on the improvement of the crop has been undertaken. Admittedly, yield enhancement has only been marginal. Since 1980, world cocoa production has been stagnant. A major part of world's cocoa production comes from genotypes that are not significantly different from their wild progenitors. This is due to restricted genetic base of the crop. The need for exploitative research in centers of diversity is highly essential to make any breakthrough in crop improvement.
2. Cocoa cultivation in many countries faces severe threat by major diseases, the control of which is not feasible by conventional methods. There is a great need to breed to develop

genetic resistance to diseases. Since the 1950s, effective measures to incorporate disease resistance have not borne fruit, and the major diseases are still a cause for worry.

3. There appears some scope if efforts in impairing disease are taken up based on national priorities. Some success in crop improvement has been achieved by recurrent selection schemes with distinct sub populations. The genotypic components of variation for all major agronomic traits are shown to be due to mainly additive gene effects, and maximum gene dispersion over subpopulations will increase the chances of detecting transgressive hybrids.
4. Clonal selection after recombination would be better for short-term progress, where clones are accepted commercially. By concentrating on development and utilization of technologies, there is need to develop stable, long-term conventional breeding work.
5. It is imperative that conventional breeding programs be maintained and expanded to develop quantitative traits such as yield and horizontal resistance. The reasons for this are (1) all desirable genes in a polygenic system cannot be assembled in a single plant in a single generation; (2) it is impractical to screen using gene markers when many genes producing small effects upon the traits are involved; and (3) quantitative traits tend to be greatly influenced by genotype/environment interactions, thus making it necessary to screen for such traits.
6. In almost all breeding programs, the breeder neglects the important trait which is flavor, as it is very difficult to quantify the same, which in turn depends on the variety and type of the material involved. Flavor improvement must find an important place in future breeding programs.
7. There should be consideration to adapt tree architecture to improve photosynthetic efficiency and harvest index. Research in cocoa based on revolutionary concepts needs to be initiated.
8. Current achievements in molecular genetics appear as if they will have a positive impact on crop improvement in the coming years. Progress in the development of new technologies in plant breeding has been tremendous since 1985. These techniques can make a useful contribution if the traditional breeding base is strong enough to support their integration.
9. Organic farming is getting increasing attention. This line should be further explored in cocoa production.
10. Physiology of flowering and fruit set in cocoa needs a detailed investigation.
11. Evolution of biological control measures against serious pests and diseases is another important area for future research.
12. Considerable attention must be bestowed on value addition at the farm level.
13. The role of "The Nutrient Buffer Power Concept," the revolutionary soil management technique developed by the author, must be exhaustively researched to improve fertilizer use efficiency and schedule economically viable fertilizer recommendations in cocoa production.

6 Coffee

Coffee is the most important nonalcoholic beverage in the world trade. It commands a turnover close to US$10 billion per annum, which makes it the second-most important commodity traded in world markets, next to petroleum. Its production forms the backbone of more than 50 developing countries (de Graaf, 1986), with its contribution to the total foreign currency earnings reaching as much as 80 percent in some countries, such as Uganda. Historically, the use of coffee has evolved from the original chewing of leaves and beans of the plant to relieve pain, hunger, and fatigue to the present-day sophisticated uses, such as espresso or decaffeinated coffee. The consumption of coffee apparently started between the fifteenth and sixteenth centuries in Arabia. The habit spread to different parts of the world, first reaching Europe through Turkey in the 17th century (Charrier and Eskes, 1997). Consumption by the producer countries has, however, remained negligible except for Brazil, which consumes about 10 percent of its production. From the center of origin in Ethiopia, *Arabica* coffee was introduced in Yemen between fifteenth and sixteenth centuries. Thereafter, the species spread to the Malabar Coast of India and Sri Lanka during the last decade of the 17th century. Coffee further spread to Java in 1706, and later to Martinique and South America via Amsterdam and Paris. Unlike *Arabica* coffee (*Coffea arabica* L.), Robusta coffee (*Coffea canephora*) has not gone through a history of extensive diaspora. First found in a state of semicultivation in the Democratic Republic of Congo (DRC), the species reached Southeast Asia in the early 20th century, where it replaced *Arabica* coffee in the lower altitude zones of Java and Sumatra due to *Hemileia vastatrix* onslaught. The species has a rather dispersed center of genetic diversity within the west and central subtropical regions of Africa. Its center of genetic diversity extends from Guinea and Liberia to Uganda and Sudan. Maximum genetic diversity is, however, apparently found in DRC. Most of the coffee spread around the world resulted from a few trees. Consequently, the majority of present-day cultivated varieties are based on a very narrow genetic base (van der Vossen, 1985). The resulting genetic uniformity is a major problem to plant breeders because it predisposes most of the varieties grown worldwide to the risk of epidemics, whenever new diseases arise. At the same time, it offers little opportunity for selection to stress factors such as insect pests and diseases. At present, the commercial production of coffee relies on two species, *Coffea arabica* and *Coffea canephora* (van der Vossen, 1985). In terms of volumes, *Arabica* coffee commands more than 75 percent of traded coffee in the world market (van der Vossen, 1985). Its predominance in world trade is attributed to the superior liquor quality inherent

The Agronomy and Economy of Important Tree Crops of the Developing World. DOI: 10.1016/B978-0-12-384677-8.00006-0
Copyright © 2010 by Elsevier Inc. All rights of reproduction in any form reserved.

in the species. The exclusive production of the species, the world over, has nevertheless been limited to the number of factors, notably susceptibility to diseases, drought susceptibility, and low tolerance to temperature fluctuations (Agwanda et al., 1997; van der Vossen, 1985). Its main competitor, *Coffea canephora*, is relatively tolerant to some of these stress factors (for instance, *Hemileia vastatrix* and high temperature, especially those prevalent in low-altitude coffee-growing zones).

Genetic Resources

Coffee belongs to the family Rubiaceae, genus *Coffea*. The genus is further organized into three sections, *Mascarocoffea*, *Eucoffea*, and *Paracoffea* (more correctly referred to as *Coffea*) of which the first two are natives to Africa, whereas *Paracoffea* is endemic to East Asia. Most of the known coffee species are found within the sections *Coffea* and *Mascarocoffea*. Based on factors such as tree height, leaf thickness, fruit color, and geographical distribution, the section *Coffea* has been subdivided into five subsections: *Nanocoffea*, *Pachycoffea*, *Erythrocoffea*, *Melanocoffea*, and *Mozambicoffea*. The cultivated species of *Coffea canephora* and *Coffea arabica* belong to the subsection *Erythrocoffea*, whereas *Coffea liberica* belongs to *Pachycoffea* subsection.

Most of the coffee species originated in Africa, with the centers of genetic diversity situated in the region covering southwest Ethiopia and southeast Sudan in the case of *Coffea arabica*, while for *Coffea canephora* and *Coffea liberia* it is the humid forest of Central and West Africa, including Uganda and Madagascar. With the exception of *Coffea arabica*, all the other naturally occurring species are diploids with the basic chromosome number of $2n = 2x = 22$. *Coffea arabica* is an allotetraploid with a chromosome constitution of $2n = 4x = 44$. The ancestry of *Coffea arabica* has been controversial for some time, but with a general consensus that *Coffea eugenioides* be considered as one of the parents. Recent molecular studies have now confirmed *Coffea canephora* to be the other parent of *Coffea arabica* (Lashermes et al., 1999).

Conservation of Germplasm

The main collections presently under conservation resulted from collection efforts made during the 1960s, which include FAO (Food and Agriculture Organization, Rome) collection mission of 1964 and the French ORSTOM (Institute Francais de la Recherche Scientifique pour les Developpemment en Cooperation) collection missions of 1966–1978. Good representations of these germplasms are conserved in Ethiopia, Costa Rica, Cameroon, Ivory Coast, Colombia, and Madagascar. In addition, a number of working collections are also conserved in various national institutes entrusted with coffee research. Traditionally, long-term conservation of the genus *Coffea* is realized through live field gene banks. This is because of the recalcitrant nature of the seeds of the species within the genus *Coffea*. Other complementary germplasm conservation techniques, such as pollen storage (Walyaro and van der Vossen, 1977), have only been limited to short-term application in breeding activities. Field collections, although most widely used at present, have a number of limitations. The collections are prone to natural hazards, such as drought and disease incidence. Wild fires are also a major threat, whereas civil unrests are increasingly becoming a major consideration in the African context. Increased pressure on the use of agricultural land due to population pressure in the developing nations has led to the encroachment of forest lands for human settlement. It is for these reasons that the recent developments in the field of cryopreservation of seeds and *in vitro* cryopreservation of zygotic embryos will be a major contribution toward securing the world's coffee collections. The methods are still not widespread, however, with the first applications being in Costa Rica.

In general, the majority of the stored germplasm has not been exploited for breeding or selection purposes. Exploitation has been limited to breeding for resistance against diseases in which only a few genotypes are involved. One reason for this scenario is the fact that most of the germplasms are, so far, not comprehensively described. The other reason is that most of the genes of interest are located in wild species for which barriers preventing crossing of such genotypes with the commercial species exist.

Breeding and Selection

Until recently, most breeding and selection efforts were directed toward enhancing yield, quality, and adaptability to agroecological conditions (van der Vossen, 1985). A number of high-yielding, good-quality varieties were developed in the early breeding efforts, which mainly involved individual plant selection. These include varieties such as "Caturra," "Mundo Novo," and "Catuai" in Brazil; "Typica" in Colombia and Central American countries; SL selections in Kenya; and N 39 in Tanzania, among others. These cultivars, usually referred to as traditional varieties, have formed the backbone of most of the coffee industries throughout the world.

The advent and subsequent spread of the diseases, such as coffee leaf rust, *Hemileia vastatrix* (Berkley, 1869, as cited by Kushalappa, 1969), and coffee berry disease (CBD), *Colletotrichum kahawae* (McDonald, 1926), have, however, made the continued use of the traditional varieties quite expensive, as this entails intensive use of fungicides to protect the crop. In an attempt to ensure that coffee production remains economical in the face of such diseases, a number of breeding programs have been given by van der Graaff (1981), van der Vossen and Walyaro (1981), van der Vossen (1985), Agwanda and Owuor (1989), Bouharmont (1995), Carvalho and Monaco (1969), Echverrie and Fernandez (1989), and others. Although varied, the present-day coffee breeding programs share one goal: to develop varieties which, under defined socioeconomic and agroecological conditions, would maximize the net earnings of the farmer. To realize this goal, higher yield, superior quality, disease resistance, and adaptability of varieties to agroecological conditions have been the main objectives of coffee breeding and are discussed briefly in the following sections.

Yield

Yield measurement as the quantum of clean coffee, that is, beans dried to 10–12 percent per unit area of land, is the most important parameter from a commercial perspective (van der Vossen, 1985). Two approaches have been used by breeders to improve yield potential. Yield was improved by selecting for vigorous trees with improved yield potential of individual trees. This approach, however, was not amenable to intensive land use and has since been replaced by the development of varieties with compact growth habits. Compact growth allows increased plant density, which in turn leads to dramatic increases in yield per unit land mass. High density is known to increase yield in the *Robusta* type coffee varieties (Browning and Fisher, 1976; Mitchell, 1976; Njoroge and Mwakha, 1983; Walyaro, 1983). To realize the full potential of compact varieties, it is, however, desirable that such varieties be resistant to the most prevalent diseases. Conditions of crowding would otherwise create microclimatic conditions favorable to development of diseases, thereby negating the benefits associated with compact varieties.

Quality

Coffee consumption globally is determined by quality. The amount and sustainability of coffee consumption depends on the desirability of the coffee brew and the pleasure derived from its consumption (Charrier, 1982). From the point of view of breeding, two levels of quality are important. The first consideration is given to the quality of beans prior to and after roasting. The bean quality, also referred to as bean grade, is important from a technological point of view, as it acts as an indicator of the suitability of a given coffee variety for roasting. Raw quality assessment results can also be used to explain liquor quality defects arising from suboptimal agronomic and processing practices. The second-quality criterion involves the organoleptic assessment of the quality of coffee brew (liquor or cup quality). Several quality traits are assessed at this stage to determine the desirability of the coffee brew for human consumption.

Bean Quality

Bean quality is judged based on shape, size, weight, and appearance. Classification of coffee beans into various grades is largely done mechanically (refer van der Vossen, 1985 for details). A study by Walyaro (1983) indicated that most of the size, shape, and weight parameters of bean quality are highly heritable and therefore easily improved through simple selection procedures, such as mass selection and backcrossing breeding. To enhance the rate of genetic gain, attention should be given to the general agronomic practices and the nature and timing of environmental factors, such as moisture stress. Optimum environmental conditions during expansion and bean-filling stages have been shown to enhance differential expression for bean grades, thereby making it possible to select superior genotypes with increased precision (Agwanda et al., 1997).

Liquor Quality

In addition to being the driving force behind coffee consumption, the cup quality of coffee determines the price offered at the world market. An important consideration when breeding for improved quality is the consistency in performance both between locations and across years, as this is of fundamental interest to the producer nations in capturing and retaining given market sectors of the world trade. Unlike bean grades, liquor quality is determined subjectively based on organoleptic procedures and by panels of experienced coffee liquors. Three traits are usually used in determining the quality of liquor, namely acidity, body, and flavor (Devonshire, 1956, Moreno et al., 1995). The traits are heavily influenced by nongenetic factors, such as edaphoclimatic conditions, cultural practices, post-harvest processing and storage, roasting, and preparation of the liquor. Nevertheless, reasonable progress from selection could still be realized by subjecting the test materials to recommended agronomic conditions and following standard sampling and liquor preparation procedures (Afnor, 1991). Efficiency of selection for overall liquor quality could also be improved by identifying the traits that could be assessed organoleptically with greater accuracy and that are strongly correlated to the overall quality of the brew (Agwanda, 1999). To enhance the differential expression between test genotypes for the various traits, it is also necessary to manipulate the moisture regimes in the selection fields in relation to the various bean growth phases (Agwanda et al., 1997).

Other Quality Criteria

The coffee market is a continuously evolving arena with new methods of coffee preparation and uses appearing in the market on short intervals. Similarly, consumer taste and preference are equally evolving, though at a slower rate. In order to keep pace with the changing circumstances, additional quality traits may have to be introduced in the coffee breeding programs. For instance, percentage of soluble solids, caffeine content, and chlorogenic acid could be used more routinely as selection criteria. Although considered minor at present, such biochemical traits are expected to continue attracting the interest of breeders, especially in *Arabica* coffee. This is particularly so because a number of coffee breeding programs are currently exploiting the Hibrido de Timor germplasm or its derivatives. Timor derivatives are known to have higher levels of these traits. Consequently, their use as donors in *Arabica* coffee breeding programs is expected to increase the three traits to levels above those characteristic of *Arabica* coffee and, hence, affect the organoleptic integrity of the newly developed varieties. Hibrido de Timor, in particular, is known to impart poor taste to its progenies (Walyaro, 1983). To realize the faster progress, studies aimed at understanding the inheritance mechanism of the traits and developed methods for relating their levels to the perceived organoleptic quality will be necessary.

Disease Resistance

Most of the coffee breeding programs concentrate on imparting disease resistance to the varieties evolved. The two main diseases, namely, CBD caused by *Colletotrichum kahawae* and coffee leaf rust caused by *Hemileia vastatrix*, are targeted in the breeding programs. The two diseases are confined to *Arabica* variety and are therefore more important to the *Arabica* coffee-producing nations. Elaborate screening techniques have been developed for the two diseases. In the case of CBD, for instance, selection for resistance is based on inoculation of 8-week-old seedlings (van der Vossen et al., 1976) or on a combination of seedling resistance and field expression (van der Graaff, 1981). Screening at the seedling stage has a number of advantages. First, it allows a large number of lines and genotypes within lines to be assessed at the same time. The ability to assess a large number of plants per cross is of significance to coffee breeders because it allows better visualization of segregation patterns. Because coffee has a long juvenile phase, the resulting long generation cycle is a major handicap to breeding and selection in the species. The use of a seedling inoculation technique greatly reduces the time required for the assessment of a given line, thereby minimizing the impact of a long generation cycle on the rate of turnover of new varieties. Breeding for rust resistance has similarly benefited from advanced selection techniques, including detached leaf, leaf disk inoculation (Eskes, 1983), and inoculation of intact leaves on young seedlings. The impact of the techniques on progress in breeding for resistance to the two diseases has been considerable, with resistant varieties being released to the farmers in countries such as Colombia (Castillo and Moreno, 1988) and Kenya (Anonymous, 1985) after a relatively short period of time. The other coffee diseases caused by pathogens, such as *Fusarium stilboides* and *Gibberella xylarioides*, are considered minor and have not attracted any attention from the breeders. Problems related to the nematodes, particularly *Meloidogyne exigua*, are now attracting some breeding attention, especially in Central America. With the increasing neglect of coffee farms, such as in the *Arabica* coffee-producing countries, and shifting weather patterns presently being experienced due to global warming and consequent climate change, the diseases hitherto considered minor may gain importance as is already the case with *Tracheomycosis* in *Coffea robusta*. Given the time taken between initiation of a breeding program and the realization of results, it may be prudent for the breeders to develop screening techniques for the minor diseases, identify genes of resistance against them, and, where possible, transfer the identified genes on genetic backgrounds that could be used for practical breeding.

Stability of Performance

Stability of performance in crop plants refers to the ability of a variety of plant to adjust its phenotypic state in response to transient environmental fluctuations in such a way as to give high and stable economic returns (Allard and Bradshaw, 1964). It can be the result of phenotypic plasticity of individual genotypes (individual plant buffering), differences of phenotypic plasticity between genotypes (population buffering), or the result of different genes being switched on and off (Allard and Bradshaw, 1964; Gallais, 1992; Yamada, 1991). Any of these mechanisms could be targeted in breeding for improved stability in *Coffea arabica*. Recent research (Agwanda et al., 1997) has demonstrated that population buffering in *Coffea arabica* could be realized by selecting for individuals that show complementary performance under different environmental conditions, but are otherwise similar in the other important agronomic trait. Because coffee plant is perennial, year-to-year fluctuations in climatic conditions could be considered as part of the transient environmental fluctuations. The combination of years and locations could thus enable the breeders to accommodate the phenomenon of biennial bearing, which is characteristic of *Coffea arabica*.

Combined Selection

To realize the goal of maximizing economic returns to the farmers, it is necessary to simultaneously improve the various economic characters that directly impact on net returns to the farmers. Toward this end, it is necessary to define, within practical limits, what could constitute an ideal variety. In principle, such a variety should contribute toward reducing the cost of production and maximize net returns to the farmer, while preserving both human and environmental health. Key factors to consider when attempting to construct such varieties include responsiveness to farm inputs, such as fertilizer use, high productivity per unit land, resistance to prevailing diseases, and the production of good bean and liquor qualities. The construction of varieties combining all these traits requires multi-criteria selection techniques, such as index selection or market-assisted selection (MAS) (Melchinger, 1990). Index selection methods have been shown to be useful in concurrent selection for yield and quality (Agwanda et al., 1997) and in predicting long-term performance (Walyaro and van der Vossen, 1979). More recent developments in molecular marker methods for coffee diseases (Agwanda et al., 1997; Lashermes et al., 1993, 1996, 1997) could also greatly improve the efficiency of selecting such ideotypes through MAS.

Insect Resistance

In the case of both *Coffea arabica* and *Coffea robusta*, only little attention has been given for developing resistance to insect attack. The main explanation to this anomaly could be that the insect pest attack is highly weather-dependent and could be easily controlled by insecticides. However, this scenario is fast changing. For instance, the berry borer menace is now a more pressing issue in Colombia and the surrounding coffee-producing countries than any other pest. Similarly, nematodes have become an economically important constraint in coffee production in Central America. The twin problems, although earlier considered minor, may now warrant a full-fledged breeding program. In the light of these considerations, and with increasing demand for organically grown coffee, insect resistance may soon be an important breeding program in coffee breeding.

Breeding Programs

Breeding for Disease Resistance

India and Brazil took up systematic coffee breeding programs as early as 1925. While the Indian program concentrated on rust resistance and yield and quality enhancement, Brazil concentrated on genetic or cytogenetic evaluations. In Africa, breeding activities commenced around 1934 in Tanzania and around 1944 in Kenya. At their inception, the programs concerned the improvement of quality, yield, and adaptation. With increased leaf rust and CBD menace, breeding priorities shifted to the development of varieties resistant to the disease. In extreme cases, total replacement of the variety with other economically viable commodities was considered, as was the case when *Hemileia vastatrix* first hit Sri Lanka in 1869. In the absence of resistant varieties, the use of fungicides to combat the disease becomes the second line of action. This can be exemplified by the elaborate spray programs developed in Kenya to control both CBD and the coffee rust. Breeding for disease resistance as the final line of action against new diseases is the development of new varieties through systematic breeding. For both CBD and leaf rust, elaborate breeding programs are now in place in a number of countries.

Coffee Leaf Rust

Currently breeding focuses on CBD and coffee leaf rust (CLR). At the global level, CLR has become the most damaging. It was in 1869 that the disease manifested its full potential when it arrived in Sri Lanka, leading to almost a total collapse of the coffee industry in the country. The disease then spread to all other coffee-growing countries, reaching Brazil in 1970 and establishing itself as the most significant coffee disease of South America. The disease was nevertheless less disastrous in these countries compared to its impact in Sri Lanka, thanks to the efficient spray programs already developed in countries such as Kenya at the time the disease reached the South American continent.

Through international collaboration, coordinated by the Centro de Investigacao das Ferrugens do Cafeeiro (IFC), Oeiras, Portugal, CLR has been studied extensively. The IFC discovered a number of genes for resistance and compatibility, which was described in detail, in the host and the pathogen, respectively. Such genes have been exploited extensively in South America and India, giving rise to resistant varieties, such as Colombia variety (Castillo and Moreno, 1988). The breeding methodology used involved initial crossing of the desired progenitor with the commercial cultivar followed by a series of selfing generations to fix the resistant genes.

Coffee Berry Disease

Unlike CLR, CBD is still restricted to the continent of Africa where it causes considerable loss in *Coffea arabica*. Reported first in Kenya (McDonald, 1926), the disease is now present virtually in all *Arabica* coffee-growing countries in the continent. Chemical methods, resistant varieties, and, to an extent, cultural practices can control the disease. Chemical control is very expensive. In Kenya, for example, CBD control is estimated at US$500 per ha in the case of susceptible varieties. With a total of 120,000 ha, the control could run up to a colossal sum of money in the country. Further, losses up to 40 percent can still occur in inclement weather, despite chemical control. Consequently, on a long-term basis, breeding for CBD resistance is the only economically viable option in *Arabica* coffee. The main coffee breeding programs dedicated to CBD resistance are found in Kenya, and to some extent in Ethiopia and Tanzania. Systematic breeding for CBD resistance in Kenya started in 1972 with the objective of introgressing CBD-resistant genes into traditional commercial cultivars, mainly SL 28 and SL 34. The program exploits the resistant genes identified in the accessions *Rume*,

Sudan, *Hibrido de Timor* and its derivatives, and one commercial cultivar K7. To combine all the resistant genes in a single cultivar, the various sources of resistance are crossed to the commercial cultivars SL 28 or SL 34. The single crosses are then crossed together in various combinations, giving rise to multiple crosses. A series of backcrosses are then conducted with commercial cultivars as the recurrent parent. The objective of the backcrosses program is to restore the genetic background of the commercial varieties, which are renowned for production of very fine, mild coffee. After every backcross, selfing of the backcross progenies is carried out to realize homozygosity at the resistance loci. Screening for CBD resistance is carried out to identify the progenies to form the next generation. So far, up to four generation backcrosses have been realized.

The Kenyan program embarked on the production of hybrid varieties as means to exploit the early generation selections for commercial use. Superior breeding lines are used as pollen donors, whereas selected *Catimor* progenies, originally from Colombia, are used as maternal parents. The use of *Catimor* as a mother population in the hybrid program has a number of advantages. First, the *Catimor* lines bring with them the rust resistance of *Hibrido de Timor*, but in a genetic background, which is commercially more acceptable than that of the original Timor hybrid. *Catimor* also possesses one of the CBD resistance genes also originating from *Hibrido de Timor*. The final advantage is the short stature, a character which is more or less fixed in the variety. The variety resulting from the hybrid program, Cultivar *Ruiru* 11, combines all these advantages with the superior quality attributes of the elite breeding lines as well as CBD resistance genes originating from *Rume*, *Sudan*, and *K7*.

Bacterial Blight of Coffee

Bacterial blight of coffee (BBC) caused by *Pseudomonas syringae* pv. *garcae* is the only bacterial disease of economic importance in coffee. The spread of the disease is highly restricted, being present mainly in Brazil and Kenya. The disease is considered inconsequential in Brazil, whereas its coverage amounts to about 3 percent of the total coffee acreage in Kenya. The disease is gaining importance in Kenya, however, as it is endemic in areas with great potential for coffee expansion. In addition, because increased area is being covered by CBD-resistant cultivars, BBC epidemiology may be expected to change as the use of fungicides, some of which are known to have bactericidal effects as well as diminishes. In addition, shifts in weather patterns and a general increase in the neglect of farms due to poor prices are becoming frequent phenomena and will undoubtedly increase the importance of the disease. Research carried out in Brazil indicates that variation for disease resistance exists in the *Arabica* coffee introductions from Ethiopia, with resistance levels ranging from complete susceptibility to immunity. These introductions could, therefore, form the basis for breeding against the disease.

Other Minor Diseases

Minor diseases in coffee can be described as those diseases that, under optimal agroeconomic conditions, can easily be put under check without having to resort to this category. They include diseases such as *Fusarium* wilt diseases and tracheomycosis. Situations in which minor diseases evolve into major problems are not uncommon, however, and may be due to a number of reasons, including deterioration in general agronomic practices, change in farming systems, and shifts in weather patterns. A recent example is the case of tracheomycosis in Uganda and the Democratic Republic of Congo, where the disease has become a major threat to economic and sustainable production of coffee in the two countries. Anticipating such problems, breeders could at least initiate activities aimed at identifying sources of resistance and, if possible, transfer the genes to genetic backgrounds where they could easily be incorporated into breeding programs whenever there arises a need.

Field Management

Environmental Requirements

Both *Coffea arabica* and *Coffea robusta* are tree crops that produce yields 2–3 years after planting with long economic life beyond 30 years, depending on local conditions and crop husbandry. Coffee is sensitive to excessive heat or cold or rapid temperature changes. Each type requires different growing conditions, with *Coffea arabica* preferring temperatures of 15–24°C, whereas *Coffea canephora* prefers warmer conditions of 24–30°C with less contrasting dry and rainy seasons. Both require an average rainfall of 1800 mm/annum for healthy growth and satisfactory productivity. In the equatorial and tropical zones, *Coffea arabica* does well at altitudes of 1200–1800 m above mean sea level (MSL), whereas it does well below 600 m above MSL in the subtropical zones. *Coffea canephora*, on the other hand, does well in the warmer zones of the plains. The ideal soils are light, deep, well drained, loamy, slightly acidic, and rich in humus and exchangeable bases, especially K.

Establishment

Seed or rooted cuttings in the nursery through tissue culture and grafts are used to raise coffee seedlings. After germination or establishment of the cuttings/tissue cultured material, they are transplanted (potted) in black polyethylene bags measuring 23 cm × 17 cm to 30 cm × 17 cm, previously filled with a mixture of top soil, manure, phosphatic fertilizer, and a suitable insecticide. They are ready for field transplantation within about a year to a year and a half after potting. Use of biological, cultural, and physical control of insects should be encouraged at both the nursery and field stages. In the field, seedlings are planted in well-prepared land free of tree stumps to avoid soil-borne diseases, such as root disease caused by *Armillaria mellea*, which affects coffee trees, and removal of difficult weeds, such as couchgrass, for good management of the young seedlings. Suitable soil conservation measures should be taken, such as bench terracing whose terrace sides should be well protected with grass or suitable trees, such as the multipurpose trees, such as *Leucaena*, *Sesbania*, *Calliandra*, and so on. Planting holes are dug 3 months before planting to minimize soil-borne diseases, such as those caused by *Fusarium* spp. They are dug along the contour as an erosion control measure. The holes are then refilled 1 month before planting with topsoil mixed with well-rotted organic manure, phosphatic fertilizer (its type depending on soil acidity), and an appropriate insecticide. Agricultural lime can also be added depending on the soil reaction (soil pH). The seedlings are planted after the onset of the monsoon. Application of mulch along the planted coffee seedlings rows or around the seedlings helps to preserve moisture, suppress weed growth, assist soil conservation, and improve soil structure. The type of mulch material is important as continued use of one type may cause soil nutrient imbalance. For example, use of Napier grass mulch can increase soil K when used for a long period, thereby disturbing the $Ca + Mg/K$ ratio. Soil amendments to rectify this anomaly then become necessary (Table 6.1).

Table 6.1 Nutrient Content of Organic Manures and Mulches

Source	Percent						ppm				
	N	P_2O_5	K_2O	CaO	MgO	SO_3	B	Cu	Fe	Mn	Zn
Boma manure	1.32	1.08	1.66	0.92	0.35	0.45	25	68	27500	916	99
Cattle manure	2.50	1.12	6.70	1.43	1.00	0.37	31	45	11200	1040	95
Coffee husks	0.48	0.07	0.40	0.31	0.08	0.15	4	96	100	32	52
Coffee pulp	3.73	0.40	6.51	0.99	0.30	0.85	18	35	880	226	18
Goat manure	2.66	3.89	4.87	1.36	1.17	0.37	45	30	1940	256	88
Maize stover	2.11	0.35	1.95	1.08	0.32	0.15	26	8	442	57	8
Napier grass	1.51	0.62	4.23	0.27	0.25	0.32	18	25	1300	157	132
Pig manure	2.34	5.27	0.96	4.23	1.55	0.52	25	211	7100	648	440
Pineapple tops	0.86	0.21	1.69	0.29	0.15	0.20	14	13	100	167	180
Poultry manure	3.54	3.24	1.55	5.06	0.98	0.57	30	190	6030	363	225
Rice husks	0.45	0.39	0.47	0.20	0.12	0.12	11	310	11500	326	142
Rice straw (immature)	1.04	0.39	1.49	0.55	0.52	1.05	19	14	4950	965	59
Rice straw (mature)	2.59	0.69	0.60	1.13	0.45	0.22	16	39	2750	902	51
Saw dust	0.36	0.11	0.16	0.24	0.03	0.10	15	70	5950	314	153
Sisal waste	1.15	0.28	1.52	5.81	0.98	0.22	36	242	1875	279	378
Sugarcane Filter mud	1.34	2.38	0.73	3.27	0.36	–	45	60	4695	969	168
Vlei grass	1.39	0.30	1.06	0.66	0.25	0.32	19	72	12950	846	85

Source: Data from Chemistry Section, Coffee Research Foundation, Kenya.

Spacing

The common coffee spacings for the traditional tall cultivars are 2.74 m × 2.74 m (1329 trees/ha), 2.74 m × 1.37 m (2658 trees/ha), and 1.37 m × 1.37 m (5320 trees/ha). For the compact cultivars, such as *Ruiru* 11 hybrid in Kenya, the adopted spacing is 2 m × 2 m (2500 trees/ha) and/or 2 m × 1.5 m (3333 trees/ha) (Njoroge, 1991), or 1 m × 1 m for *Catimor* and *Colombia* cultivars in Colombia. To facilitate efficient use of machinery, large coffee estates can practice different spacings while maintaining the above tree densities. High densities assist in soil conservation, especially through raindrop impact, control of runoff through litter fall, and weed suppression. In high-density coffee plantations, due to high foliage cover, the plants are able to maximize use of solar energy, while the well-distributed mass of roots utilize effectively the soil nutrients. During the first 2 years of coffee establishment, before the canopy closes up, nurse crops or intercrops can be introduced in the interrow spaces. These help provide soil conservation, weed suppression, and an economic return to the farmer when an economic annual crop is used apart from attaining food security. Early maturing annual crops, such as legumes, cereals, and vegetables can be sown, including non-climbing beans, peas, tomatoes, kales, carrots, Irish potatoes, soy bean, millets and sorghum, and so on, depending on the ecological zone where the food crop can perform well. The legumes and nonlegumes should be alternated seasonally or on alternate coffee interrows. Cover crops, such as *Desmodium* spp. and sweet potatoes, may strangle the young coffee seedlings, though may provide good soil cover.

Nutrient Management

Most of the soils in which coffee is grown are low in soil fertility, especially N. On average, a ton of coffee beans can remove 46 kg N, 8 kg P_2O_5, and 58 kg K_2O; parchment 2.3 kg N, 0.3 kg P_2O_5, and 1.9 kg K_2O; and pulp 15.3 kg N, 3.7 kg P_2O_5 and 27.4 kg K_2O. This, together, with losses through parchment, pulp, erosion, and leaching, leaves the soils seriously depleted of fertility. It is, therefore, necessary to return some of the losses through recycled prunings when used as mulch. Hence, fertilizers are needed for both vegetative growth of tree and production of high-quality coffee beans. To apply the correct type and rate of fertilizer and thus avoid toxicity and nutrient imbalances in the soil environment, fertilizer recommendations should best be based on soil and coffee leaf analysis results, where analytical facilities are available.

Among plant nutrients, N is the most limiting in coffee production and *Arabica* coffee has positively responded to N application rates varying from 50 to 100 kg/ha/annum in Kenya (Njoroge, 1985). Responses to as much as 300–400 kg N/ha have been obtained resulting in high yields in soils with the right kind of soil reaction (pH). The proportion of large-sized beans has been found to decrease with increased rates of N application unless balanced with phosphate fertilizer (Njoroge, 1985). Split application was found better than single ones (Njoroge, 1985). Split application of N controls leaching down of nitrates, as the fertilizer gets rapidly nitrified, especially if it happens to be of ammonium origin, better than single application. Investigations in Kenya have also shown no positive yield increases to phosphate application, despite observed low P availability (Keter, 1974). Such anomalous results have been observed in many locations with regard to P application to different crops (Nair, 1996). The primary reason could be that the routine estimations of P availability do not reflect a true picture of what happens in soils. P dynamics is extremely complex (Nair, 1996), and the author's revolutionary soil testing procedure, globally known as "The Nutrient Buffer Power Concept," provides answers in several situations in which routine soil testing do not precisely reflect soil P availability. The types of fertilizers used in coffee production include straight, compound, and foliar fertilizers. Efforts should be made to minimize excess application of plant nutrients, especially N to control leaching and ground water pollution.

Organic manures, mainly cattle manure, have been used for a long time in coffee production. Such manures increase the soil organic carbon, which improves the soil's water holding capacity (WHC) and other physical characteristics of the soil, such as soil texture and cation exchange capacity, all of which have a positive impact on soil fertility (Oruko, 1977). Its use has been reported to lead to increased coffee yields and quality of beans, especially on soils of very low soil fertility (Mitchell, 1970). As the organic manures are formed from different sources, they have varying nutrient composition, and their continuous use may lead to nutrient imbalances, adversely affecting coffee bean quality (Northmore, 1965). However, organic manures can be used as a substitute, to some extent, to inorganic ones, which will also reduce cost of production. Organic manures can be made in homestead farming by composting all plant and animal refuse. Vermicomposting is getting to be popular, but there is yet no systematic studies carried out on a global scale on its benefits. The primary benefit of organic manures is in building up soil carbon. This is especially important due to the adverse impact on soil carbon due to excessive application of inorganic fertilizers in chemical agriculture. Studies in Kenya have shown the possibility of substituting inorganic fertilizer requirements with two 13 kg tin containers of well rotted cattle manure/annum. However, this would depend largely on the source of the manure and calls for nutrient analysis of the various sources of organic manure (chicken, pig, goat, and so on). Decomposed coffee pulp, sludge from methane gas plants, and more are increasingly being used. Results of investigations from Kenya indicate the possibility of using green manures, such as lucerne (*Medicago sativa*) and multipurpose trees, such as *Leucaena* spp., *Sesbania* spp., and *Calliandra* spp., to be good sources of plant nutrients in coffee production (Kimemia et al., 2000). Green manure from permanent cover crops such as *Desmodium* spp. appear to mineralize very slowly and release the plant nutrients they contain, and so may not be of much benefit in the short run, but will act as good cover crops to control soil erosion, especially on sloppy lands. Planting multipurpose trees for this purpose on bench terraces and wastelands may also improve sustainability of the soil environment.

Weed Control

Annuals, perennials, and sedges are the weeds in coffee plantations. Weeds reduce yields by more than 50 percent and adversely impact quality (Njoroge and Kimemia, 1989). Various control measures are used, the most common being digging using a forked hoe, slashing, mulching, and, most important, using herbicides that include systemic, contact, and soil-acting herbicides (Njorge,1994). Some of those commonly used are indicated in Table 6.2.

Low rates and volumes of recommended herbicides can be made to effectively control annual weeds in coffee at the 1–4 leaf stage using low-volume nozzles (Njoroge and Kimemia, 1992). Continued use of this technology can reduce the amount of herbicides related to the environment, thereby reducing environmental pollution. The use of one type of herbicide has led to the development of herbicide tolerance by some weeds, such as tolerance of "Black Jack" (*Bidens pilosa* L.) to paraquat observed in Kenya (Njoroge, 1986). Integration of the different herbicides and other methods of control are therefore recommended. Slash and burn creates a carpet of weeds, which may help reduce soil erosion during the rains, whereas forking may encourage rain water infiltration into depths of soil, reducing surface runoff and building up soil moisture. A long-term approach to weed management through integrated weed management (IWM) would be the best option for efficient weed management and reduction in environment degradation through soil erosion, nutrient leaching, and environmental pollution.

Mulching

Mulching (covering soil surface with vegetation either dry or green) helps arrest surface runoff of both soil and water when excess rainfall is received, improving water infiltration and thereby building soil moisture. In addition, there are a number of other benefits, such as improvement of soil structure, supply of plant nutrients contained in the mulch materials on decomposition, minimizing use of inorganic fertilizers, reducing soil surface temperature, and suppressing weed growth, which

Table 6.2 Some Commonly Used Herbicides in Coffee Cultivation

Soil Acting	Contact	Systemic
Atrazine (50% and 80% wp)	Actril DS (70% EC) (a mixture of loxyil and 2,4-d)	Ametryne (80% wp)
Candex (65% wp) (a mixture of asulam and atrazine)	Amitrole (25% or 50% MI)	2,4-d amine
Diuron (48% EC, 80% wp)	Diquat	Asulam (40% SL)
Flumeturon (80% wp)	Paraquat	Dalapon (74% and 85% wp)
Linuron (50% wp)	Glufosinata-ammonium (20 and 14 SL)	Fluazifop butyl (25% EC)
Oxyfluofen (24% EC)		Glyphosate (various)
Simazine (50% and 80% wp)		Haloxyfop ethoxyethyl MCPA (various) Tardon 101 (picloran plus 2,4-d)

leads to reduced use of herbicides. It also helps in the check of thrips attack. Mulch is applied in the interrow space in alternate years. Mulching has been found to improve yield and quality of beans. Commonly applied mulch materials are, Napier grass, maize, banana stover, coffee prunings, and any other vegetative material. Use of *Desmodium* spp. may be very useful, especially on sloppy lands, but, the main shortcoming is the low biomass turnover and possible competition for moisture with the main crop coffee (Njoroge and Mwakha, 1983; Snoeck et al., 1994). The other disadvantage is that *Desmodium* mineralizes rather slowly.

Pruning

Coffee pruning involves the removal of unwanted branches and old stems. The main reasons for pruning coffee trees are to maintain a suitable crop/leaf ratio for good cropping level and maintenance of a high proportion of large beans to open the tree centers to light, to facilitate disease and pest control, and to help with harvest. This leads to efficient utilization of solar energy by the coffee plant and chemicals used to control pests, thus avoiding excess pesticides in the environment, especially the soil environment, which can have a great polluting effect as it contaminates ground water and leads to health hazards.

The pruning system depends on the type of tree training adopted. Training is the modification of the natural habit of the coffee trees to suit the particular conditions under which they are grown. Basically, there are two training systems: (1) single system and (2) multiple system, which is either capped or uncapped. The change of cycle to raise new stems is carried out after 6–7 years, especially for the multiple system. This improves coffee quality and may reduce disease problems, as the trees are healthier, leading to reduced use of pesticides. Most small coffee growers allow their coffee grow freely without capping, whereas most large coffee growers cap their plants to have ease of mechanization. The compact cultivars, such as *Ruiru* 11 hybrid, are currently recommended on uncapped coffee and require "stumping" every 5–7 years to replace the old stems with new ones (Njoroge et al., 1992). Little pruning is carried out in some countries in Southern Africa, where foliar diseases pose no major problem. Growers prefer replanting rather than raising new stems at the time of cycle change. Prunings are best left *in situ* to act as mulch, which would return nutrients to the soil. However, where fuel energy is increasingly scarce, these prunings are used as firewood. This is particularly true of the poorer African coffee growers.

Water Management

As plant nutrients are primarily transported through internal moisture movement, proper water management assumes much importance in coffee production. Good soil moisture encourages root proliferation and helps in various physiologic processes of growth. Excessive and untimely irrigation can adversely affect yield. Coffee trees require about 4–8 weeks of dry period, to build up internal water stress prior to breaking flower dormancy. The frequency of irrigation is determined by the rate of ET from the coffee field and varies according to weather conditions. Different methods of irrigation, such as overhead irrigation, need further evaluation on its efficiency and effects on pest management. These investigations, by and large, are quite sparse in published literature.

Intercropping and Shading in Coffee

Coffee is grown mainly as a monocrop in most countries; the primary reason being that quality of coffee might be adversely affected if farmers ignore coffee in preference to intercrops. These could be due to competition for plant nutrients, water, and light between the main coffee crop and the intercrop. However, coffee farmers (in particular, small farmers) have been intercropping their coffee with various food, fruit, and tuber crops, especially at the establishment and change of cycle periods and even during the production phases. Large-scale coffee estates have also been observed to move in this direction.

Because coffee occupies a substantial amount of land with high potential, available land for food crop planting is becoming limited, and hence more intercropping is expected to occur in most countries, as in Kenya. In Ethiopia, the indigenous home of coffee, the crop is grown mostly in a multistory cropping system, with trees in the upper story followed by coffee along with food crops, such as maize, sorghum, and legumes including beans, peas, and lentils. The ground floor is covered by root crops such as yam and taro, vegetables such as cabbage and peppers, and spices like ginger and cardamom (Awoke, 1997). In Kenya, preliminary results have indicated that it is possible and economical to intercrop young *Arabica* coffee with some food crops during the first 2 years after establishment (Njoroge et al., 1993). It is also possible to intercrop coffee with dry beans during the change of cycle phase (Mwakha, 1980). More studies are encouraged on this line to maximize on available land, improve food availability, and achieve higher incomes. This would also help sustain the coffee and the farmers in periods of economic severity when coffee prices are down. Intercropping also assists in protecting the soil from the vagaries of soil erosion before the coffee canopy closes up. There is also better utilization of solar energy and effective weed control.

Several tree species have been grown successfully as shade trees and wind breaks, such as *Cordia* spp., *Grevillea robusta*, *Albizia* spp., *Leucaena leucocephala*, and *Cypress* spp. In Ethiopia, indigenous trees, such as *Albizia gummifera*, *Allophylus abyssinica*, *Celtis africana*, *Cordia africana*, *Ekebergia capensis*, *Ficus sur*, *Ficus sycomorus*, *Ficus vasta*, *Millettia ferruginea*, *Macaranga kilimandscharica*, and *Croton machrostachys*, are left as shade trees (Awoke, 1997). Shade trees have been found to have a positive effect on evening out erratic yields caused by periodic overbearing, which also reduces crinkling of coffee leaves (commonly known as "hot and cold" disease) and hail damage. Shade has also been shown to reduce infection of bacterial blight in coffee due to reduced hail injury, thereby reducing pesticide use. Apart from such benefits, shade trees help to recycle soil nutrients from deep layers. Leguminous trees fix atmospheric N and assist in controlling soil erosion and weed growth. In multistory intercrop systems, most nutrients are held in the vegetative mass, which is returned to the soil with litter fall. This system also utilizes the soil mass more efficiently.

Diseases of Coffee

The economic production of coffee is influenced greatly by several major diseases whose relative importance varies depending on the locality and variety involved. The majority of these diseases attack *Coffea arabica* L, which represents 75 percent of total world coffee, both in terms of production and trade (Wrigley, 1988). All parts of the coffee tree, including foliage, berries, stems, and roots, are affected by one or the other of the various diseases. These diseases can be grouped and discussed as per the part of the coffee plant, where the pest attack occurs.

Diseases of the Foliage

Foliage diseases like the coffee leaf rust caused by *Hemileia vastatrix*, leading to leaf shedding and South American Leaf Spot caused by *Mycena citricola* and also, to some extent, Brown Eye Spot caused by *Cercospora coffeicola*, have major effects on the plant. These diseases adversely affect the production potential of the coffee plant, because they affect large areas of leaf and lead to leaf shed. Because photosynthesis provides the most essential carbohydrates for the proper metabolic functioning of the plant, any disease adversely affecting the leaves take a big toll on production. Because developing berries provide the strongest physiological sink for carbohydrates, any reduction in photosynthesis on heavily bearing trees will result in carbohydrate starvation of shoots and roots (Cannell, 1970). Because the current season's new growth carries the following season's crop, the main effect of foliage diseases is to reduce the next season's crop. Where major leaf diseases continue unchecked over a number of seasons, progressive decline in yield and plant vigor occurs. Several leaf diseases on trees carrying a crop may result in photosynthesis being unable to meet the demands of the developing crop. Carbohydrates in such cases are withdrawn from the remaining leaves and young vegetative tissue, resulting in leaf loss, overbearing stress, and dieback of young shoots and roots (Cannell, 1970). Often a large proportion of the crop on such trees fail to mature properly; the berries appear dull rather than glossy and are particularly prone to berry diseases (Waller, 1987). Yellow ripening is another characteristic symptom, and a large proportion of light and empty beans are produced with the accompanying loss of berry quality.

The CLR is, by far, the most serious among foliage diseases. It covers almost all *Arabica* coffee-growing regions. Warm and humid conditions, such as those found in equatorial regions below 1500m above MSL, are conducive to the spread of latent infection leading faster development of epidemics (Kingori and Masaba, 1994; Waller, 1972). For this reason, such regions are mostly unsuitable for growing *Arabica* coffee. During the last two decades of the 19th century, much of the *Arabica* coffee introduced into Sri Lanka and Indonesia was destroyed by CLR disease. These were replaced by the *Robusta* variety introduced from Central Africa, where they became successful, especially at lower altitudes (de Graaf, 1986). Fungicidal sprays are used to control CLR, because earlier attempts to utilize resistance were frustrated by the occurrence of many different mutations of the pathogen; until recently, resistance to all these was not available (Rodrigues et al., 1975). Copper-based fungicides have been found to be universally effective and the cheapest in terms of cost. Several systemic fungicides have also been found to be universally effective as well as low in cost. Systemic fungicides, such as Triadimephon and other Triazoles, have been used with partial success partly due to cost and variable field performance (Figueiredo et al., 1981). Currently, the results of the programs using Grade A (complete) resistance derived from "Catimor" are being used in Colombia and Kenya where commercial cultivars have been released. This resistance has so far proved durable.

Minor leaf diseases include *Ascochyta tarda* Stewart, a pathogen found at high altitudes in which young leaves are affected, leading to their death and dieback of shoot tips. Predisposing conditions, such as wounding and physiologic damage due to "hot and cold" diseases, are usually required for infection (Firman, 1965). A similar disease occurs in high-altitude coffee zones of Central and Northern Latin America, and is attributed to *Phoma costaricensis* Echandi; symptoms and conditions for infections are very similar to those of *Ascochyta tarda* (Echandi, 1958). In addition, pathogens causing leaf diseases, such as *Colletotrichum* spp., also cause leaf lesions, following damage by some other agents. Hocking (1966) has shown that *Colletotrichum* can be a primary pathogen on coffee leaves, and there are records of severe defoliation of *Coffea canephora* and *Coffea excelsa* being associated with infection by *Colletotrichum* spp. Saccas and Charpentier (1969) and Muthappa (1970) reported a stalk rot of leaves caused by *Colletotrichum* spp. in India. Both *Colletotrichum* and *Glomerella cingulata* have also been implicated in the etiology of "weak spot" (Shaw, 1977) and *Mancha mantecosa* (oil spot) in South America (Vargas and Gonzales, 1972).

Berry Diseases

Though diseases causing damage to coffee berry may not cause damage to other parts of the plant, they lead to yield loss. Several fungi have been found to affect berries. Coffee berry disease (CBD) caused by *Colletotrichum kahawae* Waller & Bridge is a particularly devastating disease, which affects developing berries, leading to berry rot and shed before the beans are formed inside. CBD occurs only in Africa, but a less virulent form of the fungus occurs worldwide, attacking only ripening berries causing "Brown blight," and the mature beans inside are not destroyed. The disease causes pulp to stick to the bean, making wet processing difficult and reducing quality (Waller, 1987).

Extensive research has been carried out on the CBD pathogen and the disease itself in the late 1960s and 1970s, reviewed by Firman and Waller (1977). The source of the pathogen's spores are acervuli on the maturing bark of young twigs and on diseased berries, and their production, dispersal, germination, and infection depend on water (Waller, 1972). Subsequent development of the disease depends on the rainfall distribution and cropping pattern of the coffee trees. In Kenya, where overlapping crops result from two rainy seasons, diseased berries of the first flowering are still present when berries from the second flowering are just developing and are therefore most susceptible. These observations resulted in the recommendation, still current in Kenya, to spray monthly in the rainy season from February to July with the aim to provide a constant protective layer of fungicide on the berry surfaces (Anonymous, 2001). Copper-based fungicides are the most popular, as they also control coffee leaf rust and reduce the intensity of bacterial blight. Other protective fungicides include Chlorothalonil, Dithianon, and Anilazine (Anonymous, 2001).

The use of systemic fungicides, such as Benomyl and Carbendazim, which were at the beginning very effective, was discontinued later due to the development of pathogen resistance over the years. The fungicide-resistant strains are very stable and easily detectable, even without selection pressure of the fungicide (Kingori and Masaba, 1991), and have been found to interfere with control of the disease by contact fungicides (Masaba et al., 1990). Chemical control of CBD and other coffee diseases account for up to 30 percent of the cost of production (Nyoro and Sprey, 1986). A majority of the small-scale farms apply these fungicides less frequently than recommended for reasons of economy. These occasional sprays induce higher levels of CBD than would occur in their total absence (Griffiths, 1972). To provide a sustainable long-term control of the disease, a breeding program to combine resistance to CBD and CLR with the high yield and high quality of Kenya coffee, was undertaken resulting in the production of *Ruiru* 11 hybrid, which has been released commercially (van der Vossen and Walyaro, 1981).

Another common but seldom important disease is berry blotch caused by *Cercospora coffeicola*. The disease is characterized by dark-brown blotchy lesions, which are often confused with lesions caused by CBD fungus (the latter are typical sunken anthracnose lesions) (van der Vossen and Cook, 1975). The normal resistance of healthy green berries is reduced when the tree is under physiological stress or when the berries are wounded and other fungi infect the immature fruit. Some yeast like fungi (*Nematospora* spp.) infect berries attacked by *Antestiopsis* spp. and can induce an internal rot of the bean.

Other fungi commonly isolated from damaged berries include *Fusarium stilboides*, *Phoma* spp., and *Colletotrichum gloeosporioides. Botrytris cinerea* may also infect coffee berries to produce "warty disease" in wet, cool conditions. Both *Cercospora coffeicola* and *Fusarium stilboides* are major secondary pathogens, attacking berries weakened by overbearing stress, and there have been several instances when true reasons for infection have been overlooked.

Dieback Diseases

When root, trunk, or foliage suffers dieback, it leads to the dying back of shoots, which eventually causes debilitation of the coffee tree. Physiological dieback is often the result of overbearing due to the tree carrying more crop than its photosynthetic capacity can provide for support. Pathological dieback diseases occur most commonly at high altitudes and are usually the result of progressive infection by the minor pathogens, for instance, *Ascochyta* and *Phoma* spp., which infect immature coffee stems. Dieback disease may not be sufficiently severe to have much effect on the crop already on the tree (which is carried on growth produced in the previous season) but can reduce the cropping potential of the tree by restricting growth of young stems so that the following season's crop may be reduced. Bacterial blight of coffee, the most severe of pathological infections, is caused by *Pseudomonas syringae* pv. *garcae*. This is the most rampant disease observed in Kenya. The bacterium is an infectant on coffee shoot, which gains entry under wet and cool situations into young tissues to produce water-soaked dark, necrotic lesions on the leaves, twigs, and berries. When terminal buds are attacked, infection spreads backward from the shoot and twigs causing dieback, which is distinctly different from overbearing dieback as leaves remain attached to the dead twigs. In extreme cases, a whole sucker or even a large part of the main stem may be infected leading to the death of the coffee plant. The disease is the most prevalent on exposed slopes as opposed to sheltered valleys. Shade has been observed to reduce the incidence of BBC (Thorald, 1945), probably by reducing wind and storm damage, which assist the entry of bacteria. Control of the disease in Kenya has been intensive, straight foliar applications of copper-based bactericides to reduce the bacterial population on the coffee trees. Kairu et al. (1991) pointed out that the disease is stimulated when organic fungicides to control CBD, which occurs together with BBC, are used. Continued use of coppers in the control of BBC is also threatened by the recent observations of presence of copper-resistant strains of *Pseudomonas syringae* on farms where copper has been routinely used (Kairu et al., 1988). The sustainability of the current control program for the disease may not be possible considering the cost of bactericides and farm machinery vis-à-vis coffee prices. For this reason, research aimed at identifying the sources of genetic resistance to the pathogen in the coffee germplasm has been initiated in Kenya.

Trunk and Branch Diseases

When translocation of photosynthates is disrupted between roots and shoots, such diseases are caused. Initially leaf wilt, leaf shed, or chlorosis is noticed, but the disease may also infect berries and leaves directly as they spread along the branches. On trees showing these early symptoms, the crop may fail to mature properly, producing many light and empty berries. Cessation of growth and shoot dieback follows, and diseased trees or large parts of them are killed. These diseases usually occur sporadically on individual trees or groups of trees and do not cause rapidly spreading epidemics. Some are greatly influenced by soil and varying climatic conditions, leading to plant stress. Others are primarily wound pathogens and become severe only on damaged trees. The most prominent are *Fusarium* bark disease caused by *Fusarium stilboides* Wollen and web blight caused by *Corticium salmonicolor or Corticium koleroga*. The former is characterized by stem cankers usually at the bases of suckers (Storey's disease) (Storey, 1932) on mature main stems, especially at pruning wounds and primary branch bases (scaly bark), and around the bases of mature stems at or just above soil level. The cankers enlarge insidiously under the bark to girdle the stem, eventually killing the affected tree. Other symptoms include berry and leaf lesions (Siddiqi and Corbett, 1963). The disease is considered a major factor limiting coffee production in Malawi, Tanzania, and Zimbabwe (Clowes and Logan, 1985; Siddiqi and Corbet, 1983). In 1970 (Baker, 1970), the disease in Kenya was largely restricted to southeastern district of Taita Taveta, which is well isolated from other coffee-growing area. The distribution has since widened progressively to cover key coffee-growing districts. Control of the disease is achieved mainly by restricting its spread to healthy trees. Pruning tools easily disseminate the spores during field operations and should be sterilized using a suitable disinfectant after using them on diseased trees. Wounds caused by boring insects and during weeding operations form common entry points for the pathogen, whereas mulch and unchecked weeds near the base of the stem create a good environment for infection and sporulation by the fungus. Earlier investigations to locate resistant varieties failed to elicit consistent results (Siddiqi and Corbett, 1965), probably due to low success of inoculation tests with pure cultures of the pathogen reported by these authors (Siddiqi and Corbett, 1963).

Wilt Diseases

The most devastating coffee wilt disease is *Fusarium* (tracheomycosis) wilt disease, which is caused by *Fusarium xylarioides* Heim & Saccas. It occurs in several African countries and has been particularly troublesome on *Coffea canephora* in West and Central Africa and Uganda (Flood and Brayford, 1997). A similar disease is caused by *Ceratocystis fimbriata* Ellis & Halsted (Waller, 1987). This disease also gains entry through wounds at the base of the trunk and causes sunken necrotic cankers, which girdle the trunk and kill the tree.

Root Diseases

The *basidiomycete* root pathogens could be troublesome in newly cleared forest land when coffee plantations are established, as is the case with many plantation crops. *Armillaria mellea* (Valhl ex Vries) Karsten is the most widely occurring root pathogen of coffee in such conditions. Moribund stumps or roots provide a food source from which the fungus can spread to infected coffee. Rapid debilitation and wilting leading to the death of the plant results. Creamy white mycelial strands can be seen beneath the bark, and clusters of characteristic mushroom-like sporophores occur on the bases of recently dead trees. Another disease is the black root rot caused by *Rosellinia* spp., particularly widespread in Latin America and the Antilles. General debilitation followed by plant death are symptoms, similar to those of *Armillaria* infection, except that the mycelia occur as broad-spreading fans rather than as rhizomorphs, and the fruiting bodies are minute spherical ascocarps (Waller, 1987). *Fusarium solani* causes a sporadic but lethal root disease of *Arabica* variety in East Africa. No visible external mycelia are present, but a distinct brown discoloration of the wood at the crown of the tree is characteristic of the infection (Baker, 1972).

Among the nematodes, a range of them attack the coffee plant, but the root-knot nematodes *Meloidogyne exigua* and *Pratylenchus* spp. have been particularly troublesome in some Latin American countries and Tanzania. General debilitation with leaf chlorosis, stunting, and dieback are the major aboveground symptoms.

Nursery Diseases

In nurseries, the most common disease is damping off, caused by *Rhizoctonia solani*, *Fusarium solani*, and *Fusarium stilboides*, either in single or in combination. Nematodes also attack the coffee seedlings in nurseries, which leads to eventual debilitation in the main plantation.

Pests of Coffee

Le Pelley (1968) has cataloged the coffee pests, most of which are with restricted distribution and of little economic importance. However, in certain seasons and in certain regions, the pests can pose a great threat to the coffee plant. Because most of them are indigenous to a region where they occur, there are also their natural enemies. Localized outbreak leads to yield loss and deterioration in berry quality. Major pests include boring beetles, scale insects, mealy and sucking bugs, leaf miners, and defoliators. A description of the major pests and their economic importance is cataloged here.

Coffee Berry Borer (Hypothenemus hampei Ferri) (Coleoptera: Scolytidae)

The berry borer is a small black beetle, which is the only serious pest of coffee. Its distribution covers East and Central Africa, Brazil, Java, Peru, Ecuador, Colombia, Central America, and Mexico (Wrigley, 1988). The first incidence in Kenya was reported in 1928 (Wilkinson, 1929) with a less than 10 percent infestation. Of late, infestation up to 80 percent has been noted. Symptoms are a single small hole, the entry point by the female beetle, at the apex of otherwise healthy-looking green or ripe berries. The beetle tunnels within the bean and lays eggs, which hatch into the larvae. These larvae continue to feed on the beans, further damaging them. Damaged beans have a distinct blue-green stain (Anonymous, 1989). When attacked, young berries usually detach and fall off to the ground. These infested beans have no commercial value. The distribution of *Hypothenemus hampei* is influenced by the altitude, tending to be more severe at lower altitudes, below 1370m above MSL. Above 1525m above MSL, the pest is rare and at and above 1680m above MSL is totally absent (Wrigley, 1988).

There are cultural methods to control the pest which primarily revolve around building up the population of natural predators of the pests. In Kenya, these include two parasitic wasps, namely *Prorops nasuta* Waterston and *Heterospilus coffeicola* Schmiedeknecht. The parasitism levels by both wasps are low (Mugo, 1994). Heavy shade from shade trees on inadequately pruned coffee bushes creates conditions unsuitable for natural predators to multiply, and hence, must be removed. To prevent build up of predator populations, berries should be picked at least once in a fortnight during peaking of fruiting and at least once in a month at other times. No ripe or dried berries should be left on the ground or on the trees. All infested berries should be destroyed by burning, deep burying, or rapid drying on trays. The old crop should, if possible, be stripped completely just prior to main flowering (Anonymous, 1989).

Insecticide sprays should only be regarded as a supplement to these cultural measures. Recommended insecticides include Dursban 48 percent emulsion concentrate and Nuvellawcl. C. (Anonymous, 1989).

*Antestia Bug (***Antestiopsis** *spp.) (***Hemiptera: Pentatomidae***)*

Antestiopsis spp. is a serious pest of *Coffea arabica* on the African continent, where the coffee tree is attacked, on average, by one or two bugs per tree, causing considerable loss in yield (Anonymous, 1989). Several species including *Antestiopsis orbitalis bechuana*, *Antestiopsis ghesquierei*, *Antestiopsis intricata*, and *Antestiopsis facetoides* occur in Kenya (Greathead, 1965). The antestia bug has a flattened body with striking black, orange, and white markings. The female bug lays egg on the underside of the coffee leaf. When hatched, the nymphs undergo five stages prior to maturing (Le Pelley, 1932). Damage by the adult bug is through its feeding on green berries, buds, and green twigs. Pierced young berries are shed, while older ones remain attached to the plant. However, the berries develop sunken, discolored patches on their surfaces due to the introduction of yeast like fungi, namely *Nematospora coryli* and *Nematospora gossypii*, vectored by the bug. On pulping, such berries yield zebra-patterned beans of low appeal to customers (Le Pelley, 1968). Such branches are difficult to prune and do not produce a good crop. When flowers are attacked, they fail to develop and turn black and die. Two antestia bug egg parasitoids, *Telenomus seychellensis* and *Hadrontus antestiae*, are common with a parasitism percentage of 80–90 percent (Le Pelley, 1968). Several nymph and adult parasitoids have also been reported (Wanjala, 1980). Coffee bushes should be kept open through regular pruning so as to make them unsuitable habitats for the pest and a suitable one for parasitism (Mugo, 1994). Antestia bugs can be controlled through pesticides (Anonymous, 1989). Spraying should be done when the average population (adults and nymphs) exceed two per tree in the drier areas or one per tree in the wet regions.

Scales and Mealy Bugs

The nymphs and adults of these insects are of considerable economic significance in coffee. These pests parasitize roots, branches, leaves, or flowers, on which they settle and suck the sap leaving the plant debilitated to die. Mealy bugs are soft-bodied insects covered with a white waxy coat, quite capable of rapid movement. Scale insects, on the other hand, remain attached to the plant and develop a tough scaly covering once they begin to feed. Their numbers increase rapidly during dry weather and tend to decline during prolonged wet spells. The most important in Kenya are the root mealy bugs (*Planococcus citri* Risso), Kenya mealy bugs (*Planococcus kenyae* Le Pelley), green scales (*Coccus viridis*), white waxy scales (*Ceroplastes brevicaudas* Hall), and fried egg scales (*Aspidiotus* spp.). Of all the mealy bugs, the most common is the root mealy bug found around coffee-growing regions worldwide (Mugo, 1995). The bug also attacks roots of citrus plants (Le Pelley, 1968). The severity of attack tends to be higher on coffee growing in soils of low fertility, deficient in Ca with low pH, around 5.0 (Baum, 1968). Affected plants yield poorly, develop yellow leaves, and eventually die. In disease-prone areas, soils should be mixed with the insecticide Furadan or Temik at the recommended rates during planting and replanting operations (Anonymous, 1989).

The origin of Kenya mealy bug is Uganda and northern Tanzania, where a complex of natural enemies kept it under control (Mugo, 1994). A parasitic wasp, *Anagyrus kivuensis*, imported from Uganda kept the pest from proliferation in Kenya (Le Pelley, 1968). Occasionally, this control was disrupted by ants, but trunk banding of infested trees using recommended insecticides offers a good solution (Anonymous, 1989). It was in Sri Lanka that green scaly bugs were first discovered, which has now spread to many coffee-growing regions worldwide. Its hosts include tea, cassava, citrus, guava, and mango (Wrigley, 1988). Severe infestation, especially in dry regions, leads to the development of stunted trees with reduced flowering. Green scales are kept under control by their many insect parasites and predators, provided ants are prevented from attending the scales (Mugo, 1994).

Coffee Leaf Miner (Leucoptera *spp.*) (Lepidoptera: Lyonetiidae)

These small moths, with their larvae mining into the leaves of the coffee plant, are widely distributed in South America and Africa (Wrigley, 1988). Symptoms are irregular brown blotches on the upper side of the leaves. Removal of the top skin of the blotch reveals fresh mines and several caterpillars, which are up to 12mm long (Anonymous, 1989). Several attacks cause major destruction of leaf tissue and serious leaf fall, reducing the amount of photosynthates badly needed by developing berries. Where this combines with coffee leaf rust, the twigs die back. The enhanced use of fertilizers and fungicidal sprays, combined with excessive mulching and leading to an upset in the ecosystem, started the pest attack in 1954 and elevated it to a major pest status (Bess, 1964). Mulching reduces predation by ants on leaf miner larvae, which drop to the ground to pupate in soil crevices. When there is better fertilization and proper insecticidal use, enhanced leaf retention results, which then forms a breeding ground for the pest, which favors old to young leaves. When the pest attack is severe, chemical control is possible using an insecticide with translaminar systemic action to get into the mines and narrow the spectrum of activity so that the pests' natural enemies will be destroyed.

Stem and Branch Feeders

Stem and branch feeders are common in the tropics and form some of the most serious coffee pests, several of which are capable of causing the death of the coffee plant. Major ones among them are the white borer (*Anthores leuconotus* Pascoe), yellow-headed borer (*Dirphya nigricornis* Olivier), and West African coffee borer (*Bixadus sierricola* White). The white borer is widely distributed in Africa, being present in Angola, Cameroon, and East Africa (Coste, 1968). It is particularly severe on *Arabica* coffee, especially at lower altitudes below 1500m above MSL. The larvae do the most damage, boring into the trunk and roots exuding woody shavings from its burrows. Young trees (2 years old) are frequently killed. Older ones wilt, and foliage become chlorotic. When they survive, yield is drastically reduced, and the trees succumb to infestation by *Fusarium stilboides* (Anonymous, 1989). Banding or spraying the stem about 18 inches from soil level using one of the recommended insecticides provides good control (Anonymous, 1989). Le Pelley (1968) reports the incidence of yellow headed borer in Senegal, Malawi, Kenya, and Tanzania. In Kenya, the pest is particularly serious in Taita Hills (Clowes, 1949) where early attempts to grow coffee were abandoned due to ravages of the borer. The larvae bore through the wood, weakening its structure and disrupting the activities of the vascular system. Affected trees therefore break easily, have low resistance to water stress, and yield poorly. Wounds caused by the borer are common entry points for bark pathogens, such as *Fusarium stilboides*.

Consumer Choice and Coffee Adaptation

Like some of the crops of great industrial importance, the uniqueness of coffee is that it is grown in the southern hemisphere, while consumption is mainly in the Northern hemisphere. The British gave a great impetus to tea cultivation in India. So it goes for coffee. The plant cannot be grown either in Europe or North America, but although consumption is very high in Europe and North America. The developed countries in the northern hemisphere comprise up to 80 percent of coffee consumption. Hence, the consumer preference and income not only determine the level of coffee consumption, but also the type of coffee grown as well as production and processing practices adopted by coffee producers.

Whether mild, Brazilian strong, or *canephora* coffee, coffee as a beverage has been traditionally consumed mainly as ground coffee, which represents 80 percent of the sales in the United States and Europe. Ground coffee is usually offered as roasted coffee in various blends, such as the instant, flavored, or decaffeinated varieties. All of these have to undergo various processing, the most important being the roasting. Current roasting techniques require exact knowledge of green coffee characteristics, not only to identify the species, varieties, and types from the different production regions, but also their particular characteristics, which depend on grading, storage, and nature of impurities and defects. These concerns of the roaster coupled with the ultimate consumer needs dictate the types and origins of coffee uses.

The consumer preferences, in terms of choice and presentation, as well as health concerns, have added a new dimension to coffee production and processing. The need for speciality or gourmet coffee has risen in Europe and the United States. The speciality concept is derived from the belief that there are certain coffees which exhibit unique crop quality and hence consumer satisfaction, as they are grown in specific regions and processed in a special manner. Such coffees include *Inter Alia Supremos* from Colombia, AA grade from Kenya, and *Blue Mountain* from Jamaica. These coffees are grown mainly at high altitudes with rich soils and high rainfall, which are ideal for bean development. Further, these coffees are processed mainly through wet method and dried in natural conditions. Due to the high premium attracted by such coffees, there is a concerted effort to develop varieties, which can do well under these highland conditions.

One of the most serious fallouts of chemical agriculture is the ever-growing concern about pesticide contamination of agricultural products. This has been the main reason for many consumers to prefer organically grown agricultural products. Coffee is no exception. Organic coffee is processed separately under natural environment without chemicals. This development, which is of a recent origin, has spurred the adaptation of minimal use of inorganic inputs on coffee crops. Consumer preferences in favor of shade-grown coffee have also been on the increase since 1975. This has necessitated the growing of coffee under all kinds of shade trees, an urgent necessity that demands varieties that do well under shade. Given the positive and negative aspects of this demand, it is a challenging task for the coffee breeder.

Research and Developmental Setup

The leading coffee-growing countries on the African, Asian, and Latin American continents (in particular, the former) have functional institutions and/or organizational set-ups, which carry out research and train scientists and those involved in developmental work in various aspects of the coffee industry. These institutions fall under different categories. They are either autonomous or function as governmental appendages. Academic institutions, such as universities, are also involved in coffee research. They often collaborate with other research organizations exclusively devoted to coffee research. An important set up is the Kenyan Coffee Research Foundation, funded mainly by coffee farmers of Kenya, the Federacion Nacional de Cafeteros de Colombia, and the Fundacao Instituto Brasiliero de Geografica e Estadistica, which are run by their respective governments. Other coffee research programs are within a wider research mandate within their respective government research centers. For instance, in India, there is a separate Coffee Research Station in Balehonnur in Karnataka State, under the administrative control of the Ministry of Commerce of the Government of India.

In addition to these research institutions, several regional and international bodies and organizations are involved in research and development in coffee. Such bodies include the European Union (EU), International Coffee Organization (ICO), World Bank, and the African Coffee Research Network (ACRN). These organizations are involved in facilitating or funding collaborative research. They also organize workshops, seminars, and conferences, both national and international, during which coffee researchers and development officials can interact and exchange ideas, share their research results, and develop a common strategy with the ultimate goal of taking coffee further on the global scale.

A Look into the Future of Coffee

It is paradoxical that while consumers keep paying more for the coffee they drink with each passing year, the benefits accrued do not reach the coffee farmers, which is a central problem of coffee industry. Rather, the coffee farmers' net income has been declining over the years. This is the biggest threat to the production of quality coffee. The main cause is the poor prices offered by the buyers at the world market. The poor prices are, in turn, the result of the large quantities of inferior quality of beans produced, which flood the global markets, from countries such as Vietnam. Here there is parallel between coffee and what is happening with black pepper. Indian black pepper, especially the "Malabar Gold" from the tiny State of Kerala, fetches a premium in world markets. But the Vietnamese pepper is of a much inferior quality. Unscrupulous traders and importers and/or exporters take advantage of this and mix the inferior quality Vietnamese pepper with the Indian variety and export it as premium Malabar pepper. But when importers detect this, the country loses heavily, not only monetarily, but in international reputation. The second contributor to the poor net earnings by the coffee producers is the increasing cost of agrochemicals, in particular, pesticides, thereby making disease and insect control quite expensive. Under the circumstances of poor prices and high cost of production, the only way to arrive at sustainable coffee production is to produce more with less input. This implies that superior varieties with good yielding capacity and quality beans need to be produced. In most cases, old varieties already combine these traits. But the main constraint is disease resistance, which is being investigated by researchers in many coffee-producing countries. Hence, the coffee breeders should focus more on improving the stability of resistance base and develop ways of rapidly responding to new diseases or new strains of the old pathogens.

Therefore, the most important and urgent task would be to focus on locating those germplasms that possess genes of economic significance, particularly those with disease and insect resistance and bean quality. Such genes should then be accumulated in genetic backgrounds, which are more attractive for breeding research through pre-breeding. The second domain that demands attention is the development of transformation procedures to be put in place as an option for rapid transfer of genes of economic interest. At present, transgenic varieties are not very popular with coffee consumers. However, by limiting coffee transformation to genes within the same family or genus, consumer acceptance could be realized with less effort.

To date, research in coffee has led to enhanced production and quality of beans. But new challenges arise in a world that is quickly changing. Some areas are still revolving around "text book knowledge." Mention should be made of soil fertility management, in particular, soil testing to quantify available plant nutrients. Coffee is a perennial crop and has similarities in nutrient absorption comparable to other perennial crops, like cardamom and black pepper. Fertilizer schedule is still rooted on the data based on classical and conventional soil testing to quantify principal nutrients like nitrogen, phosphorus, potassium, and zinc. Recent research by the author has led to a totally different and groundbreaking concept, which is now known globally, as "The Nutrient Buffer Power Concept." A detailed discussion is given (Nair, 1996) and subsequent experimental evidence has been provided with much success of the new concept in the nutrition of black pepper (Nair, 2002) and more recently on cardamom (Nair, 2006), both perennial crops. There is much scope to test the validity of this concept in scheduling precise fertilizer recommendations for coffee.

Though biotechnology might offer avenues for improving coffee production and quality, it is a totally unexplored area of research. Molecular marker-assisted selection is a powerful approach for coffee improvement. Other areas of interest would be breeding decaffeinated coffee, production of organic coffee, biocontrol of pests and diseases, and integrated pest management (IPM). Genetic modifications in pest resistance are an area that holds out much promise. Breeding coffee to suit specific tastes is another area worth exploring. Additionally, there is an imminent need to deeply research the economics of production and coffee market dynamics vis-à-vis global markets. No attempts have been carried out on these lines.

Perhaps more important, what impact has the WTO stipulations brought about on coffee market? Long ago, late Jules Nyrere, a great leader of the third world and resident of Tanzania said to those in the developed world: "Give us a better price for our products. We do not want your charity." Has this prophetic statement been lost in the winds of change? It is time to reflect on these things and rectify where things went wrong. The WTO discussions have been dragging along for almost a decade now with no end in sight that would favor the poor. Today, the fate of coffee is a classic example, where the producer is at the Mercy of market manipulators and the consumer is fleeced.

7 Oil Palm (*Elaeis guineensis* Jacquin)

The oil palm (*Elaeis guineensis* Jacquin) is economically important for its oil and has become one of the major oil crops of the world. For a long time it was the "poor man's cooking oil," but with the fossil fuel reserves fast depleting, it has become a much sought-after "green fuel." In fact, at the time of writing this book, the price of a ton of palm oil had already crossed US$900, higher than its equivalent petroleum. The oil palm yields about a threefold more oil than coconut, sevenfold more oil than rapeseed, and almost a tenfold more than soybean. To date, as an oil yielder, palm oil takes second place only to coconut and stands above rapeseed, sunflower, groundnut, and cottonseed. Under good agricultural management, a hectare of oil palm yields about 5–7 tons of oil per annum (Henson, 1991).

The global major oil production was 1.09 billion tons in 1999, of which soybean contributed 24.7 million tons (22.6 percent), followed by palm oil at 20.4 million tons (18 percent), rapeseed oil at 12.9 million tons (11.8 percent), sunflower oil at 9.3 million tons (8.5 percent), tallow at 8.1 million tons (7.4 percent), lard at 6.6 million tons (6 percent), and butter at 5.8 million tons (5.3 percent). World export of major oils and fats amounted to 34 million tons in 1999. Palm oil was the major oil traded with 13.6 million tons (40.2 percent), followed by soybean with 7.5 million tons (22.3 percent), sunflower oil with 2.9 million tons (8.7 percent), and tallow with 2.2 million tons (6.7 percent). Malaysia, world's largest producer and exporter of palm oil and its byproducts, produced 10.5 million tons (51.5 percent), followed by Indonesia, contributing 6.2 million tons (30.5 percent). Countries such as Nigeria, Colombia, Thailand, Ivory Coast, Papua New Guinea, Equador, and Costa Rica produced about 3.5 million tons of palm oil. In terms of palm oil export, Malaysia contributed 8.8 million tons (65.1 percent), while Indonesia contributed 3.1 million tons (24.2 percent), the rest of the exporting countries contributing 10.7 percent (Table 7.1).

There was a surge in the number of countries importing palm oil, which has shown an increase from 67 in 1970 to more than 120 in 1999. At least 20 major palm products are available for export from Malaysia. The major importers of palm oil from Malaysia are listed in Table 7.2. The major destinations of palm oil export from Indonesia are the Netherlands (44 percent), Germany (12 percent), Italy (9 percent), Spain (5 percent), and others (Kenya, the United States, Greece, and the United Kingdom) put together is 8 percent (Wakker, 2000).

Palm oil and palm kernel oil can be used both for edible (90 percent) and nonedible (10 percent) purposes. Palm oil can also be used for frying meat, fish, and vegetables

The Agronomy and Economy of Important Tree Crops of the Developing World. DOI: 10.1016/B978-0-12-384677-8.00007-2
Copyright © 2010 by Elsevier Inc. All rights of reproduction in any form reserved.

Table 7.1 Major World Producers of Palm Oil ('000 tons) from 1994 to 2003

Country	1994	1995	1996	1997	1998	1999	2000	2001	2002	2003
Malaysia	7403	7221	8386	9069	8319	10554	10842	11804	11909	3354
Indonesia	3421	4008	4540	5380	5361	6250	7050	8030	9200	9750
Nigeria	645	640	670	680	690	720	740	770	775	785
Colombia	323	353	410	441	424	501	524	548	528	543
Cote D'Ivoire	310	300	280	259	269	264	278	220	240	251
Thailand	297	316	375	390	475	560	525	620	600	630
Papua New Guinea	223	225	272	275	210	264	336	329	376	325
Ecuador	162	178	188	203	200	263	222	201	217	247
Costa Rica	84	90	109	119	105	122	138	138	140	144
Honduras	80	76	76	77	92	90	97	108	110	112
Brazil	54	71	80	80	89	92	108	110	118	132
Venezuela	21	34	45	54	44	60	73	80	80	79
Guatemala	16	22	36	50	47	53	65	70	81	91
Others	1265	1676	815	869	844	832	879	919	922	940
Total	14304	15210	16282	17946	17169	20625	21877	23947	25236	27383

Source: Malaysian Paklm Oil Board (www.mpob.gov.my).

and for making salad dressing, margarines, shortenings, *vanaspati* (a popular hydrogenated vegetable cooking fat popular among Indians), ice cream, confectioneries, and other emulsion-based products. As far as nonedible use is concerned, the oil is mainly used in the cosmetic industry (soap manufacturing) and also indirectly as oleo chemicals (fatty alcohols, fatty amines, amides, and nitrogen and glycerols). Carotenoids, vitamin E (tocopherols and tocotrienols), and sterols are minor components of crude palm oil that can be pretreated and encapsulated for pharmaceutical applications (Choo and Yusof, 1996).

There are other industrial uses of oil palm. The oil palm biomass can be processed to produce wood-based products (Mohamad, 2000), such as pulp and paper, particleboard of various kinds, and medium-density fiber board. Recently, there is much interest in the production of oil palm fiber plastic board as thermoplastic sheets and thermosetting boards due to their suitability for manufacturing various automobile components. The focus of this chapter is on research and development, which would ensure the sustainability of oil palm as a plantation crop. There is also an emphasis on plant breeding and selection of superior varieties and the potential of biotechnological interventions in the production of premium products, such as oleate and stearate oils for the oleochemical industry. The need to manage oil palm in adjusting to a fast-changing environment is also highlighted.

Table 7.2 Export of Palm Oil to Major Destinations ('000 tons)

Country	1990	1997
Pakistan	702	1132
China	737	1065
European Union	553	729
Australia	58	95
Bangladesh	25	141
Egypt	346	333
India	494	963
Indonesia	–	100
Japan	274	358
Jordan	45	230
Korea	215	182
Myanmar	–	158
Singapore	742	308
South Africa	–	183
Saudi Arabia	86	134
Taiwan	–	72
Turkey	191	237
United Arab Emirates	–	106
United States	143	113
Mexico	–	25
Others	1109	813
Total	5720	7477

Source: Oil World Annual.

Historical Origin and Global Distribution

The tropical rain forest of Central and West Africa is the center of origin of oil palm (Zeven, 1964). The main oil palm belt of the African continent runs through the southern latitudes of Sierra Leone, Liberia, Ivory Coast, Ghana, Togo, Nigeria, and Cameroon as well as the equatorial region of the Congo and Angola between 10° north and 10° south (Zeven, 1967). Tribal migration or intergroup exchanges spread oil palm across Africa (Smith et al., 1992). Oil palm was taken to Congo and East Africa before the arrival of Europeans there as colonizers. It was introduced to Sudan about 5000 years ago (Clark, 1976). The sporadic occurrence of oil palm on the East African coast (Uganda, Tanzania, Kenya, Rwanda, and Burundi) was probably due to Arab slave traders. The Africans probably brought oil palm to Madagascar in the tenth century (Pursglove, 1972). In the fifteenth century, slave traders brought oil palm to the New World, from where it was cultivated in Bahia, Brazil. It was grown in European conservatories until 1973. It must have reached Mauritius earlier and was found in Kolkatta (in the State of West Bengal, India) much earlier in 1836. In 1848, the Dutch imported oil palm seeds from East Africa. In 1848, the Dutch imported oil palm seeds from West Africa, and four oil palm seedlings were planted in Buitenzorg Botanical Garden (now Bogor) in Indonesia, of which two came from Amsterdam Botanic Garden and two from Bourbon Island (now Reunion) (Pursglove, 1972). This introduction laid the foundation for the oil palm industry in southern Asia, especially Malaysia and Indonesia. The Singapore Botanic Garden obtained oil palm seeds from Java around 1870 (Smith et al., 1992). Subsequently the seeds were distributed to Malaysia where they were grown as ornamental trees. The seeds were also sent to Sumatra, which later received seeds from Buitenzorg. Palms from these sources were planted in Deli, Sumatra, and evolved into the *Deli dura* populations. These are the populations currently being used as part of genetic foundation for oil palm improvement in Malaysia and Indonesia.

Taxonomy and Botany

The genus *Elaeis* belongs to the palm family *Palmae*, an important member of the monocot group under the order *Spadiciflorae*. It is included in the *Cocoineae* tribe together with the genus *Cocos* (Uhl and Dransfield, 1987). The word *Elaeis* has its origin in the Greek word *Elaion*, meaning oil, and *guineensis* points to the oil palm origin in the Guinea coast (Hartley, 1988). Within the genus *Elaeis*, two species are distinguished—the economically important oil palm *Elaeis guineensis* and the oil palm of American origin, *Elaeis oleifera*. These species hybridize readily, suggesting a close relationship in spite of their origin in two different continents (Hardon and Tan, 1969).

The oil palm has a crown of 35–60 pinnate fronds arranged on a vascular stem. It has a single bud in the base of the crown where fronds and inflorescences originate. The palm may reach a height between 15 and 30 m and can last up to 300 years. It is monoecious with male or female inflorescences occurring separately. This enforces crosspollination. Wind and insects assist pollen dispersal (Pursglove, 1972). The main pollinating insect is the *Elaeidobius* species (Syed, 1980). On an average, the fronds are produced at the rate of two per month in a regular sequence. The length of the frond is typically about 7 m, and each frond consists of a petiole, which is 150 cm long, and a rachis bearing 250–350 leaflets. Each leaflet may be about 130 cm long. The leaflets are arranged on two lateral planes. An inflorescence primordium forms in each axil of the frond. Male and female flowers are produced at different times in separate inflorescences. The oil palm flower primordium has both male and female organs (Beirnaert, 1953). In a potentially female flower primordium, the two accompanying male flowers are suppressed and remain rudimentary, while in a potentially male primordium, the female organ is suppressed. Very rarely, both the gynoecium (female) and androecium may develop to give a hermaphrodite flower. The process of sexual differentiation occurs at 28 months stage before anthesis (Wood, 1984). The male inflorescence has an approximately 40 cm long stalk, with 100–300 finger-like spikelets. Each spikelet has about 600–1500 yellow flowers. Pollen production ranges from 10 to 30 g per inflorescence. The basic structure of the female inflorescence is similar to that of the male, but the spikelets are shorter. Each spikelet bears 5–30 flowers that are receptive for 2–3 days. Fertilized female flowers produce fruits that grow and ripen over about 6 months. Oil palm fruits are sessile drupes borne on a large compact bunch. Each fruit consists of a hard kernel inside a shell (endocarp) that is surrounded by a fleshy mesocarp. The mesocarp contains about 49 percent palm oil and the kernel about 50 percent palm kernel oil. A mature bunch contains a few hundred to a few thousand fruits and its weight ranges from 5 to 50 kg depending on the age of the palm and the genetic traits and the environmental conditions in which the palm grows.

The palm has differing external appearance. The most common type, known as *nigrescens*, is deep violet or black in color prior to ripening. An uncommon type is green in color and is called *virescens*, which is the dominant type. Another type, *albescens*, characterized by the small amount of carotene in the mesocarp, is a low yielder. *Albescens* fruit can be of the *nigrescens* or *virescens* type and is referred to as *albo-nigrescens* and *albo-virescens*, respectively (Pursglove, 1972). In internal structure, the most important difference is thickness of the shell. A single gene controls the shell thickness of the oil palm (Beirnaert and Vanderweyen, 1941). The recessive homozygote, the pisifera (sh –, sh –), is shell-less, whereas the dominant homozygote, the dura (sh +, sh +), has a thick shell. When the dura crosses with the pisifera, a heterozygote, the tenera (sh +, sh –), which has a thin shell surrounded by a ring of fibers in the mesocarp, is produced. The most commonly cultivated fruit form is the high yielding tenera.

Growing Conditions

Oil palm is a crop of the tropics, such as those found in Southeast Asia, like Malaysia and Indonesia, on the African continent (Western and Central) and South America (Hartley, 1988). From as far as 16° north in Senegal to 13° south in Malawi, and 20° south in Madagascar, one can find isolated oil palm groves (Piggot, 1990). However, it is grown commercially in more than 20 countries with most areas within 10° north and 10° south of the equator (Goh, 2000). In this respect, Malaysia and Indonesia are the world's main producers and also exporters of palm oil. In general, oil palm can be grown in the tropics with assured irrigation.

Climatic Requirements

High rainfall (between 200 and 300 cm/year) is required for good growth of oil palm. At least 15 cm of rainfall is required each month without distinct drought season or months with less than 10 cm rain (Goh, 2000; Hartley, 1988; Piggot, 1990). A water deficit greater than 30–40 cm/annum will significantly reduce fresh fruit bunch (FFB) yield (Mohd Haniff, 2000). Optimal temperatures range between 22°C and 33°C, with the lowest temperature supporting the plant close to 20°C (Goh, 2000; Hartley, 1988). However, the growth rate of young seedlings is inhibited at temperatures of 15°C or lower, but the maximum tolerable temperature is 32°C, which did not affect plant growth adversely (Goh, 2000). The daily requirement of sunlight is between 5 and 7 daylight hours and at least 2000 hours of sunshine annually (Hartley, 1988; Piggot, 1990). In Malaysia, the high rate of annual growth is the result of high levels of annual light interception (Mohd Haniff, 2000). Relative humidity should be between 75 and 100 percent (Piggot, 1990).

Soil Requirements

Most soils are suitable and the crop does not demand high fertility, except that the soils should not be heavy with large amounts of clay, which during the monsoon season leads to waterlogging due to impeded drainage (Piggot, 1990). The suitable soil texture is sandy loam of more than 75 cm depth. Lateritic, sandy, or peat soils are problematic soils that need proper manuring and maintenance for optimum palm growth (Hartley, 1988). Ideally, oil palm should be grown in flat areas. The elevation and slope of an area proposed for oil palm cultivation are important factors that determine its suitability. In general, oil palms are not recommended for planting in areas with an elevation of more than 200 m above MSL (Paramananthan, 2000). In undulating areas, cost of maintenance, harvesting, and transportation would be higher. The planting density varies according to soil type. For inland soils (sedentary), planting is done in a triangular form, with a distance of 8.8 m, giving 148 palms/ha. However, with coastal alluvial soil, the recommended density is 136 palms/ha, and for peat soils, 160 palms/ha (Hartley, 1988).

Nutrient Requirements

When palms are planted on inland soils, one-quarter to one-half of N, P, and K have to be supplied through inorganic fertilizers; however, in fertile soils there is no need for this (Tarmizi, 2000). Malaysian soils need between 0.5 and 1.1 kg of N, 1.1 kg P_2O_5, and 0.5 and 2.0 kg K_2O per palm annually (Tarmizi, 2000). The nutrients are required mainly to offset and/or compensate losses through soil immobilization, as for instance in the case of P_2O_5, leaching, as in the case of N, and to offset recycling losses from pruned fronds and palm oil mill effluents. Response to high K application on inland soils showed that the ratio of oil to bunch is decreased. N increased the number of bunches, their weight, and total oil produced, whereas P increased only bunch weight, but not other bunch determinant yield parameters.

Fertilizer scheduling in oil palm is still rooted in classical textbook knowledge. Recent advances in soil fertility management with perennial crops like black pepper (Nair, 2002) and cardamom (Nair, 2006), based on the path-breaking "The Nutrient Buffer Power Concept" (Nair, 1996), hold out much promise in the case of oil palm, especially in the case of P and K.

Collection of Oil Palm Germplasm

In the past, a number of expeditions were made to collect oil palm genetic materials from its center of origin. Some oil palm researchers were in Belgian Congo after the Second World War to initiate a collection there (Vanderweyen, 1952). In the early 1960s, Nigerian oil palm breeders collected 72 open-pollinated progenies from the eastern part of the country. These materials were planted and evaluated at the National Institute for Oil Palm Research (NIFOR) main station. Some outstanding palms were selected and introduced into their current breeding program (Okwuagwu, 1985). In Cameroon, Blaak (1967) sampled oil palm materials from Bamenda highlands of the country. Some of the palms were planted at Lohe, in Cameroon, while the rest were distributed elsewhere. Between 1974 and 1975, the Institute de Researches pour les Huiles Oleagineux (IRHO) prospected *Elaeis guineensis* materials from the western region of Cameroon. The French oil palm workers systematically evaluated natural stands of oil palms in Ivory Coast, and the selected ones were progeny tested and utilized as new foundation material in their breeding schemes (Meunier, 1969). It was reported that IRHO also selected four palms in Malaysia and another 38 palms in Benin and used them as their original *tenera* stock.

Recent Collections

Four *Deli dura* palms were initially utilized as parental stock in breeding program. As such, the genetic base of oil palm breeding in the Far East was extremely narrow. A number of systematic expeditions to collect oil palm genetic materials were mounted by researchers in Malaysia. The objective of the exploration was not only to broaden the genetic base of the current oil palm breeding materials, but also to ensure conservation of oil palm genetic resources for posterity. The first attempt was made in Nigeria (Rajanaidu, 1985). This effort was followed by collections from other countries on the African continent, namely the Republic of Cameroon in 1983, Democratic Republic of Congo in 1984, Madagascar and Tanzania in 1986, Angola in 1991, Senegal in 1993, Gambia in 1993, Guinea and Sierra Leone in 1994, and Ghana in 1996. During the prospection in each country, the sites and palms were chosen at random. One bunch was harvested from each of the sampled palm and the fruits were kept separately. The mean coefficient of variation percentage (CV%) of the traits scored in each country was computed (Rajanaidu, 1985). The following details pertain to the collections from each country, gathered by the Malaysia Palm Oil Board (MPOB).

Nigeria

Genetic materials were collected from 45 sites distributed throughout the country from the oil palm growing regions in 1973 (Rajanaidu, 1985). Five to ten palms were sampled from each site depending on the rainfall pattern, soil type, and density of oil palm groves. The total number of palms sampled during the prospection was 919, which consisted of 595 *duras* and 324 *teneras*. The *pisifera* was virtually absent in all populations (Rajanaidu, 1985).

Republic of Cameroon

Pamol, a subsidiary of Unilever, carried out the prospection in 1984. Samples were collected from 32 sites. One to 15 palms were sampled at each site. A total of 95 palms (58 *duras* and 37 *teneras*) were sampled with an average of 3 palms from each site (Rajanaidu, 1985). About 19,000 seeds were brought to Malaysia (Rajanaidu, 1985).

Democratic Republic of Congo

The germplasms were collected from 54 sites with the collaboration of Plantation Lever Zaire (PLZ) between April and July 1984 (Rajanaidu, 1985). Five to ten palms were sampled at most sites. In all, 369 bunches were collected, consisting of 283 *duras* and 85 *teneras*. A total of 73,800 seeds were dispatched to Malaysia after the intermediate quarantine (Rajanaidu and Jalani, 1994c).

Tanzania

With the cooperation of the Ministry of Agriculture, Tanzania, and financially supported by the International Board of Plant Genetic Resources (IBPGR), collection of oil palm germplasm was initiated in Tanzania in 1986. Half of the samples collected were deposited with the Ministry of Agriculture, Kigoma, Tanzania. The materials were sampled from 13 sites near Kigoma, along Lake Tanganyika. At each site, one–seven palms were chosen randomly, and a total of 60 samples (42 *duras* and 18 *teneras*) were collected (Rajanaidu, 1986a).

Madagascar

Because the distribution of oil palm in Madagascar was very sparse, only 17 palms were sampled from four sites. One to six palms were sampled from each site. This expedition was carried out in 1986 with the collaboration of the Ministry of Agriculture in Madagascar and sponsored by the International Board of Plant Genetic Resources (IBPGR) (Rajanaidu, 1986a). The palms observed were very poor in growth compared to those found in Nigeria, Cameroon, DRC, and Tanzania (Rajanaidu and Jalani, 1994c).

Angola

Oil palm genetic materials in Angola were taken from eight sites in 1991. Only the coastal areas were covered. At each site, 2–14 samples were collected. A total of 54 bunches (42 *duras* and 12 *teneras*) were collected (Rajanaidu et al., 1991).

Senegal

Collection of genetic materials in Senegal was carried out in July and August 1993, with the Ministry of Agriculture, Senegal, assisting in the efforts. Palms were sampled from 13 sites. Five to ten palms were selected from each site, and a total of 104 accessions (all of them *duras*) were collected (Rajanaidu and Jalani, 1994b).

Gambia

The Ministry of Agriculture and Forestry, Gambia, was involved in the collection. A total of 45 palms were sampled from six sites. At each site, 5–10 plants were sampled. Only *dura* palms were encountered in Gambia (Rajanaidu and Jalani, 1994a).

Sierra Leone

The samples were collected from April to May 1994 with the cooperation of the Ministry of Agriculture. A total of 56 samples (52 *duras*, 3 *teneras*, and 1 *pisifera*) were obtained from 14 sites. At each site, two–six palms were sampled. In terms of fruit color, 54 were *nigrescens*, 1 *virescens*, and 1 *albescens*. The seeds were equally divided between the Ministry of Agriculture and MPOB (Rajanaidu and Jalani, 1994c).

Guinea

The palms were collected in May 1994 from 14 sites. The prospection was carried out jointly by MPOB and the Ministry of Agriculture, Guinea. At each site, three–five palms were sampled. A total of 61 samples (58 *duras*, and 3 *teneras*) were collected. All the fruits were *nigrescens* (Rajanaidu and Jalani, 1994c).

A general comparison was made between the populations collected from the different countries mentioned, and data on bunch weight, single fruit weight, and mesocarp to fruit weight among *duras* and *teneras* have been summarized (Table 7.3).

Table 7.3 A Summary of Bunch Characters of Oil Palm Genetic Materials Collected from Africa

Country	Dura			Tenera		
	Bunch Weight (kg)	**Single Fruit**	**Mesocarp to Fruit**	**Bunch Weight (kg)**	**Single Fruit Weight (g)**	**Mesocarp to fruit (%) (g)**
Nigeria	11.8	7.9	47.3	10.9	8.5	70.9
Cameroon	16.8	10.3	39.7	17.3	8.6	62.4
DRC	17.6	14.2	43.9	17.4	12.6	64.1
Tanzania	18.4	16.9	46.7	13.7	15.5	70.8
Angola	21.4	14.2	48.9	16.0	11.7	70.9
Senegal	5.9	2.6	35.1	–	–	–
Gambia	5.7	2.3	33.4	–	–	–
Sierra Leone	21.4	14.2	48.9	–	–	–
Guinea	11.4	6.4	35.0	–	–	–

Source: Rajanaidu et al. (2000).

The data in Table 7.3 reveal that the mean bunch and fruit weight recorded for the Cameroonian populations was higher than that observed in the case of the Nigerian population. The samples from Tanzania, which were obtained from a fringe population, showed bunch and fruit qualities comparable to samples obtained from Nigeria, Cameroon, DRC, and Angola. These traits tended to increase among Nigerian and Tanzanian populations. For *dura* palms collected from Angola, the percent mesocarp to fruit and bunch weight was higher compared to the populations from Nigeria, Cameroon, and DRC. The mean fruit weight was similar to those obtained from DRC. As for *teneras*, the percent mesocarp to fruit recorded for Angola was similar to the Nigerian and Tanzanian populations and higher than those obtained in the case of the populations from Cameroon and DRC. The *dura* bunch weight and mesocarp to fruit of the Gambian populations were the lowest compared to other materials. The values of these characters increased in the populations from Angola to Senegal, whereas the mean fruit weight increased in the populations from Senegal to Tanzania. In the case of Gambian populations, the mean fruit bunch weight and mesocarp to fruit weight ratio were similar to that observed as in the case of populations from Senegal (Rajanaidu et al., 2000).

Elaeis oleifera

Elaeis oleifera genetic materials were collected from six countries, namely, Colombia, Panama, Costa Rica, Honduras, Brazil, and Suriname in 1981–82 (Rajanaidu, 1986b). This American species is attractive to the oil palm breeders because it possesses a number of desirable traits such as slow increase in height, high iodine value, and resistance to diseases, such as *Fusarium* wilt.

Characterization of Germplasm

Morphological Variation

The collected germplasms were planted and maintained as open-pollinated families at Kluang MPOB Research Station, Johore, Malaysia. The palms were planted in several experimental designs, such as cubic lattice, randomized complete block design (RCBD), and completely randomized design (CRD), to study their phenotypic characters. Data on yield, oil and kernel content, fatty acid composition of oil, physiological parameters, and flower characters were recorded and analyzed. The palm oil was also screened for iodine value and carotene content. The performance of the Nigerian germplasms was examined and a total of five populations (12, 13, 16, and 19) were identified. These populations were from the east central part of Nigeria. The traits of interest found in these populations are as follows:

1. *High yield and dwarfness*: Some palms (*teneras*) gave high yield, more than 10 tons/ha/annum. In addition, they were short with annual height increment of 15–25 cm only compared to 45–75 cm of the commercial planting materials. Palms that possessed these traits were sampled from populations 12, 13, and 14 (Table 7.4).

Table 7.4 Performance of Dwarf Nigerian *Duras* and *Teneras*

Trial Number	Family	Palm Number	Oil/Palm/year (kg)	Oil/ha/year (tons)	Height Increment (cm/year)
Duras					
0.149	14.07	2703	38.4	5.6	21.0
0.149	14.07	2705	40.2	5.9	17.5
0.149	12.01	14483	43.9	6.5	19.1
0.149	12.01	14676	40.1	5.9	24.0
Teneras					
0.149	28.17	12724	83.8	12.1	23.1
0.149	19.11	12279	75.9	11.2	21.5
0.149	13.05	12094	76.2	11.2	24.0
0.150	16.21	4352	70.3	10.4	24.9
0.150	19.13	3759	71.5	10.5	22.5
0.151	14.03	128	59.0	8.7	25.7
0.149	12.01	2577	62.4	9.2	19.0
0.149	12.01	1704	62.7	9.2	17.5
0.149	12.01	11525	64.3	9.5	14.0
Current Planting Material				5.0	45–75

Source: Rajanaidu et al. (2000).

2. *High iodine value*: Positive correlation exists between iodine value (I.V.) and the level of unsaturation of palm oil. Palms having high iodine value will have highly unsaturated oil and subsequently produce more liquid oil. A number of Nigerian palms had iodine value more than 60, which is higher than that observed in the D × P commercial (I.V. values of 52–53). The prospects of marketing such palm oil in temperate countries are bright.
3. *High kernel content*: A few Nigerian families had mean kernel to bunch ratios of more than 15, which is higher than that of current breeding populations (4–8) (Rajanaidu and Jalani, 1994d). It was shown that by maximizing the level of kernel per bunch in oil palm, it is possible to get better economic returns.

The Cameroonian population showed a yield variation from 12 to 116kg/palm/annum. The bunch number varied from 2.5 to 32kg, and the average bunch weight ranged from 2.2 to 11.6kg. High-yielding palms were from populations 22 and 28 (Rajanaidu et al., 2000). The yield recorded for populations from DRC was slightly better than that of the Cameroonian populations, but lower than that of the Nigerian populations. Analysis of yield parameters showed that the populations had yielded between 11.5 and 183.4kg/palm/annum. The bunch number varied from 2.3 to 28.8, and the average bunch weight from 2.8 to 15.4kg. Populations 30 and 31 had higher yields and bunch numbers (Rajanaidu et al., 2000).

Isozyme Variation

Eighteen populations of *Elaeis guineensis* germplasm from nine countries in Africa (Angola, Cameroon, Guinea, Madagascar, Nigeria, Senegal, Sierra Leone, Tanzania, and DRC) and one *Deli dura* population were studied using seven enzyme isozyme systems. On average, 75 percent of the loci were polymorphic. The mean expected heterozygosity was 0.177. The genetic differentiation between populations was high (F ST = 0.384) indicating that only 62 percent of the isozyme variation as among progenies within populations. The dendrogram constructed with the data obtained showed that the *Deli dura* population was closely related to Sierra Leone populations (Hayati et al., 2000).

Molecular Variation

The extent of genetic variability between and within palm populations was estimated using restriction fragment length polymorphism (RFLP) technique. Oil palm germplasm materials collected from Nigeria, Cameroon, DRC, Tanzania, Madagascar, Angola, Senegal, Sierra Leone, Gambia, and Guinea were screened (Maizura, 1999). An advanced breeding population, *Deli dura*, was included as a reference. DNA extracted from each sample was digested with five restriction enzymes and hybridized with four oil palm cDNA probes. A total of 111 bands were observed and 58 loci were identified. Based on percentage polymorphism, a considerable difference was found between population 12 (high-yielding dwarfs) and population 39 (low-yielding talls) of the Nigerian origin. Correlation analysis showed that a high degree of interrelationship exists between level of polymorphism and mean yield ($r = 0.77$). This pointed to the possibility of using this parameter to select genetically divergent populations. Among the collections, the Nigerian populations exhibited the highest value for all genetic variability parameters with the most number of rare alleles, suggesting that Nigeria may likely be the center of diversity for oil palm. Compared to *Deli dura*, six oil palm germplasm collections exhibited higher level of polymorphism. These collections are from Nigeria, Cameroon, DRC, Tanzania, Angola, and Senegal. These collections should be given priority for characterization and utilization. Oil palm populations from Tanzania showed the closest genetic relationship in *Deli dura*, indicating that the breeding population could have originated from there.

In all, a total of 687 accessions belonging to 11 African countries were screened with eight primer combinations of amplified fragment length polymorphisms (AFLPs). Nigerian materials recorded the highest number of polymorphic bands. The estimated mean genetic diversity showed lower genetic variability in *Deli dura*; the population from Cameroon had the highest variability. Overall results proved that Nigeria, more specifically the east-central part, is the center of origin of oil palm. A considerable amount of mixing of natural population was noticed in the investigation. This was probably due to human influence (Kularatne et al., 2000).

Characterization of Elaeis oleifera Germplasm Collection

As in the case of *Elaeis guineensis* germplasms, the same parameters were used in the case of *Elaeis oleifera* germplasms as well. Significant differences existed among the germplasms obtained from Panama, Costa Rica, Colombia, Honduras, Brazil, and Suriname for yield ratio of oil to bunch, height, total economic product, and iodine value. In general, populations from Suriname showed the lowest value for all the yield attributes, except the ratio of oil to bunch (Mohd Din and Rajanaidu, 2000). Hardon (1969) pointed out that *Elaeis oleifera* has more unsaturated oil than *Elaeis guineensis*. Fatty acid composition of the two species has been investigated (Rajanaidu et al., 1984). It was observed that the oil characteristics of *Elaeis oleifera* were quite close to olive oil. The collections from Colombia, Panama, and Costa Rica had I.V. (iodine value) of more than 90. The C 18:1 level ranged from 52 to 66 percent and the C 18:2 level varied between 15 and 23 percent. The I.V. for the populations from Brazil ranged from 76 to 81, and the level of C 18:1 was lower than in accessions from Colombia, Panama, Costa Rica, and Honduras. The fatty acid composition of the Suriname genetic materials was rather unique. It has the highest level of C18:1 and the lowest of C 18:2 when compared to materials from other countries (Rajanaidu et al., 1994).

Utilization of Oil Palm Germplasm

Following are the ways in which oil palm germplasm is being currently utilized (Rajanaidu et al., 2000):

1. *Direct selection of individuals*: Individuals that possessed interesting yield attributes and acceptable yield were selected and introgressed into another advanced material. In the Nigerian population, about 3 percent of the *teneras* had oil yields comparable to that of the current planting materials. One third of these had annual increments significantly less than the commercials. These elite palms were cloned by tissue culture techniques.
2. *Progeny testing*: Some of the most promising Nigerian *teneras* were progeny tested with a range of *Deli duras* available in the Malaysian industry and MPOB. The progenies and parents were selfed and would be used for seed production following the reciprocal recurrent selection procedure (Jacquemard et al., 1981).
3. *Broadening the genetic base of* Deli duras *and* teneras: Crossing the *Deli duras* with Nigerian *duras* could broaden the overall genetic variability of current *Deli duras*. Such crosses would provide the basis for further selection and breeding. The selected Nigerian *teneras* could also be introgressed into the current *tenera* breeding populations.
4. *Foundation Breeding Program*: The outstanding *dura* and *pisifera* populations from the Nigerian collection are currently being used to initiate an entirely new breeding program aimed at producing superior alternatives to the current *Deli duras* and modern *pisiferas*.
5. *Development of elite planting materials*: Using the selected Nigerian palms, MPOB developed several commercial oil palm planting materials. These materials possessed high yield, dwarfness, high I.V., and high kernel content.

Genetic Improvement Through Breeding and Selection

The oil palm improvement project has the following objectives:

1. Enhance yield on an area basis
2. Enhance oil quality

3. Reduce annual height increment
4. Develop palms resistant to the ravages of pests and diseases.

Additionally, the oil palm improvement project also aims at the development of planting material with other attributes, such as enhanced early yield, high kernel yield, carotene, and vitamin E levels. The project also aims to explore the most profitable environment × genotype interaction to sustain the yield potential.

The various breeding programs can be classified into the following categories.

Dwarfing Program (PSI)

Primarily, the breeding program focused on yield enhancement. Between 1962 and 1988, the oil yield increased from an average of 5.0–9.6 tons/ha/annum, representing a jump of almost 100 percent (93.2 percent, Table 7.5), which works to an annual increment of 3.6 percent or 0.2 tons/ha annually (Lee and Toh, 1991). This increase was based on the *Deli dura*, which had its origin in the Bogor palms planted in 1848. They have a narrow genetic base and additive variation left on the *Deli dura* after generations of selection is low (Thomas et al., 1969). Crossing with other *dura*, such as those from the Nigerian population, would lead to enhanced genetic variability. Current oil palm materials grow at the rate of 40–75 cm/annum. After 20 years, the palm becomes very tall, resulting in very difficult harvest of the fruit bunches. Hence, there arose a need to dwarf the palms. Efforts were then initiated in this direction. Sources of dwarf genes are Dumpy E 206 palm (selected in 1920–21 at Elmina Estate, Kuala Selangor, Malaysia), Dumpy-AVROS *pisifera* (created by HRU and Golden Hope), *Elaeis oleifera*, and, recently, palms from the Nigerian collection. The availability of these sources led to the development of PSI (high-yield dwarf) planting material.

The dwarf mutant E206 selected by Jagoe (1952) had low yield compared to other materials, making it unattractive for exploitation. Planting materials produced from *Deli dura* × Dumpy-AVROS are 20 percent shorter with yields comparable to the current planting materials (Kushairi et al., 1999).

Table 7.5 The Yield Performance of Oil Palm Planting Materials

Planting Materials	Year Planted	Number of Progenies	Fresh Fruit Bunch (tons/ha/year)	Oil to Bunch (%)	Projected Oil Yield (tons/ha/year)
DD × CI	1962	32	22.0	22.2	4.9
DD × UAC	1962	15	24.6	20.6	5.1
DD × AVROS	1964	22	31.0	23.5	7.3
DD × AVROS	1968	16	31.1	22.1	6.9
DD × AVROS	1970	29	31.6	24.2	7.6
DD × AVROS	1979	5	34.5	25.8	8.9
DD × Yagambi	1988	66	34.9	25.9	9.6

Source: Lee and Toh (1991); Lee (1996) Mukesh and Tan (1996).

Elaeis oleifera was crossed to *Elaeis guineensis* to produce short-stem hybrids. However, due to poor fruit set (thick shell and thin mesocarp), the oil yield of these hybrids was disappointing. Selected *teneras* from the Nigerian population were progeny tested with the industry *Deli duras*. Simultaneously, these parents were selfed, and based on the progeny testing results, the respective selfs could be used for commercial seed production. Progeny test results of these crosses showed an increase of oil-to-bunch yield and mean yield from 25 to 32.9 percent and 180 to 245 kg, respectively. In general, the progenies are 15–20 percent shorter than the current planting materials (Rajanaidu et al., 2000). The Nigerian *duras* were introgressed into *Deli dura* populations to broaden the genetic base of the latter. These intercrosses are shorter and have oil yields comparable to or higher than the commercial *teneras*. Further exploitation of the Nigerian population involved population 12. This population is unique—high yielding, short, and compact. Various families of population 12 were progeny tested as D × T to create a new foundation stock. Some of the crosses are high yielding and shorter compared to the control (commercial D × P). T × T crosses within population 12 were also made to develop dwarf *pisiferas* (Rajanaidu et al., 2000).

High Unsaturation Oil (PS2)

An increase in unsaturation oil is desirable due to consumer demands for monounsaturated and polyunsaturated dietary oils and fats. A higher unsaturation level in palm oil enables penetration into the liquid oil market. Currently, palm oil is semisolid at room temperature (28°C) and can be fractionated into 70 percent liquid olein and 30 percent solid stearin. The palm oil quality can be improved by reducing the stearin and increasing the olein levels. Three families of the Nigerian *duras* were chosen for high yield and high I.V. (unsaturation). The palms were selfed, and early observations showed that I.V. is transmitted from parents to the progenies. These palms were progeny tested with industry *pisiferas*. The progenies had low yield and oil-to-bunch yield but high I.V. (57–59) compared to the control (53). These results were based on crosses in which one parent had a higher I.V. (more than 60), which can be expected if both parents were selected for the trait. Along this line, the high I.V. *duras* are being progeny tested with the high I.V. Nigerian *tenera*. Selected palm of the same collection was selfed to generate *pisifera* for the next generation of high I.V. planting materials known as PS 2 (Kaushari et al., 1999).

The iodine value for the *Ealeis oleifera* germplasm collections is about 80, much higher than that of the African palm. The interspecific hybrids revealed I.V. values of 65. However, the yield of the hybrids was too low for commercial exploitation (Rajanaidu et al., 2000).

High Lauric Acid (PS3)

The principal sources of lauric acid are coconut and oil palm. An economic analysis showed that breeding for higher levels of kernel content realizes the highest return from oil palm. An increase of 5–10 percent of kernel to bunch would bring about a net profit of US$300 from every 100 tons of fresh fruit bunch (Rajanaidu et al.,

1996). Cultivation of high kernel planting materials (PS3) could be a gainful venture. The current planting materials have 5–10 percent kernel to bunch. Populations taken from the northern part of Nigeria have higher levels of kernel to bunch (more than 10 percent). Selected Nigerian *duras* are being progeny tested with AVROS *pisifera*. Nigerian *teneras* and *pisiferas* and Serdang *pisiferas* are other sources of high kernel genes that could be exploited for production of PS3.

High Carotene, Vitamin E, and Sterol

The carotene content for the Nigerian palms ranged from 273 to 3512 ppm, whereas the current Deli *duras* had 500 ppm. A special breeding program was developed to exploit the Nigerian population with high carotene content. *Elaeis oleifera* has low oil yield (less than 0.5 tons/ha/annum) but has much higher carotene (4300–4600 ppm), vitamin E (700–1500 ppm), and sterol (3500–4000 ppm) contents compared to *Elaeis guineensis*, which has these yield attributes in the range of 500–700 ppm, 600–1600 ppm, and 1100–1250 ppm, respectively (Jalani et al., 1997).

Disease Resistance

Breeding for tolerance to vascular wilt caused by *Fusarium oxysporum* is an important program in West Africa. This disease does not occur in the Far East. It causes the death of young and adult palms. Tolerant palms and screening techniques for identifying tolerance in a population are available (Meunier et al., 1979). Basal stem rot caused by *Ganoderma boninense* is a serious problem for older palms planted in the Far East. A screening technique is being developed (Ariffin et al., 1995) to breed for disease resistance. The American oil palm *Elaeis oleifera* was utilized to develop palms tolerant to "sudden wilt" and "little bud rot," which adversely affected oil palm plantation in Central and South America. A few hybrids exhibited tolerance to this disease (Meunier, 1989).

Prospects for Biotechnological Interventions

The physical and chemical properties and their applications of vegetable oils are determined by their fatty acid composition. The wide range of applications of vegetable oils, 90 percent edible and 10 percent nonedible, reflect their diversity in fatty acid composition (Stobart et al., 1993). Vegetable oils are renewable resources that can serve as feedstock to produce environment-friendly industrial products, such as lubricants, paints, detergents, body care products, but, most important, the "green fuel," as is being demonstrated by recent developments. Palm oil has become the oil much sought after by the automobile industry as a substitute for fossil fuels, and that is precisely the reason for its skyrocketing price. Demand for these oils is likely to escalate in the nonfood sector to augment depleting mineral oils (Murphy, 1994). Soybean (*Glycine max*), oil palm (*Elaeis guineensis*), rapeseed (*Brassica* spp.), and groundnut (*Arachis hypogaea*) are the major crops which provide limited number of fatty acids. These include unsaturated oleic (18.1 percent) and linoleic (18.2 percent) acids and generally lower levels of alfa-linolenic acid (18.3 percent), saturated stearic acid (18 percent), and palmitic acid (16 percent) (Stumpf, 1987). Hundreds of different fatty acids have been identified and characterized in the plant kingdom (Hilditch and Williams, 1964), but most are of no economic value. Genetic engineering provides the means to tap these vast resources by producing fatty acids of economic importance in storage lipids of oil crops (Murphy, 1994). Achievements through breeding alone or in combination with mutagenesis, such as in the development of rapeseed oil with low content of erucic acid (22.1 percent) (Stefansson, 1983) and sunflower oil with high levels of oleic acid (Garces and Mancha, 1990), indicate that plants can tolerate wide variations in fatty acid composition of storage lipids.

Genetic engineering in the past decade of the 20th century has led to breakthrough results in altering fatty acid composition of storage oils in oil seed crops. These significant results include modifications involving chain length (introduction of medium fatty acids (Voelker et al., 1992), level of unsaturation, increase in saturated fatty acids (Knutzon et al., 1992), and positioning on the glycerol backbone of fatty acids (increase in erucic acid at the sn-2 position, Brough et al., 1996). In several cases, the analysis of fatty acids and other relevant investigations carried out on the transgenic plants have provided new insights into the regulation of fatty acids in plants.

The Production of Oil in Oil Palm Fruits

The storage oils obtained from oil palm fruits are of two kinds—palm oil from the mesocarp and kernel oil from the kernel. They are different in composition, physical properties, and usability (Hartley, 1988). The kernel oil is rich in medium-chain saturated fatty acid: 51 percent lauric acid (12.0) and 18 percent myristic acid (14.0), which serve as important feedstock for the oleochemical industry (Pantzaris and Yusof, 1990). Palm oil, which contains about 50 percent saturated fatty acids, 40 percent monosaturated fatty acids, and 10 percent polyunsaturated acids, is a semisolid fat at room temperature. Its fatty acid composition consists of 44 percent palmitic acid (16:0), 5 percent stearic acid (18:0), 39 percent oleic acid (18:1), and 10 percent linoleic acid (18:2). The main applications of palm oil are in the edible oil food industry field, mainly as solid fat for margarine, shortening, and cooking oil. Nonedible or technical applications, however, are substantial and enlarging day after day. These applications include soaps manufacture, oleochemical production, and automobile energy sources, the last use having been moved up in priority because of the depleting fossil fuel reserves.

During early (about 10 weeks old) development, young fruits contain very little lipids (about 5–8 percent) per fresh weight. Storage oil synthesis in the oil palm mesocarp can be detected as early as 12 weeks after anthesis. High rate of oil accumulation begins at 16 weeks of growth and stops when the fruits are about 20 weeks old after anthesis (Oo et al., 1986). Oil is stored in oil bodies found in the cytoplasm of mesocarp cells of ripe fruits. Studies on mesocarp cell morphology by Aziz et al. (1990) showed that small-size oil bodies can already be observed at around 13 weeks after anthesis when oil synthesis begins. Oil accumulation in the kernel starts at around 12 weeks after anthesis and stops at 14 weeks of growth. During this period, the kernel gradually solidifies (Hartley, 1988). The fatty acid in the mesocarp of young fruits consists mainly of polyunsaturated linolenic acid (18:3) and linoleic acid (18:2), which are components of membrane lipids. When rapid oil accumulation begins, the level of linolenic acid drops to insignificant value. The level of linoleic acid also drops but is maintained with stability at the level of 10 percent in mature fruits. The highest increase observed is in the level of oleic acid from 22 percent at 13 weeks to 39 percent in ripe fruits (Aziz et al., 1986).

A Blue Print to Modify Palm Oil Composition

During the past years, various strategies to modify the oil composition of palm oil have been employed with the objective to produce more liquid oil. This is considered of immense importance, because it provides the means to diversify the use of palm oil. The objective is achievable by increasing the level of unsaturation or iodine value, thus changing the physical properties of palm oil (Rajanaidu et al., 1993). Physical and chemical processes involving two levels of fractionation and transesterification showed promising results in changing oil composition, but were not viable for commercialization. In addition, such approaches could produce triglycerides different from the ones obtained from natural processes (Wong et al., 1991).

By selecting yield attributes, classical breeding programs have led to remarkable success (Rajanaidu, 1987). Even though much emphasis is placed on these lines, research attention is also being directed to study sustainability factors, arresting tall growth, adaptability to varying soil and climatic conditions, resistance to pests and diseases, slow vertical growth to assist harvesting, and attaining high levels of fatty acid unsaturation (Yong and Chan, 1996).

The South American species *Elaeis oleifera* produces mesocarp oil with I.V. between 77 and 88, which is much higher than that obtained in *Elaeis guineensis*, which has an I.V. of about 53 (Sambanthamurthi et al., 1996). The interspecific hybridization of *Elaeis oleifera* and *Elaeis guineensis* was carried out by various groups to introgress the high oil unsaturation and slow vertical growth of *Elaeis oleifera* into the high-yielding commercially grown *Elaeis guineensis*. The FI hybrids are backcrossed to the *Elaeis guineensis* parent to improve yield and vegetative traits. The genomic *in situ* hybridization (GISH) technique, which can differentiate the genomes of the two species, was developed for the oil palm (Madon et al., 1999). This molecular technique is a valuable aid to plant breeders to determine the genomic composition of progenies so that those having a high proportion of the *Elaeis guineensis* parents can be selected. Another breeding strategy employed was to introgress the high I.V. characteristics of *Elaeis guineensis* genetic material from Nigeria into commercial breeding material. Restriction fragment length polymorphism analysis detected high levels of genetic variability in the normal population within *Elaeis guineensis* species, which can be further exploited to improve existing planting material (Maizura et al., 1996). However, for oil palm, which has a long generation time, it is believed that the achievement of goals through these breeding programs will take a long time to accomplish and require a lot of space and manpower (Rajanaidu et al., 1993).

Production of novel high-value products by genetic engineering provides the opportunity to diversify use and increase the economic value of palm oil. Production of speciality oils for industrial applications would be a very attractive proposition because oil palm is the most productive oil crop. Achievements in manipulating fatty acid biosynthetic pathways using recombinant DNA technology leading to the production of transgenic oil crops, especially rapeseed with modified oil compositions, have made this technology very attractive for oil palm. Reported success in raising the level of lauric acid (Voelker et al., 1992) and stearic acid (Knutzon et al., 1992) in rapeseed oil proved that both fatty acid chain length and the level of fatty acid unsaturation can be modified. Field investigations indicated that characteristics of the transgenic plants were similar to normal plants, and the oil production was not affected by substantial modification in fatty acid composition (Kridl et al., 1993). The main advantage for a dicotyledon such as rapeseed is that it has a simple and efficient *Agrobacterium*-mediated transformation method. The successful application of the microprojectile bombardment method in transforming monocotyledons such as rice (Christou et al., 1986) and wheat (Vasil et al., 1992) provides hope for applications of this technique in the case of oil palm. There are various projects under the genetic engineering program of the oil palm crop. It has been envisaged that the production of oil containing high amounts of monounsaturated oleic acid will have industrial significance in producing chemical derivatives that can serve as alternatives to fossil fuels. The strategy is to manipulate the fatty acid biosynthesis pathway in the mesocarp to reduce production of palmitate and channel toward increasing oleate (Cheah et al., 1995). The pathway for palmitic and oleic acid synthesis in oil palm is similar to other plants (Oo, 1988). An understanding of the biochemical pathway is critical in order to choose the most suitable gene(s) that should be manipulated in order to modify the pathway to produce the desired oil composition. In plants, *de novo* fatty acid synthesis occurs in the chloroplasts and the nongreen plastids in nonphotosynthetic tissues.

The first committed step is catalyzed by acetyl CoA carboxylase, which produces malonyl CoA from acetyl CoA and bicarbonate. A group of dissociable enzymes that make up the fatty acid synthase (FAS) complex catalyze the sequential addition of 2-carbon units derived from malonyl CoA to a growing acyl chain esterified to acyl carrier protein (ACP). The primary products of this type II synthase are 16:0-ACP and 18:0-ACP. Elongation of palmitoyl-ACP (16:0) to stearoyl-ACP (18:0) is catalyzed by beta-ketoacyl-ACP synthase II (KAS II), one of the condensing enzymes from the FAS complex. Stearoyl-ACP desaturase inserts the first double bond into stearoyl-ACP to produce oleoyl-ACP. The acyl chains are then released into the cytoplasm by soluble acyl-ACP thioesterases. With the exception of plastidial desaturation, further elongation and desaturation of fatty acids occur in the cytoplasm.

That the palmitic and oleic acids in the oil palm mesocarp are controlled by the enzymes acyl-ACP thioesterase and KAS II is proved by biochemical investigations. Thioesterase activity, both in the crude extract (Sambanthamurthi and Oo, 1991) and in a distinct fraction after purification by anion exchange chromatography, showed a strong preference for palmitoyl-ACP as a substrate (Abrizah, 1995). The studies showed that the activity of type synthase II increased with ripening, and it correlates positively with the level of unsaturation in the crude oil extract

(Cheah et al., 1995). It was observed that the oil palm has an active oleoyl CoA desaturase that readily converts oleate (18:1) to linoleate (18:2). It may be necessary to downregulate the expression of the gene encoding this enzyme for increasing oleic acid content (Sambanthamurthi et al., 1999). About seventy-fold purification of glycerol 3-phosphate acyl-transferase was achieved from oil palm tissue cultures and mesocarp acetone powder (Arif and Harwood, 2000). This was an arduous task since glycerol 3-phosphate acyltransferase is a membrane-bound enzyme. Biochemical studies of acyl transferases are important because they determine the species of fatty acid incorporated into the glycerol backbone during triacylglycerol assembly and hence the fatty acid composition of vegetable oils. Full-length or nearly full-length cDNA clones for genes encoding key enzymes in oil synthesis including stearoyl-ACP desaturases (Siti Nor Akmar et al., 1999), palmitoyl-ACP thioesterase (Abrizah et al., 1999), and acyl-carrier proteins (Rasid et al., 1999) have been isolated and characterized. The cDNAs encode precursor proteins containing N-terminal leader peptide for targeting the plastid, which is the site of *de novo* fatty acid synthesis. Down regulation either by co-suppression or antisensing of oil palm palmitoyl-ACP thioesterase and stearoyl-ACP desaturase are strategies that will be taken for producing high oleic and stearic oils, respectively. The identity and specificity of the encoded product of the isolated thioesterase gene is being confirmed biochemically using enzymes obtained by over-expressing the genes in bacterial systems. Consistent with the observation in other plant species, more than one gene coding for acyl carrier protein was found in oil palm. Two differently regulated stearoyl-ACP desaturase genes were identified. One is constitutively expressed, suggesting a housekeeping role in membrane lipid biosynthesis. Expression of the second is induced in mesocarp and kernel tissues in phase with oil synthesis indicating its direct involvement in oil synthesis. Polyclonal antibodies were raised against the oil palm stearoyl-ACP desaturase and used to study post-transcriptional regulation using Western blot analysis. Together, these provide useful background information for mesocarp oil modification by genetic engineering.

The use of microprojectile bombardment technique as a major breakthrough in oil palm transformation has been reported. The production of transgenic oil palms containing herbicide-resistant genes was reported by Parveez (1998). Regulatory sequences controlling tissue and temporal-specific gene expression are essential for the genetic engineering effort. Comparison of protein synthesized *in vitro* from mRNA of mesocarp at different ages of fruit detected differential gene expression during development of this tissue (Siti Nor Akmar et al., 1994). Differential screening and subtractive hybridization techniques have been employed for the isolation of cDNA clones specifically expressed in the mesocarp during the period of oil synthesis (Siti Nor Akmar, 1999). These clones will be used for obtaining the desired promoters to be used in modifying mesocarp oil composition by genetic engineering (Siti Nor Akmar et al., 1995).

The Place of Tissue Culturing in Oil Palm

It was in mid-1970s that results of successful investigations in tissue culturing oil palm were reported (Jones, 1974). Currently, about 20 oil palm laboratories around the world use the technique to produce seedlings, with capacities ranging from 10,000 to 200,000 plantlets per annum. Tissue culturing is superior to conventional breeding, because the technique allows rapid multiplication of plantlets with uniformity having desired characteristics. It provides opportunities to improve planting materials using existing individuals that have all or most of the desired qualities, such as good oil yield. It also opens new avenues for oil palm biotechnological research, because tissue culture is the means for regeneration of tissues transformed with genes of specific traits of interest.

There are several reasons for carrying out extensive investigations in tissue culture in oil palm. For instance, it is used as a means to produce good *tenera* palms for commercial planting. High demand exists for high-quality planting material not only in Malaysia, but also in other countries. Based on current demand for oil palm seeds, it is estimated that a ready market exists for more than 100 million tissue cultured plantlets annually (Zamzuri et al., 1999). Next, tissue culture is made to multiply good parents (both *dura* and *pisifera*) for seed production. Furthermore, tissue culture is also used as a means to expedite the exploitation of genetic potential of progenies from interspecific *Elaeis oleifera* × *Elaeis guineensis* crosses. Interspecific hybridization with South American species of *Elaeis oleifera* was carried out by various breeding groups to introgress the high oil unsaturation and slow vertical growth characters into high-yielding, commercially grown *Elaeis guineensis*. The F 1 hybrid was shown to have an intermediate I.V., but was not viable commercially due to its low yield and undesirable excessive vegetative vigor. Even after using various strategies of backcrossing with *Ealeis guineensis*, progenies with an improved I.V. suitable for commercialization have not been obtained (Yong and Cahn, 1996). Finally, in Costa Rica, for instance, tissue culture is also used to salvage diseased palms (Guzman, 2000).

Tissue Culture in Oil Palm

The ortets that are selected from the field should be superior in quality. Tissue culture laboratories are linked to effective oil palm breeding and improvement programs that can supply the large reservoir of required explants. Ideally, an ortet should possess several desirable heritable traits, such as high oil yield, low height increment, and high quality. The selected ortets are supported by at least 4 years of field data showing good performance on yield and vegetative and physiological traits, such as bunch index and oil characteristics (Rohani et al., 2000). Oil yield is determined by the oil extraction rate (OER) or oil-to-bunch ratio (O:B) and weight of fresh fruit bunch (FFB). Since O:B ratio has been demonstrated to be highly heritable and transmitted from ortets to ramets, it is given emphasis in ortet selection. Leaves, inflorescences, and roots have been used as explants for oil palm tissue culture. Young leaf spears are preferred in most laboratories. Leaf explants can be surface sterilized easily and give higher clonability rates (Rajanaidu et al., 1997). The process of oil palm tissue culture can be divided into different stages. Callus is initiated from the explant, followed by embryogenesis, shoot and root regeneration, hardening of ramets for nursery, and field evaluation. The regeneration process through oil palm tissue culture takes 2–4 years, depending on the genotype. Growth conditions for the different stages are typically at 28°C, plus or minus 2°, with equal light and dark photoperiods except during callogenesis, where the culture is maintained in complete darkness. The explants are placed in media containing either 2, 4–8 (2,4-dichlorophenoxy acetic acid) or NAA (naphthalene acetic acid) for 12–20 weeks for callus formation. The calli are maintained in media containing lower concentrations of 2,4-D or NAA for up to 12 months for multiplication and embryogenesis (Rohani et al., 2000). The rates of callusing and embryogenesis in oil palm have been demonstrated to be genotype dependent (Ginting and Fatmawati, 1995). For example, tissues of variety *La Me* origin produced as high as 60 percent callus, whereas tissues from *Yagambi* produced only 5–20 percent callus. The average rate of embryogenesis is about 36 percent for leaf-derived calli (Rajanaidu et al., 1997; Wooi, 1995). A few oil palm clones produce embryoids after 1 month, whereas others may take as much as 2 years (Ginting and Fatmawati, 1995). It was shown that the propensity of callusing and the rate of embryogenesis are not influenced by ortet age (Wooi, 1995). Small clumps of polyembryoids are transferred onto a basal nutrient medium and kept for at least 3 months for induction of the shoot. Shoots obtained are separated from the polyembryoids and placed onto solid shoot development media containing low concentrations of NAA in culture tubes (it takes two–three shoots) or in flasks, which take 15 shoots. When the shoots are 5 cm in height, they are transferred into liquid root initiation media. Shoot development and rooting that involve many culture transfers are labor intensive. Based on the experience of *in vitro* culture in ornamentals and banana, Zamzuri (1998) introduced the double-layer rooting technique in oil palm. In this technique, the solid shoot development medium is overlayed with liquid root initiation media. This has an effect of improving worker efficiency by eighteen-fold and reduces the cost of rooting by about 94 percent. The plantlets are transplanted into small polybags containing 1:1 ratio of soil and sand and kept for 3–4 months under shade with relative humidity of more than 70 percent for acclimatization before transferring to field nursery. In the nursery, the plantlets are treated like seed-derived plant.

Clonal Abnormality

Fruit and floral abnormalities in clonal palms were reported in the late 1980s (Hartley, 1988). The abnormality referred to as mantling affects flower development involving feminization of the male parts of the flowers; the vegetative parts appear normal. Considerable variation exists in the severity of abnormality and such variation may influence oil yield because of the effect on fruit set. It was suggested that the abnormalities may be due to physiological factors, whereby an increase in the concentration of hormones in the palms could cause biochemical disorders leading to an expression of abnormality (Obasola et al., 1978). Studies in other plants showed correlation of increased 2,4-D levels with increased ploidy levels (Sogeke, 1998) and increased DNA methylation, hence affecting genome expression (Rohani et al., 2000). Sogeke (1998) used NAA, which is a milder auxin, in oil palm tissue culture media instead of 2,4-D. This technique, which was able to reduce callusing time and hasten production of embryoids, successfully produced normal plants in the field.

Field Performance

There is an improvement of 20–30 percent oil yield in clonal material over seedling planting (Soh, 1986). The difficulty in selection for resistance to *Ganoderma*, *Fusarium*, and *Blastobacter* lies in achieving true-to-type reproduction of plants, selected as ortets, especially with the incidence of mantled abnormality. The percentage of abnormality in the field has generally been reduced to less than 5 percent and maintained at this tolerable level (Maheran et al., 1995; Khaw and Ng, 1997). The same authors suggested the need to ensure prudent selection of good ramets at both the *in vitro* and in the nursery stages. Their observations indicated that clones appearing as normal in both stages gave low levels of abnormality in the field (about 2.2 percent). Economic analyses reported by Maheran and Chan (1993) indicated that the higher productivity of clonal materials will not only compensate the high initial investment, but will also give higher returns than for conventional planting materials. It was estimated that the initial investment will be covered in the sixth year, after which returns from the clonal materials will be much higher than normal D × P planting materials. Zamzuri et al. (1999) investigated the commercial feasibility of producing clonal palm planting material using four different production systems working under single or double shifts to produce 90,000–700,000 plantlets for more than 30 years. For all the models, it was concluded that the venture would be attractive if the plantlets are sold at RM 20 per piece. Since an oil palm tree with productive life of more than 20 years that will produce about 3 tons of FFB in total and was estimated to be worth RM 1200, it would be worthwhile to put this initial investment on oil palm cloning.

Pests and Diseases

All of the insect pests of oil palm in Malaysia are of local origin; they have adapted to the crop ever since its introduction close to a century ago. In general, oil palm pests can be classified into insects and vertebrates. The insect pests consist of leaf defoliators, bag worms, nettle caterpillars, crown attacker, rhinoceros beetle, and the bunch moth. Other insect pests that attack the nursery stages are cockchafers and grasshoppers. Vertebrate pests include rodents, wild boar, porcupines, and elephants (Chung, 2000).

Insect pests are generally kept in balance by biotic and abiotic factors. The former include natural enemies (parasitoids, predators, and diseases caused by bacteria, fungi, and viruses). The latter are mainly environmental factors, such as rainfall, temperature, and humidity. Nevertheless, the balance may sometimes be disrupted by injudicious use of insecticides, which often trigger pest outbreak. It is now imperative that chemical use is minimized to avoid the problems of insect resistance, persistence in the environment, resurgence of secondary pests, and detrimental effects on natural enemies, beneficial organisms, and nontargets. However, when the need arises, especially in controlling outbreaks, it is important to ensure that pest numbers are kept below a certain economic threshold but not to eliminate the pest completely. This is in line with the concept of integrated pest management (IPM) where natural control is integrated with chemicals (Wood, 1976). However, despite this approach, outbreaks of some pests, such as bagworms and nettle caterpillars, are quite common, causing yield loss of up to 40 percent (Basri, 1993). It is important to find factors governing outbreaks because, with this information, a pest can be more effectively and efficiently managed in an environmentally friendly manner.

Fundamental research in biological control approaches, that is, parasitoids, predators, *Bacillus thuringiensis*, and fungal and viral pathogens, has intensified because of their role in substantially reducing the use of chemical insecticides (Basri et al., 1995). In controlling leaf-eating pests, the propagation of natural enemies by planting beneficial plants is widely recommended. *Cassia cobanensis* was found to be the most suitable in sustaining the longevity of parasitoids in the laboratory (Basri et al., 1999).

Attacks by rhinoceros beetle are now common because of the extensive replanting programs being undertaken in Malaysia. Other agronomic practices such as application of empty fruit bunches (EFB) as mulch, underplanting of oil palm, and "zero burning" have also aggravated the problem (Basri and Norman, 2000). Earlier research efforts were directed at the use of chemical insecticides (Norman et al., 1999; Toh and Brown, 1978) for integrated control of this pest. Information on the ecology of the rhinoceros beetle, especially in the zero-burning environment (Samsudin et al., 1993), will further enhance IPM approaches for *Oryctes rhinoceros*. Greater emphasis is placed on the development and use of microbial pathogens, such as *Metarhizium anisopliae*, *Bacillus thuringiensis*, and so on, for the control of *Oryctes rhinoceros* (Ho, 1996). For general review on insect pests of oil palm, Wood (1968, 1976) is a good reference.

In addition to attack by insect pests, the oil palm plant is also susceptible to the onset of diseases from the time the seed germinates until planting in the field. Fungi cause the most damage to oil palm. In general, there are five major diseases: vascular wilt, basal stem rot (BSR), bud and spear rot, red ring disease, and quick wilt. In Malaysia, the most significant disease is BSR caused by *Ganoderma* sp. The other diseases mentioned are not present in Malaysia, but assume serious proportion in other countries on the African and Latin American continents.

BSR symptoms are retarded growth, one-sided yellowing of the lower fronds, and pale green foliage (Ariffin, 2000). In older palms, the canopies of infected palms are pale green with multiple unopened spears. This is typical resulting from decreased water intake due to the rotting stem. With the progression of the disease, fruiting bodies will start to appear, initially as small white buttons, subsequently developing into bracket-shaped sporophores (Ariffin, 2000). When the fruiting bodies appear, the palm is already in an advanced stage of decay and near to death.

The symptoms of BSR disease usually manifest after the palms are 10–12 years of age (Gurmit, 1991; Benjamin and Chee, 1995; Khairuddin, 1990). In areas where oil palm is planted after coconut, the disease symptoms occur much earlier, within 1–2 years after planting. Serious incidence of BSR occurs in coastal clay soils compared to inland soils (Gurmit, 1991). Control measures include injecting systemic fungicides into diseased palms (Chung, 1990), though chemical control has not proved economically viable (Ariffin, 2000). The best method to control the disease is by clean clearing, completely removing the infected palms, and destroying the bole and root masses that harbor the pathogen. This is particularly important, as the detached infected root segments can still transmit the disease and cause outbreak (Gurmit, 1991).

The latest breakthrough on the BSR is the identification of four species of *Ganoderma: Ganoderma boninense, Ganoderma zonatum, Ganoderma miniatocinctum*, and *Ganoderma tornatum*. Although the first three species are pathogenic, *Ganoderma boninense* is the most aggressive. *Ganoderma tornatum* is nonpathogenic (Idris et al., 2000).

Oil Palm Plantations and the Environment

To ensure sustainability of the environment, one must conserve it properly. Among the countries that grow oil palm, Malaysia is at the forefront (51.9 percent). The country has undertaken intensive research and development activities since 1975, and the findings have been implemented. Under the Environment and Quality Act of 1974 and the Environment Quality Act of 1978 (Clean Act Regulations), open burning of felled palms during replanting is prohibited in Malaysia, and zero burning has been in operation virtually by all plantations in the country. Zero burning involves shredding of oil palm trunks with an excavator so they are 5–10 cm thick and stacking the shredded trunks in the interrows (Mohd Hashim et al., 1993). Decomposition of the shredded biomass takes place within 2 years (Norman and Basri, 1997), after which it does not remain a suitable breeding substrate for the rhinoceros beetles. In addition, this technique has merit in terms of recycling large quantities of plant nutrients through decomposition and improving soil physical properties. The quantities of available nutrients from oil palm trunks were 219.6 kg/ha N, 21.2 kg/ha P, 314 kg/ha K, and 52.6 kg/ha Mg (Mohd Hashim et al., 1993). Chan et al. (1981) reported the presence of 146 kg/ha of Ca in the felled palms. Theoretically, these reserves could provide to the palms N, K, and Mg for 6–7 years and P for about 2 years. Thus, by employing zero-burning practices, the requirement for inorganic fertilizers can be reduced by 20–30 percent during the first 4 years. However, considerable research is needed to optimize fertilizer inputs during replanting with zero burning (Zin Zawawi, 2000).

Traditionally, empty fruit bunches (EFB) were incinerated to produce bunch ash, which is a good source of K fertilizer. Under the two acts mentioned earlier, incineration is prohibited because it causes air pollution. To circumvent the problem, EFB are applied in the plantations as mulch within palm circles and interrows as partial sources of nutrients. Chan et al. (1981) estimated that the quantities of available nutrients from EFB were 5.4 kg/ha of N, 0.4 kg/ha P, 35.2 kg/ha K, 2.7 kg/ha Mg, and 2.3 kg/ha Ca. This has become a standard agronomic practice within the plantations.

The total economic life span of an oil palm plant is about 25 years. During replanting, the plant produces considerable biomass. In Malaysia, which is the lead country in oil palm production on account of massive replanting programs (yielding 4 million tons of trunks and 2.9 million tons of EFB, annually, during 2001–03), opportunities other than agronomic applications have been considered. Intensive research has been undertaken since 1980, and technologies are currently available in converting oil palm biomass into value-added products, such as pulp, paper, particle board, medium-density fiber board, and thermoplastic (Mohamad, 2000). However, further investigations are required to determine the balance between the quantum of biomass removed to manufacture value-added products and the quantum that needs to remain in the plantation to maintain organic matter content and a good soil fertility base. Although the usage of pesticides in oil palm plantations has been minimal, research has been intensified in the use of microbial pathogens and naturally occurring predators to control major insect pests, such as rhinoceros beetle and bag worm. Basri and Norman (2000) reported that of the three strains of *Beauveria bassiana*, isolated from the bag worm *Metisa plana*, one strain was highly effective against the bag worm, yet safe against the pollinator, *Elaeidobius kamerunicus* (Ramle et al., 1995). Nuclear polyhedrosis virus (NPV) was also isolated from the bag worm. Since the infection was only tertiary, there is but little scope of using the virus to control this pest biologically (Siti Ramlah et al., 1996). An effective strain *Metarhizium anisopliae* to control the rhinoceros beetle has been field evaluated and current efforts are focused on its mass production for application in oil palm plantations. Based on the investigations of Basri et al. (1999), a number of major plantation companies in Malaysia are planting beneficial plants (in particular, *Cassia cobanensis*) within their plantations. These plants provide a good source of nectar to the parasitoids of the bag worm and other pests, thereby extending the life span of the natural enemies in the oil palm ecosystem. It is anticipated that with the use of microbial pathogens and parasitoids, the use of chemicals will be reduced in terms of insect pest control.

There is also a growing awareness among planters that palms which meet the minimum environmental standards alone be grown. A number of plantations have been accorded the ISO 14001 certification. This ensures production in a clean environment. This has important implications on the trade of palm oil in the future in the sense that developed countries would prefer importing palm oil from companies that address environmental concerns.

It is imperative to enhance the value of palm by many folds. Genetic engineering is one avenue to achieve this, which will hopefully ensure high-value products, such as oleate oil, speciality oil, bio plastics, pharmaceuticals, and nutraceuticals. The derivatives of oleate oil can be used as industrial feedstock for the oleochemical industry. As mentioned earlier, major breakthroughs have been made in producing oleate oil, and research in this area is currently being prioritized. Intensive collaborative research is also being undertaken between the Malaysian Palm Oil Board (MPOB) and the Massachusetts Institute of Technology (MIT) in the production of bio plastics. The ultimate aim would be in the production of palms that will produce different premier products to meet specific market needs. Instances can arise whereby the price of palm oil can decline below the cost of production. These instances involved low demand by importing countries and oversupply of the world's oil and fats brought about by good growing conditions in other oil-producing countries. Under these situations, it would be profitable to convert part of the stock to produce palm diesel fuel for automobiles. The technology is available and it has been proven on the road by automobiles made to run on palm diesel for a few hundred thousand kilometers European Auto Industry (2009). Even if the price is not favorable in relation to the cost of production, the project is still economically viable because it also produces by products of high value, in particular, carotene and glycerol. The oil palm needs to be adapted to the presence of other crops during its growth, either during its period of immaturity or throughout its life span, depending on the types of crops chosen. This essentially involves crop and animal integration. In essence, the project is aimed at increasing total income to the grower without affecting the overall production of palm oil. The types of economical plants being investigated by MPOB include yam, timber trees, banana, pineapple, and other fruit trees. The animals under investigation include cattle, sheep, poultry, sheep, and so on. These investigations are still in the preliminary stage.

A Look into the Future of Oil Palm

The steadily increasing world population will make a huge demand on oils and fats. According to the Malaysian Palm Oil Board (MPOB), the projected growth must be around 24.6 percent by 2020. Palm oil is expected to contribute 4.4 percent annually to account for this increase. As such, palm oil has a very important role to play in the world's oils and fats sector.

There are other factors as well that contribute to this escalating demand. First, changes in planting policies of other types of oil crops, such as a reduction in the total area planted, brought about by low prices, would enhance the demand for palm oil. Second, depleting fossil world stock of fuels will also be in favor of palm oil derivatives, which can be substituted as industrial feed stocks. The recent emergence of palm oil as a "green fuel," to replace fossil fuel, is an important indicator in which direction the future development will taker place. In 2008, palm oil outpriced crude oil in the world market. This has important consequences—first, the adverse impact on cooking oil. Palm oil is still the cooking medium for the vast majority of people on the Asian continent. Palm oil can replace other oils, such as coconut oil. This was a very recent development in the State of Kerala, India, where coconut oil was the preferred cooking medium for millennia. Cheaper palm oil has not only found its way into the kitchen, but is also being increasingly used in industry, as in the case of manufacture of toiletries, such as soap manufacture. However, it must also be pointed out in this context, an "anticoconut lobby" has been in operation for decades to project it as a health hazard, in the wake of its "alleged" role in increase in cholesterol in those consuming coconut oil, leading to cardiac problems. This has been contested by medical science, and common sense tells why there should not be a "coconut oil scare" which has been in use for ages in the State of Kerala, with no direct relationship between increased cardiac problems in the people and coconut oil consumption.

These reasons support the development of a strong research and development effort in palm oil industry. The R&D has to ensure that the industry remains well sustainable and competitive. Furthermore, the R&D has to be aligned to a new direction that delivers products and processes that serve the ever-changing fancies and needs of the end users.

The availability of novel genes from MPOB germplasm collections has made it possible for the development of new oil palm planting materials (PSI: high yield and dwarf; PS2: high yield and high iodine value). Such availability of genetic variability could also be exploited for the production of planting materials with high carotene content, vitamin E, and sterols. By 2020, major changes are likely to occur in the types of planting materials used by commercial plantations.

There is an urgent need to fully exploit the potential for genetic engineering in oil palm. The production of the first transgenic oil palm containing herbicide tolerant genes was a landmark in oil palm breeding. Currently, research is geared to production of varieties with high oleic oil: Following the field evaluation of the materials, production of other new and value-added products will be easily achieved, because, in essence, the same tools and techniques will be applied.

Other areas of interest in the sphere of genetic engineering include industrial oils, such as palmitoleic and ricinoleic acids, thermoplastics, and nutraceuticals. The oil palm has two storage sites, the kernel and the mesocarp, which can be the targets for storing genetically engineered products. The substrates and intermediates in these tissues may be channeled to alter the levels of existing products or to produce novel value-added products without deleterious effects on the plants. In comparison to other oil crops, the application of transgenic technology to oil palm provides specific advantages. First, it is conceivable that genetically engineered products can be produced at high levels because oil palm is highly productive. Compared to most other oil crops, its productive index per unit area of land is manyfold. Second, being perennial in nature, the palm will not face similar problems of instability of transgenic inheritance as with annual oil crops, because the same crops remain in the field for a very lengthy period of 25–30 years. Even the coconut tree does not have this long a life span.

Admittedly, the early reports on investigations on clonal abnormality slightly dampened the research interest in oil palm tissue culture. But there is now a renewed interest in this technology for the mass production of planting materials. Unfolding research evidence indicates that clonal materials from elite explants are more uniform and superior in performance compared to D × P seedlings. Heritable traits, such as fruit and vegetative characters, were shown to be inheritable from ortets to clones. Good household practices by prudent selection at various stages, minimizing the time in culture, and reducing hormone levels in culture media are found to substantially reduce the rate of abnormality. In addition, investigations are being carried out on abnormality to unravel the underlying mechanism that causes somaclonal variation and to develop markers for early detection. The present system, which requires repeated subculturing of embryonic tissues, is labor intensive, expensive, and time consuming. Media and culture protocols that are applicable to a wide range of genotypes are being developed so that large-scale production will be cost effective, opening up the possibility of automation. Some laboratories have initiated liquid cultures, which offer several advantages, including the ability of embryonic cells to proliferate rapidly, requiring far less subculture and greater savings in labor and space requirements.

Other areas of research interest will be focused in the future—in particular, those which have an important bearing on the environment. These areas include the utilization of beneficial plants for parasitoids in the field, precision agriculture with a focus on optimizing nutrient and water requirements, the use of geographic information systems (GIS), which is gaining rapid recognition in the United States and Canada, the management of oil palm residues, biological control of *Ganoderma*, and enhancement of field mechanization.

8 Rubber (*Hevea brasiliensis*)

Almost 99 percent of world's natural rubber is produced by *Hevea brasiliensis*, the "Para Rubber" of international commerce. The term *rubber* has, indeed, a very interesting origin. The British scientist Joseph Priestly coined the word, because of the ability of rubber to rub out pencil marks. In all of the Indo-European languages, barring English, the term is derived from the Amerindian name for rubber trees: *cachuchu* (weeping wood). The Spanish name *caucho* indicates the ecological origin of the majority of rubber-bearing plants, because Spain was the principal colonial power in South America at the time rubber started to become known in Europe. French interest continued throughout the eighteenth century, in particular, the significant contribution made by the French botanist, Jean Baptiste Fusée Aublet, who published the first taxonomic description of *Hevea* in 1775. The taxonomy of the genus has undergone considerable changes since then. The name itself is a Latinized version of the Ecuadorian Indian name, *Hheve*, and there was some earlier competition with other possible names such as *Siphonia* and *Caoutchoua* (Jones and Allen, 1992).

The first attempt at vulcanization of rubber was done in the nineteenth century, which also saw the first development of specialized machinery and techniques to manufacture rubber goods. This led to commercial trade in rubber and also the commercial cultivation of rubber plant. The rubber boom in Southeast Asia in 1910 led to the great first surge in commercial planting on a very large scale, after which the crop developed remarkably, from a wild jungle tree to a major domesticated crop within a span of about four decades. The growth of plantations in Southeast Asia was favored by rapid developments in the transportation sector, such as railways and steamships, and the opening of the Suez Canal. By the end of the nineteenth century, natural rubber became one of the major plantation crops. The evolution and adaptation of natural rubber, an important ingredient of many modern facilities, thus illustrates the evolution of modern civilization. Natural rubber is a microcosm of the world history, embodying the change from colonization to independence and from parochialism to internationalism. The industry has withstood the many shocks that it has been subjected to by the vicissitudes of the world economy (Jones and Allen, 1992).

The Importance of Natural Rubber

Millions around the world cultivate rubber as the principal means of livelihood. Natural rubber is the raw material for the manufacture of a very large number of finished

The Agronomy and Economy of Important Tree Crops of the Developing World. DOI: 10.1016/B978-0-12-384677-8.00008-4
Copyright © 2010 by Elsevier Inc. All rights of reproduction in any form reserved.

products. Rubber is used to fabricate hard and strong structural materials, resilient elastic materials, conductors and nonconductors of electricity, shock absorbers, mountings for motors, and other machinery, such as transmission belts and gaskets, hoses to transport gases and liquids, transparent and translucent materials, clothing articles to protect from rain, sports goods, cements, paints, plastics, pharmaceuticals, and, most important, the manufacture of tires, the chief outlet for rubber and having a worldwide use.

Origin and Distribution of Rubber

Natural rubber, an industrial raw material of strategic importance, is considered to be one of the most versatile agricultural products, finding its use in about 50,000 products the world over. In India, around 35,000 products are made from natural rubber, and it supports an industry with a turnover of about 140,000 million people annually. From an agroecological point of view, rubber is an ecofriendly tree species.

The genus *Hevea* occupies the whole of Amazon River basin in Brazil, also extending south and north to parts of Brazil, Bolivia, Peru, Colombia, Ecuador, Venezuela, French Guiana, Suriname, and Guayana. All the ten species are grown in Brazil, whereas seven are found in Colombia, five in Venezuela, and four in Peru (Wycherley, 1992). *Hevea brasiliensis* extends about half of the range of the genus, mainly occupying the region south of the Amazon, extending to the states of Acre, Mato Grosso, and Parana in Brazil, parts of Bolivia, Peru, north of the Amazon to the west of Manaus, as far as the south of Colombia.

Currently, the rubber species is grown mainly in the tropical regions of Asia, Africa, and South America, in countries, such as Malaysia, Indonesia, India, Sri Lanka, Thailand, China, Philippines, Vietnam, Cambodia, Myanmar, Bangladesh, Singapore, Nigeria, The Republic of Cameroon, Democratic Republic of Congo, Ivory Coast, Ghana, Zaire, Liberia, Brazil, and Mexico. However, the major share of the total production comes from tropical Asia.

The Rubber Industry—A Historical Perspective

Even before the time of Columbus, the various uses of rubber had evidently been well established, as methods of tapping the rubber tree and processes for making crude articles from the latex were well known. The earliest European visitors to the new continent recorded the use of articles made of rubber, such as protective garments, balls for playing games, and syringes. This was an illustration of the advanced civilization that European explorers encountered in the Western hemisphere. Schurer (1957) traced the use of rubber in Peru, in the Mayan civilization in the Yucatan, and in the ancient Mexican civilization, where it was an important element of religious rites. In the Yucatan and Mexico City, rubber was preserved and used in the liquid form and in ceremonies; it symbolized the blood in human sacrifices (Baulkwill, 1989). However, the literature on rubber dates only from the first visit of Columbus to America, when the use of rubber was already well established in the Western hemisphere.

Rubber as an Important Commercial Commodity—The Evolutionary Cycle between 1700 and 1870

There was a commercial lull between the sixteenth and seventeenth centuries in Europe for rubber. By the middle of eighteenth century, rubber was rediscovered by the French scientists. The astronomer-geographer Charles-Marie de la Condamine in his book on the Amazon in 1745 described how Amerindians in Ecuador and Brazil made torches, boots, bottles, and syringes employing an elastic resin or gum called *caoutchoua* from the sap of a tree called *Hheve*. During the period 1750–1800 French scientists discovered solvents for rubber, after which rubber solutions were used to make flexible tubes—catheters for medical purposes. Rubber solutions were applied to the silk fabric of the hydrogen balloons that made successful ascents in France in 1783. By 1790, European scientists reinvented the rubber syringe, a device long known to the Amazonian Indians.

The nineteenth century saw rubber transformed from a curiosity—used chiefly as an eraser—into an important commercial and industrial product. Thomas Hancock in 1821 developed the revolutionary masticator, which shredded and compressed solid rubbers and scraps to a warm, homogenous, moldable, or rapidly soluble mass, which was a remarkable technical advance of that period that made rubber processing a possibility. Charles Mackintosh in 1823 made waterproof fabrics using his unique fabric—rubber fabric sandwich and coal tar derived naphtha as the solvent in Scotland. The introduction of vulcanization methods during 1839–1844 by the two giants of rubber industry, Thomas Hancock and Charles Goodyear, created a new demand for rubber by virtue of the properties, namely resistance to hot and cold air and melting. The process of cold vulcanization by dipping thin rubber articles in sulfur chloride solution was first patented by Alexander Parkes in England in 1846.

It is the rapid growth of the automotive industry in the nineteenth century that propelled demand for rubber astronomically. Annual consumption expanded from a negligible amount to thousands of tons. The nineteenth century also witnessed remarkable advancements in scientific knowledge leading to technical advancement. The chemical structure of the rubber molecule was revealed, and the first "rubber-like" material, the "synthetic rubber," was manufactured artificially. The three most important contributions to the new rubber plantation industry include (1) the discovery of vulcanization, (2) transfer of *Hevea* from the west to the east in 1876, and (3) the introduction of new methods of tapping the latex coagulation.

Early History of Domestication of Rubber Industry

The successful transfer of *Hevea brasiliensis* to the east and the subsequent establishment of rubber plantations in response to the rising demand for raw material were the results of the political forethought and practical abilities of a few people combined with favorable circumstances. The domestication of *Hevea* in the east is the most spectacular story in the history of rubber, as between 1870 and 1914 a novel plantation industry of some 900,000 ha was established to meet the new industrial demand (Baulkwill, 1989).

The real initiative of the historic domestication of natural rubber in the Far East was taken up by Sir Clements Robert Markham of the India Office in London.

Similarly leading roles were played by Joseph D. Hooker, Director of Kew Gardens, London; Henry Wickham, planter, rubber trader, and naturalist.; Henry Nicolas Ridley, protégé of Hooker and from 1888 onwards by the Director of the Singapore Royal Botanic Gardens (Baulkwill, 1989).

The entire development of rubber plantations in the east is attributed to the famous Wickham collection (Lane, 1953). In 1876, Wickham collected 70,000 seeds of *Hevea brasiliensis* from Brazil, near Boim on the Rio Tapajoz and from the well-drained undulating country toward the Rio Madeira. This area produced excellent wild rubber. About 2700 of these seedlings were raised in Kew Garden, England. In 1876, 1919 of them were dispatched to mainly Sri Lanka and a few went to Malaysia, Singapore, and Indonesia (Baulkwill, 1989). In India, rubber was first received in 1878 from Sri Lanka.

A few initial unsuccessful attempts were made to establish *Hevea* in Kolkata (the capital of West Bengal State) Botanical Gardens, after which Sri Lanka was selected as the most suitable place for acclimatization, and thus rubber cultivation moved outside the Indian mainland and was established in Sri Lanka (Thomas and Panikkar, 2000). Sri Lanka was thus the center of early activities, with the Heneratgoda Botanic Gardens in Colombo becoming a major source of rubber seeds for domestic use for exports (Jones and Allen, 1992). Early development research on diseases (Petch, 1911), latex flow, and the use of acetic acid to coagulate latex (Parkin, 1910) were also carried out in Sri Lanka.

In India, cultivation of rubber started in 1878 in the town of Nilambur, in Kerala State, South India, as a forest crop, using the planting materials brought from Royal Botanic Gardens in Heneratgoda in Sri Lanka (Petch, 1914). With the help of state administrators, British planters initiated commercial rubber plantations in India, and the product was exported to London. Gradually, a rubber-planting community was developed in Kerala, and by 1947, 73 percent of the cultivated area (115,304 ha) under rubber was controlled by Indian companies (Sarma, 1947; Thomas and Panikkar, 2000). In the twentieth century, rubber experienced a tremendous boost in demand on account of the development in the automotive industry, mainly propelled by the escalating demand for tires. This intensified the search for new sources of rubber, encouraged efforts to cultivate rubber yielding plants, and motivated chemical research into the phenomenon of elasticity of natural rubber. The response of rubber growers to the escalating demand was very quick, and rubber succeeded equally on highly capitalized estates and tiny family holdings. In the two world wars of the twentieth century, rubber was found to be a necessity for national survival. It became essential to the enjoyment of the conveniences and amenities of modern life.

Area, Production, and Productivity

Trends in Planted Area and Production

The extent of planted area and production from the 1970s to late 1900s in natural rubber-producing countries reveals the fluctuations in area and production during this period (Table 8.1). Declines, notably in Malaysia and Sri Lanka in Asia and Nigeria

Table 8.1 Estimates of Natural Rubber (NR) and NR Production in the Mid-1970s and Mid-1990s

Country	NR Planted Area ('000 ha)		NR Production ('000 tons)	
	1970s	**1990s**	**1970s**	**1990s**
Bangladesh	0	47	–	2
Cambodia	45	60	19	40
China	300	610	65	441
India	235	533	137	566
Indonesia	2322	2322	842	1594
Malaysia	1981	1635	1532	980
Myanmar	63	133	15	26
Papua New Guinea	13	18	6	6
Philippines	45	92	29	64
Sri Lanka	228	161	144	105
Thailand	1400	1960	382	2067
Vietnam	42	300	22	203
Asia Total	6674	7871	3193	6094
Cameroon	20	44	16	58
Ivory Coast	20	72	16	102
Gabon	0	12	–	9
Ghana	6	17	3	10
Guinea	0	6	–	3
Liberia	118	120	84	59
Malawi	0	2	2	–
Nigeria	240	200	66	58
Congo	90	40	30	11
Africa Total	494	513	215	312
Brazil	20	180	19	56
Guatemala	12	36	7	35
Mexico	6	15	–	–
Other Latin American Countries	8	15	7	23
Latin American Total	46	246	33	114
World Total	7214	8630	3441	6520

Source: IRSG (2000).

and Zaire in Africa, were more than offset by the expansion of more than 1 million ha in Indonesia, with an increase of 500,000 ha in Thailand, China, and India. In relative terms, many smaller producing countries, such as Brazil, Vietnam, Guatemala, Ivory Coast, and Cameroon, also performed well in terms of area expansion. A few newcomers to the industry include Bangladesh and the most recent entrants to the industry, Gabon and Guinea in Africa, where production started in the 1990s. A number of producers have reentered the industry since the early 1970s, for example GREL in Ghana and Mandala in Malawi. Cambodia and Vietnam, whose industries were virtually destroyed in the war, have recovered their rubber area, very slowly in the case of Cambodia, where political conflict continues, and quite spectacularly in Vietnam, where most of the present 300,000 ha were planted since reunification in 1975. There has been a rapid expansion of planted area in China, which now has the world's fourth-largest rubber base with 610,000 ha.

Area Under Rubber

In the late 1900s, Asian countries held the maximum area under rubber cultivation, constituting about 93 percent of the total cultivated area, with African and Latin American countries contributing 5 percent and 2 percent, respectively. Indonesia is the largest cultivator of rubber with an area of 3,329,000 ha, followed by Malaysia (1,315,000 ha), Thailand (1,968,000 ha), China (618,000 ha), and India (573,000 ha, see Table 8.1).

Production and Productivity

Natural rubber (NR) is produced largely by small farmers in the developing countries of Asia, Africa, and Latin America. Despite increasing output from other countries, the share of production of the top three rubber-producing countries, Thailand, Indonesia, and Malaysia, was still quite high, but falling from its peak of 80.5 percent in 1969 to 74 percent in 2003. The influence of Malaysia, the leader in rubber production for many years, had an output of 1.6 million tons in 1986. But the production level has been declining since then. Many reasons could be attributed to this. Economic development of the country, leading to labor shortage, as rubber tappers moved from rubber plantations to urban jobs, is one of the main causes. Malaysia currently produces less than 1 million tons annually, whereas Thailand and Indonesia, trailing Malaysia, have an annual output of 2.9 million tons and 1.8 million tons, respectively (Table 8.2).

Most of the rubber-producing countries have increased their production and productivity substantially since 1980. The increase in rubber yields in India has been substantial, given the small size of the rubber plantations and the overall size of the rubber industry in the country. The production increased from a very low level of 23,730 tons in 1955–56 to a high of 707,100 tons in 2003. The average yield per hectare increased from 353 kg in 1955–56 to 1592 kg in 2002–03, the highest productivity among rubber-producing countries. Thailand had a remarkable increase in NR production, where the productivity that had been as low as 350 kg/ha in 1975 shot up to 1130 kg/ha in 1990s. As per the recent trends in production, Thailand contributes 36 percent of total global production, followed by Indonesia and

Table 8.2 Area Under NR in the Main Rubber-Producing Countries

Country	Year	Area ('000 ha)
Thailand	1998	1972.0
Indonesia	1998	3344.0
Malaysia	1998	1568.0
India	1998	553.0
China	1996	618.0
Sri Lanka	1998	158.0
Brazil	1998	180.0
Nigeria	1999	150.0
Vietnam	1997	275.0
Ivory Coast	1998	96.0
Cameroon	1997	42.0
Philippines	1999	92.0
Others	1999	658.5
World Total	1999	9706.5

Source: IRSG (2001).

Malaysia at 22.4 and 12.3 percent, respectively. India contributes 8.9 percent with other countries contributing the balance of 20.4 percent (see Table 8.3).

World Trade in Natural Rubber

In 2003, world consumption of natural rubber touched almost 7,910,000 tons. The United States has been the largest consumer among the world rubber-consuming countries during the past two decades with a steep increase in consumption from 666,000 tons in 1975 to a high of 1,078,500 tons in 2003, an increase of 62 percent. China now ranks first with an annual consumption of 1,485,000 tons (18.8 percent), followed by Japan with 784,200 tons (9.9 percent), and India consuming 717,100 tons (9.1 percent) annually as per 2003 statistics (Table 8.3).

The Botany of *Hevea* Plant

Cytotaxonomic Background

Hevea brasiliensis belongs to the family *Euphorbiaceae*. At present, a total of 10 species are recognized within the genus *Hevea*:

- *Hevea benthamiana*
- *Hevea brasiliensis*

Table 8.3 Leading NR Producing and Consuming Countries in the World in 2000

NR Production			NR Consumption		
Country	**Tons**	**World Share (%)**	**Country**	**Tons**	**World Share (%)**
Thailand	2,873,100	36.0	United States	1,078,500	13.6
Indonesia	1,792,200	22.4	China	1,485,000	18.8
Malaysia	985,600	12.3	Japan	784,200	9.9
India	707,100	8.9	India	717,100	9.1
China	480,000	6.0	Malaysia	420,700	5.3
Vietnam	384,000	4.8	Korea	332,800	4.2
Ivory Coast	127,000	1.6	Germany	251,000	3.2
Sri Lanka	92,100	1.2	France	300,200	3.8
Liberia	110,000	1.4	Thailand	298,600	3.8
Philippines	84,000	1.1	Brazil	256,000	3.2
Others	344,900	4.3	Others	1,985,900	25.1
World Total	7,980,000	100.0	World Total	7,910,000	100.0

Source: IRSG (2004).

- *Hevea camergoana*
- *Hevea camporum*
- *Hevea guianensis*
- *Hevea microphylla*
- *Hevea nitida*
- *Hevea pauciflora*
- *Hevea rigidifolia*
- *Hevea spruceana.*

Hevea brasiliensis, along with three other species originally described under the genus *Siphonia*, was brought under the genus *Hevea* by J. Mueller Argoviensis in 1865 (Wycherley, 1992). Murca Pires reported the last species, *Hevea camergoana*, in 1981. All of the species except *Hevea microphylla* were placed under the subgenus *Heveae*, whereas *Hevea microphylla* was placed under the subgenus *Microphyllae* (Wycherley, 1992).

The Morphology of the *Hevea* plant

The rubber tree is a sturdy, quick growing, erect perennial, growing to a height of about 30m with an economic life span of more than 30 years in plantations. It has a straight trunk with light gray bark, and the branches develop to form an open, leafy crown. The young plants show characteristic growth patterns of alternating periods of rapid elongation and consolidated development. The leaves are arranged in groups or storeys, with each storey having a cluster of spirally arranged trifoliate glabrous leaves and extra floral nectaries present in the region of insertion of the leaflets. *Hevea* is a deciduous tree that sheds its leaves during December through February (wintering-partial or complete) followed by refoliation and flowering. The plant is monoecious with unisexual flowers produced in pyramid-shaped panicles in the axils of leaves having numerous small male flowers and fewer female flowers of bigger size. The female flowers are confined to the tip of the panicles and their branchlets. The ovary is tricarpellary syncarpous, which on pollination develops into a three-lobed dehiscent capsule (regma) with three large, mottled seeds. Pollination is by insects, and fruits ripen in 5–6 months after fertilization. Seeds contain an oily endosperm.

Cytogenetics of the *Hevea* plant

Hevea brasiliensis and its congeners are uniformly diploids with $2n = 2x = 36$ (Majumder, 1964). The 36 chromosome genera are, therefore, probably old tetraploids based on $x = 9$. A few putative hybrids between several different species of *Hevea* have been collected in the wild, and interspecific hybrids have also been produced in breeding programs. Indian scientists have reported development of an experimental tetraploid (Saraswathy Amma et al., 1984) and synthesis of a triploid (Saraswathy Amma et al., 1980) in the clone RHII 105. A spontaneous triploid (Nazeer and Saraswathy Amma, 1987) and a genetic variant with dwarf stature (Markose et al., 1981) have also been identified.

Commercial Adoption of Vegetative Propagation

Credit for the widespread adoption of vegetative propagation by budding goes to the horticulturist Van Helten in 1916 in collaboration with two planters, Bodde and Tass, in Indonesia (Dijkman, 1951). This was one of the contributing factors for the rapid growth of the rubber plantations with genetically improved clones. Depending on the color and age of the buds (scion) and the age of the stock seedlings, there are two types of budding: brown and green budding. In brown budding, which was traditionally adopted in *Hevea*, vigorously growing seedlings of about 1 year of age are used as stock plants with brown buds collected from mature shoots of the same age. In the green budding technique developed in North Borneo during 1958–60 (Hurov, 1960), 3- to 5-month-old, healthy, vigorous seedlings were used as stock, and tender green-colored buds were used as scion. Stock-scion union takes place in about 3 weeks. A budding success of 90–100 percent is possible if quality stock and scions are used and budding is done in favorable seasons. Brown-budded stumps are hardier than the green-budded ones. In general, green budding results in higher budding success than brown budding during summer (Marattukalam and Premakumari, 1982), which is of advantage since the green budded plants can be raised in polyethylene bags in summer and can be planted in the field during the ensuing planting season itself. Another type of propagation of young budded plants, called young budding, is produced by budding very young stock plants of 7–8 weeks of age (Seneviratne, 1995) with green buds from shoots of comparable girth.

The possibility of forming a three-part tree with improved rootstock, trunk, and crown components originally envisaged by J.S. Crammer in Java in 1926 (De Vries, 1926) was further developed into the crown-budding technique. This method was first adapted to develop trees resistant to South American leaf blight (SALB). However, due to the practical difficulties of developing crown-budded plants on a large scale, this technique is adapted only to a limited extent. The feasibility of using the bench grafting technique, namely indoor grafting of stock plants (Marattukalam and Varghese, 1998), was also established in rubber. This type of budding can be adapted under adverse climatic conditions, such as severe cold, extreme heat, and heavy rains, when outdoor grafting will not be successful.

To cut short the immaturity period, different types of advanced planting materials were developed. Among the advanced planting materials, such as mini or maxi stumps and poly bag plants, the poly bag plants are being used on a commercial scale. Uniform growth, fewer casualities, early establishment, cost reduction, less weed growth, and so on are some of the advantages possible through the use of appropriate advanced planting materials.

Bark Anatomy

The economic product from rubber trees, latex, a specialized cytoplasm, is contained in a laticiferous system in the bark differentiated by the activity of the vascular cambium. The early researchers during the first quarter of twentieth century realized the importance of studies on the anatomy of the laticiferous system. Our basic understanding of all aspects of the anatomy of laticifers of *Hevea* originated from the contributions of Bobilioff (1919) and Gomez (1974) and Premakumari and Panikkar (1992) subsequently provided further detailed information of *Hevea* laticifers.

Hevea bark has two distinct zones—the outer hard bast, which is more protective in function with mostly discontinuous and nonfunctional laticifers, and the inner soft bast, containing productive and continuous latex vessels differentiated from the vascular cambium. The latex vessels are of the articulated anastomizing type with vessels of the same ring interconnected tangentially. Latex vessels are more concentrated in the region near the cambium in the soft bast. The laticifer differentiation from the cambial derivatives is a rhythmic process, and a ring of laticifers is produced each time, forming concentric rings alternating with layers of other phloem tissues. The laticifers are generally oriented 3–5° to the vertical, in a counterclockwise direction.

The quantity of laticiferous tissue in a tree is determined by factors such as the number of latex vessel rows, density of vessels within rows, distribution pattern of and distance between the vessel rings, as well as bark thickness and girth of the tree. Highly significant clonal difference has been reported for these characters

(Premakumari and Panikkar, 1992). During tapping, only a thin slice of bark of 1–1.5 mm thickness is shaved off to cut open the latex vessels, leaving the cambium intact for bark regeneration. Since the renewed bark is exploited after the consumption of virgin bark, bark regeneration is of great significance.

Although investigations on the structural aspects of laticifer anatomy have helped the conceptual development of scientifically correct tapping systems, studies at the ultracytological level (Dickenson, 1965; Gomez, 1974) have contributed significantly in furthering our understanding of the physiology and biochemistry of latex.

The Physiology of Latex Flow

When there is no tapping, no movement of latex occurs inside the laticiferous systems once it is synthesized. Latex flow from a tapped tree is an abnormal physiological phenomenon inducted by tapping. The cytoplasmic nature of latex, both in terms of structure and function, was elucidated by high-speed centrifugation (Moir, 1959) and later by electron microscopy (Dickenson, 1965; Gomez and Moir, 1979).

It is the very high turgor pressure in the latex vessels and elastic contraction of walls after a sudden release of turgor resultant from cutting the vessels open which trigger the latex flow. After a few hours, capillary forces regulate the flow until it ceases as the latex coagulates and plugs the vessels. An inherent clotting mechanism is present within the latex vessels, which is responsible for the cessation of latex flow (Southorn and Yip, 1968) consequent to the plugging of the opened ends of laticifers. Milford et al. (1969) proposed an index—the plugging index—which is a time-flow constant expressed as the ratio of the flow for the initial 5 minutes to the total volume yield for measuring the extent of plugging. Sethuraj (1981), through a theoretical analysis of yield components of *Hevea brasiliensis*, represented the effect of the major yield components by the following formula:

$$Y = \frac{FlC_r}{p}$$

where
Y = Yield
F = Initial flow rate per unit length of the tapping
l = Length of the tapping cut
Cr = Percentage of rubber content
p = Plugging index.

Of these four major components, the length of the tapping cut is determined by the girth, and thus the growth vigor, and is influenced by the total biomass production and partitioning between growth and latex production. The inherent characteristics, exploitation systems, and the environment also influence the other three components. Yield is a manifestation of the major yield components as well as a number of anatomical, physiological, and biochemical subcomponents influencing the main components (Sethuraj, 1992).

Tapping Process

The rubber tree is exploited by tapping the bark—a process of controlled wounding of bark to cut open the latex vessels by regularly removing a thin shaving of bark from the surface of the tapping cut at previously specified intervals. The native method of crop harvest from wild rubber trees involved a crude method of latex extraction, which led to the formation of irregular secondary growth, poor yield, and early abandonment of trees (Cook, 1928). Henry Ridley, Director of the Singapore Royal Botanic Gardens during 1888–1911 and a very prominent personality during this period, made significant contributions to the basic method of present-day tapping. This excision method of tapping, known as Ridley's method, involves the removal of a thin layer of bark from the cut end at each tapping, thus permitting a smooth flow of latex and allowing the bark to regenerate (Ridley, 1897). Ridley made other significant contributions, such as establishing the importance of tapping in the early morning, the effects of daily and alternate days tapping, and the most optimum age to begin tapping.

A budded tree is opened for tapping when it attains a girth of 50 cm at a height of 125 cm from the bud union, whereas trees of seedling origin are tapped on attaining 55 cm girth at 50 cm height. One tapping panel covers half the circumference of the tree. The tapping cut of a budded tree should have a slope of about 30° to the horizontal. Since the latex vessels in the bark run at an angle of 3–5° to the right, a cut from high left to low right will open a maximum number of vessels. Tapping is a highly skilled work where the depth of tapping should be within 0.5 mm of the cambium to get optimum yield without injuring the cambium. Tapping should commence in the early morning, as late tapping will reduce the latex yield due to increased transpiration leading to lower turgor pressure.

Ethephon—A Chemical Yield Stimulant

Ethephon (2-chloro-ethyl-phosphonic acid) (Abraham et al., 1968) revolutionized the ability to regulate commercial latex yield at will, based on physiologic and economic requirements. Judicious use of ethephon-like stimulants helps to sustain rubber plantations during times of low rubber prices. Detailed field studies in different institutes have led to specific commercial recommendations based on different situations. The different methods of application include employment on the bark, groove, lace, and also in multiple bands, tape, and soil.

It is the ethylene production, either endogenously or through hydrolytic decomposition (Abraham et al., 1968; Gomez, 1983), that the yield stimulants act. Stimulation delays latex vessel plugging and prolongs duration of latex flow (Boatman, 1966). Tupy and Primot (1976) observed that the increase in pH as a result of stimulation enhanced the rate of latex regeneration through increased invertase activity. Stimulation with ethephon results in a 20 to 100 percent increase in latex yield. The stimulation effect has been reported to last at least for 3 months (Chapman, 1951). However, the yield stimulation will lead to 2–7 percent reduction in the dry rubber content, even though it has no effect on rubber properties (Sumarno-Kertowardjono et al., 1976).

Tapping Notations

Tapping notations are a set of internationally approved signs and symbols describing the method of tapping and its frequency. The description of notations for tapping systems was first introduced by Guest (1939) and later revised by Lukman (1983). The notations for tapping method include symbols for length of the tapping cut, direction, frequency, and so on. The type of cut given is denoted by a capital letter "S," which signifies spiral cut, "V," which means V-cut, and "C" indicates circumference (used for two or more unspecified cuts on a tree tapped on the same day), and "Mc" indicates minicut (5 cm or less in length). Length of the cut denotes the proportion of the circumference of the trunk that is cut for tapping and is represented by a fraction preceding the symbol of cut, except in minicut, where the actual length is given in cm. Examples are S for full spiral cut, V for one full V-cut, C for one full circumference (unspecified cut), ½ S for one-half spiral cut, ¼ S for one-fourth spiral cut, ⅓ V for one-third V-cut, ¾ S for three-fourths spiral cut, ½ C for one-half circumference cut, and Mc2 denotes a minicut 2 cm in length. Frequency of tapping is denoted as a fraction, which explains the interval between tappings. The numerator of the fraction is d, which denotes the tapping period (day), whereas the denominator denotes the actual interval between tappings in days or in a fraction of a day. Examples are d/1 for daily tapping, d/2 for tapping once in 2 days, and d/3 for tapping once in 3 days, and so forth.

The notation for yield stimulation using chemical stimulants consists of three parts. The first part indicates the stimulant and its concentration; the second, its place of application quantity of stimulant used and method of application; and third, the number of application and periodicity. The notation ET 2.5% Pa2 (1) 16/y (2w) means the that the tree is stimulated with ethephon (ET) at 2.5% concentration applied on panel (Pa) with 2 g of stimulant per application on 1 cm band in 16 applications per annum at fortnightly intervals.

Tapping Systems

In rubber, the response to different tapping systems varies from clone to clone. Budded trees are tapped on half-spiral alternate days (½ S d/2), and seedlings are tapped on half-spiral every third day (½ S d/3). However, in high-yielding budded clones prone to tapping-panel dryness (TPD), reduced tapping intensity of once in 3 days is recommended though, in general, small growers prefer higher frequencies. The latex yield from trees varies from clone to clone, the age of the tree, soil fertility, climatic conditions, tapping system followed, and the professional skill of the tapper. Intensive tapping is done for maximum exploitation of the tree before felling it for replanting. The methods adopted include increased tapping frequency, extension of tapping cut, opening of double cuts, and use of yield stimulants. Upper-level tapping, such as controlled upward tapping (CUT), is now gaining popularity for longer exploitation of the virgin bark above the lower panel (Vijayakumar et al., 2000).

Biochemistry of Latex and Rubber Production

Latex is a specialized form of cytoplasm containing a suspension of rubber and non-rubber particles in an aqueous serum. Besides rubber and water, the other components of latex are carbohydrates, proteins and lipids, inorganic salts, and other minor substances (Archer et al., 1963). Rubber particles are the most important constituents in the latex, which make up to 35–40 percent of the volume. Rubber particles are usually spherical, but are sometimes oval or pear-shaped (Dickenson, 1965) and are strongly protected in suspension by a film of adsorbed protein and phospholipids (Archer, 1964). This protein-phospholipid layer imparts a net negative charge to the rubber particle, contributing to colloidal stability (Bowler, 1953).

Lutoid particles are next, in abundance, making up 10–20 percent of the latex volume. These are subcellular membrane-bound bodies (Southorn and Yip, 1968) that enclose a fluid serum known as lutoid serum or B serum, which is a destabilizer of rubber hydrocarbon. One to three percent of the latex volume is occupied by Frey-Wyssling particles, which are yellow and spherical (Dickenson, 1965). Quebrachitol is the most concentrated soluble carbohydrate in the latex along with sucrose and glucose (Low, 1978). The total protein content in the fresh latex is about 1 percent, of which about 20 percent is adsorbed on rubber particles, an equal quantity found in the bottom fraction, and the remainder in the serum phase (Archer et al., 1963). Lipids constituting about 1.6 percent of latex play a vital role in the stability and colloidal behavior of latex. Other minor components are thiols (main reducing agents), various amino acids, organic acids, and inorganic ions, such as K, Mg, Cu, Fe, Na, Ca, and phosphate (Archer et al., 1963).

The biosynthesis of rubber is initiated with the generation of acetyl coenzyme A, which is converted to isopentanyl pyrophosphate (IPP) and then polymerized into rubber. The sustained production of rubber is largely dependent on the latex biosynthetic capacity of a clone for *in situ* latex regeneration between two successive tappings. Biochemical components, namely total solid content (TSC), thiols, inorganic phosphorus (Pi), Mg, and sucrose, play an important role in latex regeneration (Jacob et al, 1986). The quantity of rubber per unit of sucrose present in the latex, termed as the sucrose conversion efficiency, is an indirect estimate of the overall regulation of the metabolic conversion of sucrose into rubber (Dey et al., 1995).

Genetic Improvement of Rubber

Genetic improvement of *Hevea*, despite being a very elaborate and time-consuming procedure, has paid rich dividends in increasing yield by several folds and making available several high-yielding clones for commercial planting.

Breeding and Selection

Rubber breeding in Southeast Asia is one of the outstanding success stories of plant breeding. The first steps in breeding were taken by the Dutch workers in Java and Sumatra. They demonstrated that yields per tree were variable, that clonal propagation was possible, and that individual clones differed from one another. With ordinary seeds as planting material in the initial years of rubber research, productivity was around only 300 kg/ha/annum. Subsequent selection of high-yielding mother trees and multiplication by budding resulted in early primary clones with improved yield potential. From the 1920s onwards, budded clones and superior seedling populations progressively supplemented the random open pollinated seedlings. Since the Second World War, clones have been dominant and rubber breeding has long been a matter of selecting and testing clones from genetically variable populations. Subsequently, breeding new clones through hybridization resulted in a series of high-yielding hybrid clones.

The Objectives of Breeding

The principal objective of breeding in *Hevea* is the synthesis of ideal clones with high-production potential combined with desirable secondary attributes, such as initial vigor, smooth thick bark with good latex vessel system, good bark renewal, high growth rate after opening, tolerance to major diseases and pests, wind havoc, TPD, good response to stimulation, and low frequency of tapping. Clones with early attainability of tapping girth and high initial yields are preferred to clones with higher yields in later phases of exploitation.

Specific objectives, however, vary depending on agroclimatic and socioeconomic requirements. Marginal and nontraditional areas demand priority for development of clones resistant to prolonged drought, high summer and low winter temperatures, strong winds, altitudes, and so on. Such situations also demand genotypes responding well to high-density planting, poor soil fertility, and low input agriculture based on sustainable farming systems. In places where labor is relatively cheap, clones suitable for high-intensity tapping are economic, whereas under labor shortage, low-intensity tapping is preferred. Since rubber planters are predominantly small holders, breeding objectives are to be streamlined to take care of their specific needs also.

Breeding Methods

The conventional breeding methods in *Hevea* can be classified as:

1. Introduction
2. Ortet selection
3. Hybridization
4. Clonal selection.

Introduction of Clones

Introduction or exchange of available clones among the rubber-growing countries in the early years constituted the original breeding pool in each country. Recent introductions under bilateral and multilateral clone exchange programs organized by the International Rubber Research and Development Board are confined to potential clones of good performance. Thus, popular clones evolved in different countries are being introduced to member countries and evaluated under the local agroclimatic conditions, and promising selections are being recommended for large-scale planting. In India, so far, 127 Wickham clones evolved in countries such as Malaysia, Indonesia, Sri Lanka, Thailand, China, and Ivory Coast formed the exotic component of the gene pool, in addition to the recent introduction of around 7000 genotypes of the Wild Brazilian germplasm.

Ortet Selection

Mother tree selection, which is also known as ortet selection, is the oldest breeding method aimed at systematic screening for outstanding seedling genotypes resultant of natural genetic recombination. Clones developed through ortet selection are called primary clones. Screening of extensive seedling plantations in Indonesia, Malaysia, and Sri Lanka resulted in a good number of early primary clones of importance, such as Tjirl 1, PR 107, GT 1, BD 10, AVROS 255, Gl 1, PB 28/59, Mil 3/2, and Hil 28, of which clones such as GT 1 and PB 28/59 are still widely planted. In India, earlier mother tree selections include 46 clones, of which RRII 1, RRII 4, RRII 5, RRII 6, RRII 33, RRII 43, and RRII 44 are among the promising ones (Marattukalam et al., 1980).

Hybridization

Hybridization and clonal selection, the most important conventional method in *Hevea*, offers scope for exploitation of heterosis in hybrid progeny of potential parent clones. Desirable recombinants once selected can be fixed easily through vegetative multiplication. As a result of hybridization and selection, a good number of hybrid clones of commercial significance have evolved. The early primary clones as parents of the first hybridization series resulted in early hybrid clones of commercial significance, such as RRIM 500 and RRIM 600 series, the yield levels of which were much superior to those of the parent clones. The best clones in each series were further used as parents in subsequent series. Rubber breeders followed this type of cyclical generation-wise assortative mating (GAM) over the past 50 years. In Malaysia, the Rubber Research Institute developed clones of RRIM 500 to RRIM 1000 series, whereas the Prang Basar Institute in the private sector selected certain PB clones of commercial significance. The Indonesian Research Institute for Estate Crops in Java and Sumatra (Balai Penelitian Purkeburan Medan, BPPM) evolved Profestation voor Rubber (PR), Algemene Verneiging Rubber Planters Cost Kust Sumatra (AVROS), BPM, S'Lands Caoutchoue Bedrijven (LCB), PPN, and Rubber Research (RR) clones. The Rubber Research Institute of Ceylon (now Sri Lanka) (RRIC) clones originate from Sri Lanka, Kohong Rubber Estates (KRS) clones from Thailand and Haiken, YRITC, and South China Academy of Tropical Crops (SCATC) clones from China.

Inasmuch as India is concerned, rubber improvement programs were initiated in 1954 and a large number of hybrid progeny was developed and evaluated for potential recombinants. The early hybrid clones developed by Rubber Research Institute of India (RRII) in Kottayam, Kerala State, include RRII 100, RRII 200, and RRII 300 series (Annamma et al., 1990). Among the clones of RRII 100 series, RRII 105 is a highly successful and popular clone (Nair and George, 1969). The clones RRII 116 and RRII 118 are outstanding in terms of plant vigor, but are of only medium yield. Clones RRII 203 and RRII 208 (Saraswathy Amma et al., 1987) and RRII 300 and RRII 308 (Premakumari et al., 1984) are the best selections in the 200 and 300 series, respectively. In another set of hybrid series, nine clones revealed marked heterotic increases for yield over the first 3 years of tapping (Licy et al., 1998). Of these, five clones designated as RRII 400 series, namely RRII 444, RRII 417, RRII 422, RRI 429, and RRII 430, have been included in Category III for planting recommendations of the Rubber Board of India, situated in Kottayam, Kerala State, since 2001. These clones have shown significant improvement in yield over the clone RRII 105, to the tune of 23–46 percent during the first 8 years of tapping in the small-scale evaluation trial (Saraswathy Amma, 2003). Another set of 50 hybrid clones were identified as having better potential for genetic advancement, based on early growth and yield (Varghese and Mydin, 2000).

How to Overcome the Constraints in Breeding?

The primary constraints in breeding and quick release of cultivars include the narrow genetic base, heterozygous nature of the crop, seasonal and nonsynchronous flowering pattern, low fruit set, long breeding and selection cycle, and lack of fully reliable early selection parameters (Varghese and Mydin, 2000). Pollen storage and induction of off-season flowering have been suggested to overcome the limitation of nonsynchronous flowering. Low average fruit set of less than 5 percent limits the size of legitimate families available for selection. Treatments such as applying boric acid sucrose solution to the stigma and enclosing the panicles in butter paper (instead of sealing individual flowers) yielded relatively higher fruit set (Mydin et al., 1989), but the extent of fruit set was still low and inadequate demanding further investigation.

The lack of fully reliable early selection parameters hampers quick release of clones. The widely adopted method for juvenile selection, namely a modified Hamaker–Morri–Mann method (Tan and Subramaniam, 1976) where 2- to 3-year-old plants are tapped successively to quantify the latex yield, identifies a fair proportion of high yielders. Making an incision at age 1 of the tree to determine latex yield was suggested (Annamma et al., 1989). A study on early evaluation incorporating clones of high-, medium-, and low-yield potential revealed performance index at an age of 2 years to be good enough for selection of a fair proportion of high-yielding clones at an early age (Varghese et al., 1993). Biochemical components, such as sucrose, total solid content, thiols, and organic phosphorus at immature phase, were suggested to provide more precision for early selection (Licy et al., 1998). However, with the available early prediction methods, nursery yield can be considered as only a fair indicator of mature yield. Detailed investigations at major yield components and stable subcomponent levels are required to develop fully reliable parameters for early prediction of yield.

Genetic Studies

In *Hevea*, the characters of economic importance are, in general, controlled polygenetically. For effective selection and utilization of genotypes for breeding programs, a clear understanding of the magnitude of genetic diversity for the desired traits as well as estimates of genetic parameters, such as heritability and genetic advance, is essential. Successful attempts have been made at RRII in this direction (Premakumari and Panikkar, 1992). Similar studies revealed the preponderance of additive gene action in the statement of yield and yield components suggesting better scope to select based on specific traits.

An understanding of the genetic divergence of the parent clones used in crossbreeding as well as component clones of polyclonal seed gardens is of paramount importance to properly exploit gene recombinants. Studies indicate the genetic divergence of the parent clones used in crossbreeding as well as component clones of polyclonal seed gardens, which is of paramount importance to properly exploit gene recombinants. Studies indicate that in spite of the narrow genetic base, sufficient genetic diversity existed in the original Wickham material (Markose, 1984). In studies on genetic diversity of Wickham clones, based on yield and various yield components, two sets of 40 and 35 Wickham clones were grouped into eight (Mydin et al., 1992) and nine (Abraham et al., 1997) genetically divergent clusters. In another recent study, 80 wild accessions were grouped into nine different clusters (Abraham, 2000). Selection of genetically unrelated parent clones offers greater chances of heterosis and superiority of resultant progeny. Use of molecular markers has simplified such studies considerably.

Genetic Base and Germplasm Resources

Is There a Need to Broaden Genetic Base?

The genetic base of *Hevea* in the east is very narrow, limited to a few seedlings originally collected from a minuscule of the genetic range in Brazil referred to as the "Wickham base" (Simmonds, 1989). These few seedlings were collected from a minuscule sample of the genetic range of the species in Boim, near the Tapajos River in Brazil (Wycherley, 1968). Using this gene pool, substantial improvement in productivity has already been achieved over the past five decades. The original narrow base has further narrowed down through the unidirectional selection for yield, a cyclical generation-wise assortative mating pattern and a wider adaptation of clonal propagation by budding. A cyclical breeding pattern with the best genotypes in one breeding cycle used as parents for the next has led to the selection and release of clones that are more or less related. The parentage of popular clones bred in various rubber-growing countries can be traced to a handful of parent clones (Tan, 1987). The main emphasis given on productivity alone has resulted in a certain amount of genetic erosion with respect to characters such as disease tolerance.

The dreadful South American leaf blight (SALB), caused by *Microcyclus ulei*, a fungus prevalent in the American hemisphere (Chee and Holliday, 1986; Edathil,

1986), is the most damaging to rubber trees. None of the Wickham clones have been reported to have tolerance to SALB. There are also indications of erosion of genes controlling resistance to *Oidium* and *Gloeosporium* in the original Wickham material (Wycherley, 1977) and several other minor diseases assuming epidemic proportions. Apart from the problem of disease susceptibility, variability for resistance/tolerance to various abiotic stress situations, such as drought, cold, high elevations, and high-velocity winds, assumes significance in the present-day context of extension of rubber cultivation to marginal and nontraditional areas. In order to select clones adaptable to such specific locations, the base material should contain ample genetic variability and a broad genetic base (Varghese and Abraham, 1999).

Fresh Germplasms: The 1981 Collection

Considering the urgent need to broaden the genetic base, fresh, wild germplasm was collected from the center of diversity in Brazil, as a result of the joint expedition of the International Rubber Research and Development Board and the Government of Brazil in 1981. The team collected 64,736 seeds (Ong et al., 1983) and budwood from 194 presumably high-yielding mother trees from the states of Acre, Mato Grosso, and Rondonia (Ong et al., 1983). The Malaysian and Ivory Coast germplasm centers established this germplasm in nurseries. Varying proportions of the wild germplasm were established in different rubber research institutes. In India, a total of 7562 genotypes, including 126 ortet clones, were received from the Malaysian center and have been established in traditional and nontraditional areas (George, 1989). Their conservation, evaluation, and utilization in breeding programs are in progress.

Characterization, Evaluation, and Utilization

Studies of the 1981 collection conserved in nurseries or field evaluation gardens are underway in different institutes across the world, where the genotypes have been introduced and established. The wild genotypes are first characterized using a set of descriptors and subjected to a preliminary evaluation in source bush nurseries. Selections for desirable characters are then planted in field evaluation trials laid out in augmented design. Wild germplasm, in a phased manner, is subjected to characterization, where it is documented based on a set of descriptors that consists of passport data, plant type data at the juvenile phase, followed by data on relevant qualitative and quantitative characters on interest in the later stages of growth (Varghese and Abraham, 1999). Data on morphological characterization of a set of wild germplasm at the juvenile stage revealed interesting variability (Abraham et al., 1994). Interesting morphological variants for growth form, flower color, and fruit shape were observed among the wild genotypes (Madhavan et al., 1997).

Preliminary evaluation of the Brazilian genotypes established in the traditional areas in India revealed wide variability in the early growth phase with respect to morphological parameters and juvenile yield (Annamma et al., 1989; Abraham et al., 1992). Several individual accessions superior to the control clone RRII 105, for certain yield traits, were identified in the immature phase for yield and other characteristics. Genotype MT 999 had a higher number of latex vessel rows, with higher diameter and cross-sectional area for the latex vessels (Abraham et al., 1992; Reghu et al, 1996). A few genotypes with higher test-tap yield in the immature phase than that of the popular clone RRII 105 were identified (Abraham et al., 1992). Provenance-wise, comparison of genotypes for various morphological and anatomical characters and test-tap yield showed the genotypes from Mato Grosso to be superior to those from Acre and Rondonia (Abraham et al., 1992; Reghu et al., 1996 in India, and Lam et al., 1997 in Vietnam). In Malaysia, however, Rondonian genotypes recorded superior yield compared to those from Acre and Mato Grosso (IRRDB, 1996).

Investigations on screening wild germplasms for resistance to major diseases and to drought and cold conditions have been initiated. Field observations have identified a general field tolerance of the wild germplasm belonging to Mato Grosso to shoot disease (Mercy et al., 1995). In general, the wild genotypes recorded poor yield compared to the domesticated control clones. Similar results have also been reported from Malaysia (IRRDB, 1996), Ivory Coast (Clement-Demange et al., 1997), and Vietnam (Lam et al., 1997). Searches for superior timber characters in wild germplasm have also been initiated in India and other countries.

Incorporation of superior wild genotypes as one of the parents in Wickham × Amazonian hybridization programs has been initiated in various rubber research institutes worldwide, including those in India. In India, among 12 cross combinations involving wild genotypes and popular Wickham clones, hybrids of the cross RRII 105 and RO/JP/3/6 had superior test-tap yield at 2 years of age (RRII, 1994). Another set of nine superior genotypes has been used in crosses with popular Wickham clones, such as RRII 105 and RRIM 600, and the progenies are under evaluation. However, there is a long way to travel to realize the expected benefits from such a large collection of wild germplasm. Utilizing powerful molecular tools, such useful genes should be identified and utilized in the production of transgenic rubber plants for specific stress situations.

Biotechnological Interventions

In *Hevea*, biotechnological advances offer immense possibilities, such as propagation of elite planting materials, eliminating stock-scion interaction, conservation and characterization of genetic resources, construction of linkage maps, development of transgenic plants incorporated with agronomically important genes, and marker-assisted selection (MAS) (Varghese et al., 2001).

In Vitro *Methods*

Micropropagation

The first successful studies on micropropagation by Carron and Enjarlic (1982) were followed by subsequent work on the *in vitro* multiplication of *Hevea brasiliensis* using seedling explants by Gunatilleke and Samaranayake (1988), Seneviratne and Flegmann (1996), Sereviratne and Wigeskare (1997), Sobhana et al. (1986), Asokan et al. (1988), and Lardet et al. (1998). These investigators used explants derived from mature clonal trees for the micropropagation work. In the RRII, protocols were developed for the *in vitro* generation of *Hevea* plants using shoot tips derived from mature field grown trees (Asokan et al., 1988; Sobhana et al., 1986).

Protoplast Culture

Only very limited studies related to protoplast culture in *Hevea* are known. Nurhaimi et al. (1993) successfully isolated protoplasts from calli and cell suspensions from anther tissues. During *Hevea* protoplast culture, a rapid decrease in viability was associated with an increase in ethylene production. These reactions are common to stress conditions. A number of physiological phenomena associated with lack of miotic division of stem protoplasts of rubber were observed. However, after optimizing different parameters, Sushmakumari et al. (1998) developed an efficient pathway for callus induction from protoplasts.

Somatic Embryogenesis

Attempts to develop somatic embryogenesis as an *in vitro* propagation technique were first made in the 1970s. The first plantlet from somatic embryogenesis through anther culture was reported from China by Wang et al. (1980). This was followed by reports of Wan et al. (1982) and Asokan et al. (1992). Carron and Enjarlic (1982) reported somatic embryogenesis and plantlet regeneration from inner integumental tissues and fruits in France. Asokan et al. (1992) reported somatic embryogenesis and plantlet formation from some commercial clones using inner integumental tissues. Carron et al. (1998) observed clonal variation in the micropropagation efficiency through somatic embryogenesis. In their study, the *in vitro* plantlets of clone PB 260 recorded better growth in the field than those of PR 107, and were also superior to other controls. Kumari Jayasree et al. (1999) reported high-frequency somatic embryogenesis and plantlet regeneration from immature anthers of commercial clones.

In Vitro *Conservation*

Cryopreservation is a viable option in *in vitro* preservation of *Hevea* for long-term storage of germplasm in field gene banks, although attempts in this direction have been rather scanty. In the late 1990s, two efficient cryopreservation protocols, one using a classical freezing process and the other using a simplified freezing process, were developed for embryogenic calli of a commercial clone of *Hevea*, PB 260 (Engelmann et al., 1977). High survival and rapid regrowth, as well as production of somatic embryos, were obtained with calli cryopreserved using both protocols. The simple freezing protocol was used successfully with a second commercial clone, PR 107. However, long-term studies are required to establish the feasibility of this *in vitro* technique as a convenient and secure alternative to the present *ex situ* method of conservation.

Genetic Transformation

Gene transfer in *Hevea* has been successfully established using genes particle gun and by *Agrobacterium* mediation (Arokiaraj and Wan, 1991). Anther-derived calli were transformed with vectors carrying the beta-glucuronidase (GUS) and the neomycin phosphotransferase (npt II) genes (Arokiaraj et al., 1996). Arokiaraj et al. (1998) reported the establishment of a gene transfer system for *Hevea* with *Agrobacterium tumefaciens* GV 2260 (p35SGUSINT) and LBA 4404 (pAL 4404/pMON9793) and that the CaMV 35S promoter directs glucuronidase expression in the laticiferous systems of transgenic plants. Attempts toward incorporation of tolerance to drought and tapping-panel dryness have been initiated at the Rubber Research Institute of India. Calli derived from immature anthers of the high-yielding clone RRII 105 were transformed with the gene coding for sorbitol 6-phosphate dehydrogenase, isopentanyl transferase, superoxide dismutase, and 1-aminocyclopropane 1-carboxylic acid (ACC) synthase genes in the antisense orientation under the control of the CaMV 35S promoter. The vectors containing the npt II and GUS genes were used for selection and scoring of the transformants.

Biochemical and Molecular Markers

Advances in biochemical markers, namely isozymes, and DNA markers, such as restriction fragment length polymorphism (RFLP), random amplified polymorphic DNA (RAPD), and simple sequence repeats (SSR) or microsatellites, offer rapid and attractive adjuncts to the elaborate conventional genetic analysis in perennial species, such as *Hevea* (Varghese et al., 2001). These molecular markers are potential tools in characterizing diversity and analyzing the *Hevea* genome. Seguin et al. (1996) were able to group *Hevea* germplasm into six genetically divergent clusters utilizing data on isozyme and RFLP markers. A comparison of genetic variability in the wild and Wickham germplasm using isozymes (Chevallier, 1988) and RFLPs by Besse et al. (1994) revealed significantly higher variability in the former. The applicability of RAPD markers for genetic analysis in *Hevea* was evaluated using 42 informative primers in a set of 24 clones from the breeding pool of the Rubber Research Institute of India. Estimation of genetic distance among the tested clones facilitated identification of genetically divergent clusters (Varghese et al., 1997). Polymerase chain reaction (PCR)-based markers, such as RAPDs, have wide application for routine analysis in a breeder's laboratory in comparison to RFLPs since they are easier and faster to screen large numbers of genotypes and assays of genetic variability of the whole genome level (Varghese, 1998). More recently, Lespinasse et al. (2000) reported development of a saturated genetic linkage map of *Hevea* sp. based on RFLP, amplified fragment length polymorphism (AFLP), microsatellites, and isozyme markers. Molecular aspects of resistance to diseases and drought are areas that deserve attention. Rozeboom et al. (1990) isolated and purified hevamine, an enzyme (molecular weight 29,000) with both lysozyme and chitinase activity from *Hevea brasiliensis* latex, which was homologous to certain pathogenesis-related proteins from plants. Variations in RAPD profiles between *Phytophthora* resistant and susceptible genotypes were reported (Jacob, 1996). Thulaseedharan et al. (1994) observed variation in the RAPD profile of TPD-resistant and susceptible seedling trees and identified two random polymorphic DNA markers tolerant to tapping-panel dryness. Molecular approaches for identification and incorporation of resistance to TPD assume much importance since many high yielders are susceptible to this syndrome, which has not yet proven to be a disease.

Crop Management Under Different Cropping Systems

Rubber is grown under a wide range of management conditions. Management practices in rubber have been modified and improved over the years to suit the different cropping systems adopted. In the Amazonian forests of South America, where wild rubber was exploited for latex, field management was limited to maintaining access paths to the highly scattered trees in the jungle. Here the trees after they reach tappable size are exploited for 20–30 years until the bark reserves are exhausted. Intercropping in small holdings raised in newly cleared areas was a very old practice. Food crops, such as rice, maize, cassava, bananas, and other crops, were raised for the first 2–3 years. The management practices were kept at a minimum, with hand slashing to control weed growth (Watson, 1989). In most of the private sector plantings, and in some government-funded small holder plantings and large development schemes, rubber is usually raised as a monocrop. Here, sensitive management practices are used, as the main objective is to get maximum economic returns from the plantations by reducing the immaturity phase and maximizing the yield during the mature phase. Over the years, the cropping cycle has been reduced with the existing plantations being replanted with more high-yielding clones as they become available.

Crop Management Practices

Rubber trees face more competition in the immature phase than in the mature phase. As rubber is usually grown in plantations in association with ground cover, and with intercropping becoming a common practice, in the modern cropping system, proper field upkeep is essential during this phase so as to provide the best environment for the young rubber plants to survive the competition. Important management practices during this phase include establishment of ground cover, intercropping, weed management, mulching, induction of branching, pruning, and thinning. These practices are meant to reduce the immaturity period and maintain optimum plant density (Punnoose et al., 2000).

Cover Crop Establishment

Leguminous creepers were found to be the best cover crops as they were superior in terms of high nitrogen fixing capacity, fast growth, tolerance to shade, drought, pests, and diseases, and provided minimum competition for nutrients with rubber plants. Leguminous cover crops reduce the cost of cultivation by substantially reducing the amount of costly nitrogenous fertilizers, which are supplemented by the cover crop. They reduce soil erosion to a great extent from hilly rubber-growing tracts in addition to reducing expenses on weeding, increasing soil moisture, and organic matter. Some of the important leguminous cover crops grown in rubber plantations include *Pueraria phaseoloides*, *Mucuna bracteata*, *Calapogonium muconoides*, and *Centrosema pubescens*.

Weed Management

Traditionally, weed control in rubber plantations has been via hand labor by slashing the planting strips in the first 4 years until the canopy closes. Application of herbicides for weed control in rubber plantations is limited to mostly dry seasons. The two main types of herbicides are pre-emergent (applied to soil) and post-emergent (applied on the weeds). Some of the common herbicides used in rubber plantations are paraquat (Gramoxone), 2,4–8 (Fernoxone), which controls broad-leaved weeds, and Glyphosate (Glycel, Roundup, or Weedoff), which controls grass weeds. To achieve weed management most economically and efficiently, an integrated approach is practiced in which a combination of manual and chemical controls are used along with the establishment of cover crops.

Intercropping

In general, rubber will best grow in a monoculture, with interrow areas protected by leguminous cover crops during the immaturity period. On the other hand, the land in such a system is underutilized with regard to the interrow space. In the first 2–3 years after planting, before the tree canopy closes, it is possible to cultivate a variety of suitable crops in the interspaces available in the young plantations. This brings income to the grower during the immaturity period of 6–7 years of the rubber trees. When properly done, intercropping has been reported to enhance the growth of rubber (Jessy et al., 1996). Legume cover can also be incorporated in the intercropping system by growing legumes in the alternate interrows or available interspaces.

Intercrops should not be planted too close to the young rubber plants in order to minimize competition for nutrients. Since intercropping necessitates soil tilling at different degrees, it is desirable to restrict the practice to level lands and gentle slopes so as to avoid soil erosion (Punnoose et al., 2000). Commonly cultivated intercrops are pineapple, banana, ginger, turmeric, vegetables, coffee, some medicinal plants, and fodder crops. Yogaratnam et al. (1995) observed intercropping of tea with rubber to be successful in Sri Lanka.

Irrigation

Irrigation in rubber is usually restricted to nontraditional rubber-growing areas, where drought is more pronounced and the plants are under severe water stress. Irrigation during the summer can enhance growth and reduce the immaturity period (Jessy et al., 1994; Pushparajah and Haridas, 1977). Irrigation in rubber nurseries is also practiced in traditional areas for good growth of the seedlings and poly bag plants.

In the traditional areas, summer irrigation during the young phase can reduce the immaturity period by 6 months to 1 year (Jessy et al., 1994). Philip (1997) reported the beneficial effect of irrigation in reducing the effect of cold weather in north-eastern parts of India, where the winter temperature can drop to −10°C. In some drought prone areas or nontraditional rubber-growing areas in India, irrigation up to 50 percent of the estimated crop water requirement could reduce the immature period from 10 to 6 years (Vijayakumar et al., 1998).

Ideal Soils for Rubber and Manuring Schedule

Rubber grows on a wide variety of soils, mostly acidic, of the humid tropics. Deep, well-drained soils with a pH below 6.5 and free from underlying sheet rocks are well suited for rubber. In the traditional rubber tract in India, the soils are mostly laterite and lateritic types; red and alluvial soils are also seen in some areas. In general, the laterite and lateritic soils are very porous, well drained, moderate to highly acidic, deficient in available P, and variable with regard to available K and Mg. However, adverse physical, chemical, and physiographic features of soils influence growth and productivity of rubber to a considerable extent (Krishnakumar and Potty, 1992).

Rubber plants have been found to respond well to systematic manuring. Proper manuring during the immature stage accelerates growth, thereby reducing the unproductive phase and optimizing productivity in the mature phase. The major nutrients N, P, K, and Mg have positive effects on growth and yield, and application of NPK increases the yield substantially (George, 1962). Fertilizer recommendations have been formulated for different regions based on soil types. A system for routine discriminatory fertilizer recommendation based on soil and leaf analysis has been set up in the different rubber-growing regions of India. This facility has the advantage of avoiding unnecessary use of fertilizers resulting in monetary benefit and reduced environmental pollution. A diagnosis and recommendation integrated system can be used to improve diagnosis and fertilizer recommendation (DFR).

Diseases and Pests of Rubber

The climatic conditions and cultural practices followed in cultivation in a region decide the susceptibility of the rubber plant to various diseases. Significant leaf diseases include abnormal leaf fall (ALF) caused by the common fungus *Phytophthora* sp. (Jacob, 1996) or secondary leaf fall (SLF) caused by *Oidium heveae*, *Colletotrichum gloeosporioides*, and *Corynespora cassicola.*

Until recently, ALF disease caused by *Phytophthora* sp. and powdery mildew caused by *Oidium heveae* were the two major leaf fall diseases of rubber in India. However, the threat of minor diseases caused by such fungus as *Corynespora* has increased significantly. As a result of the severe incidence of *Corynespora* leaf spot disease in Sri Lanka, RRIC 103, one of the most popular high yielders planted extensively, had to be withdrawn from the planting recommendation, necessitating replanting of vast areas under this clone (Liyanage et al., 1991). In India, reports (RRII, 1998) on the incidence of *Corynespora* leaf spot observed in the popular clones RRII 105 and RRIM 600 in the rubber-growing tracts of Karnataka State in South India and certain parts of Kerala State, have caused serious concern over the possibility of the disease attaining epidemic proportions.

South American leaf blight caused by *Microcyclus ulei*, fortunately still confined to the tropical Americas, is the most devastating disease of the rubber plant. However, rubber plantations in the east are also under the potential threat of SALB due to the favorable climatic conditions. Hence, breeding and selection of SALB-resistant clones assumes significance for the Asian countries as well. Available reports indicate that disease resistance in *Hevea brasiliensis* is inherited polygenetically, implying that the possibility exists of obtaining horizontal resistance, which is more stable and durable (Varghese, 1992).

Another significant disease in rubber that causes drying of the main stems is pink disease, caused by the fungus *Corticium salmonicolor*. The disease causes drying of the main stems and branches of 3- to 7-year-old immature plants, particularly in the fork region. Dry rot disease caused by *Ustilina deusta* and patch canker caused by *Phytophthora palmivora* do only occasional damage. Among the panel diseases, black stripe caused by *Phytophthora* sp. cause damage to the tapping panels. Three root diseases observed are white root disease caused by *Rigidoporus lignosus*, brown root disease caused by *Phellinus noxius*, and red root disease caused by *Ganoderma philippii*.

The damage by insect pests is only minimal in rubber. Termites, cockchafers, mites, and thrips are some of the pests found in rubber plantations. Control schedule for infestation by most of the diseases and pests have been developed.

Tapping-Panel Dryness (Brown Bast)

Brown bast or tapping-panel dryness is a syndrome characterized by prolonged dripping of latex with a gradual decline in volume yield, precoagulation latex, and partial or complete drying of the tapping area. In some instances, browning and thickening as well as cracking and deformation of the bark takes place. High-yielding clones are most susceptible to the syndrome, resulting in drying up of the tapping panels in 10–25 percent of the trees. Reduced tapping intensity or tapping rest for 3–12 months are considered curative measures for this syndrome. It is considered to be a stress-related physiological disorder.

Technological Developments

Natural rubber (NR) harvested from the tree as latex accounts for 70–80 percent of the crop, the rest being the field coagulum. Latex can be processed into ribbed smoked sheets (RSS), pale latex crepe (PLC), technically specified rubber (TSR), speciality rubbers, preserved field latex, and latex concentrate, whereas field coagulum is processed and marketed as either block or crepe rubber. Conversion of fresh latex into ribbed smoked sheets is the oldest method of processing and has been widely adopted, especially by small growers, due to its simplicity and low cost. At present, sheet rubber accounts for around 75 percent in the Indian NR market (Kuriakose and Thomas, 2000).

The need for production of technically specified rubber (TSR) arose when the traditional visually graded sheet and crepe rubbers were not considered adequate enough to face stiff competition from the synthetic rubbers. Consequently, new methods of processing were developed to produce and market TSR in compact medium sized blocks, wrapped in polyethylene, and graded adopting technical standards. The implementation of new manufacturing practices and international standards has resulted in an increased use of TSR compared to other marketable forms, which on a global scale accounts for around 50 percent of the NR processed (George et al., 2000). In block rubber processing, the coagulum is made into crepe by passing it through macerators, granulating it in hammer mills, and pressing it into blocks.

Concentrated latex is produced by various processes, namely evaporation, electrodecantation, creaming, and centrifugation, of which centrifugation has been developed on a large scale and is currently the most widely accepted method accounting for more than 90 percent of the concentrated latex produced (Mathew and Claramma, 2000). Since a number of products, such as foam, elastic threads, carpet backings, adhesives, gloves, balloons, rubber bands, and other dipped goods, are made directly from latex, latex preservation with appropriate chemicals for long-term storage and concentrations are essential. Although ammonia is being used conventionally as an ideal preservative, low ammonia preservation system have been developed to rectify some of its inherent defects.

Modified Forms of Rubber

Latex harvested from the tree contains around 94 percent hydrocarbon and nonrubber substances, such as proteins, fats, fatty acids, and carbohydrates, which influence the chemical and physical properties of the hydrocarbon polymer. Studies on physical and chemical modifications of NR have resulted in several modified forms suitable for specific processes and applications. These include viscosity-stabilized rubber, thermoplastic NR, oil-extended natural rubber (OENR), epoxidized NR (EONR), graft polymers (GP), chlorinated rubber (CR), liquid natural rubber (LNR), and so on.

An increase in viscosity of NR during primary processing and subsequent storage under ambient conditions leads to storage hardening. Therefore, constant viscosity (CV) rubbers have been developed by adding small amounts of chemicals, such as hydroxylamine neutral sulfate, which help to preserve the original Mooney viscosity for a long time. The controlled and stable Mooney viscosity provides easy and uniform processing and also minimizes the premastication time, resulting in energy saving. In comparison with NR vulcanizates, thermoplastic natural rubber (TPNR) blends are remarkably resistant to heat aging and ozone (Elliot, 1982). Chlorinated rubber with 65 percent chlorine is highly resistant to chemicals and is used in anti-corrosive and heat-resistant paints, coatings, adhesives, printing inks, textile finishers, and so on. Epoxidized natural rubber (EONR), a chemically modified form, has qualities such as improved resistance to hydrocarbon oils and low air permeability, while retaining the high-strength properties of NR (Baker et al., 1985).

Oil-extended natural rubber has reduced tensile strength and resilience, but has a good resistance to tear and wear when blended with polybutadiene rubber. OENR has definite advantages when used in the manufacture of tire treads, which have better grip on snow and ice and have gained considerable acceptance in European countries. Deproteinized natural rubber (DPNR), a highly purified form of NR, is suitable for electrical engineering applications (for reduced creep and stress relaxations, superior dynamic properties, and consistency to stiffness).

Natural rubber can be modified chemically to graft copolymers by polymerizing vinyl monomers, such as methyl methacrylate, styrene, and acrylonitrile (Claramma et al., 1984). Methyl methacrylate (MMA)-grafted NR and styrene-grafted NR are two graft copolymers developed. The major use of the former is in adhesives for the shoe and tire industries, whereas the latter finds application in microcellular solings in place of the high styrene resin grade of solid black rubber (SBR).

Extensive size reduction of molecular chains of NR by depolymerization leads to the formation of liquid natural rubber. This serves as a reactive plasticizer in compounding and also as a substitute to synthetic liquid elastomers, such as fluid silicones used in elastic molds for various industrial and art work (George et al., 2000). A wide range of engineering applications of rubber have been developed in almost every industrial sector, namely engine and suspension mounts, bridge bearings, earthquake protection of buildings, and many other uses.

Discovery of Vulcanization

Credit of the historic discovery of vulcanization, a process of heating rubber with sulfur at a high temperature to improve its strength properties, goes to Charles Goodyear (1800–1860), an American rubber manufacturer. All articles of rubber were observed to become sticky in hot weather and brittle in cold weather. Goodyear worked on the problem for making rubber more stable and less susceptible to heat, cold, and light. The fundamental changes in the properties of NR through vulcanization removed most of its susceptibilities to climatic conditions and its limitations as a raw material for mechanical applications.

Natural rubber is basically a high molecular weight polymer, *cis*-1,4-polyisoprene with viscoelastic properties. The elasticity of rubber depends on the predominant *cis* configuration of the polymer. In vulcanization, the *cis*-isoprene units from intermolecular crosslinks with sulfur, makes it hard, resistant to abrasion, heat, light, and oils, while retaining its unique elasticity, which makes rubber useful in manufacturing a variety of products.

Product Development

The different forms of natural rubber and its chemical modifications are being used in the manufacture of thousands of products. The major share of the dry forms of natural rubber finds applications in the tire sector. Other products include belts and hoses, footwear, and molded items, such as mountings, bushes, seals, mats, and pharmaceutical closures. Concentrated latex is the raw material used in the production of foam, condoms, gloves, latex threads, and adhesives.

Technology has been developed for the rubberization of roads, which is becoming popular. Rubberized roads are more durable compared to asphalt roads, and the service life increases by 50 percent or even more. Moreover, the surface bears heavy traffic, provides skid resistance, and withstands both extreme cold and hot weather conditions. The additional cost for seal coats using 2 percent rubber in asphalt is 12–15 percent, whereas the additional cost for rolled asphalt with 4 percent rubber is 16 percent (Haridasan and Gopalakrishnan, 1980; Gopalakrishnan,1994).

Ancillary Products

The major ancillary products from rubber plantation, especially in the mature phase, are rubber honey, seeds, and rubber wood. Commercial exploitation of these ancillary products adds to the efforts to maximize returns from rubber, especially in the case of small farmers.

Rubber Honey

Rubber plantations have been identified as an important source of honey. During the refoliation and flowering period, honey bees collect large quantities of nectar from the extrafloral nectarines in the leaflets. Lack of honey flow in the rest of the year in rubber plantations is managed by alternate bee flora.

There has been a drastic reduction in the production of rubber honey in India from a peak of 2750 tons in 1990–91 to an all-time low of 550 tons in 1993–94, due to the outbreak of sacbrood disease in bees (George and Joseph, 1994). The rehabilitation measures by the introduction of *Apis mellifera* have resulted in the rise of honey production to 1500 tons in 1997–98 (RRII, 1998). *Apis mellifera* had a reported average yield of 60 kg/hive/year compared to just 19.46 kg/hive/year for the popular Indian honey bee, *Apis indica* (Haridasan et al., 1987). On an average, 15–20 *Apis indica* hives can be placed per hectare, and the results of a recent survey showed an average yield of 12.1 kg/hive/year for the Indian honey bee (Chandy et al., 1998). Mature plantations in India, thus, have the potential to produce 67,886 tons of rubber honey annually, though only 2 percent of this potential was exploited in 1996–97 (George et al., 2000).

Rubber Seed Oil

A rubber seed has an average weight of 3–5 g, of which about 40 percent is kernel, 35 percent shell, and 25 percent moisture. Oil content in the kernel ranges from

35 to 38 percent, and the recovery rate of seed cake is in the range of 57–62 percent. The two major products processed from rubber seeds are rubber oil and rubber seed cake. In India, the seeds mature between July and September. The estimated production potential is about 150 kg/ha production of rubber seed oil and cake in India for 1997–98 was 2890 and 4710 tons, respectively (RRII, 1998). Among the three methods of extracting rubber seed oil, solvent, expeller, and rotary extraction, the rotary extraction method is commonly used. The recovery of oil and cake depends on the quality of the kernel, the extent of drying, and quantity of molasses used for processing (George et al., 2000).

Rubber seed oil is used in the soap manufacture industry (Hardjosuwito and Hoesnan, 1976). Epoxidized rubber seed oil is used in the formulation of anticorrosive coatings and polymer additives, and in alkyl resin casting (Vijayagopalan and Gopalakrishnan, 1971). It also serves as a substitute to linseed oil in the paint industry. Rubber seed cake is rich in protein and is a source for cattle and poultry feeds (Amritkumar et al., 1985).

Rubber Wood

One of the recent developments is that rubber plantations have become a major source of industrial timber in the rubber-growing countries of South and Southeast Asia. Since the 1980s, rubber wood has been internationally accepted as an eco-friendly source of timber for manufacture of household furniture. The rubber wood has excellent physical properties; because of this, rubber wood has become an important source of raw material for the manufacture of panel products, such as particle board, block board, medium-density fiber board, and more (Yusof, 1994). The current size of the world market for rubber wood-based furniture and allied products is to the tune of US$1.5 billion globally (George et al., 2000). The current estimated average production of rubber wood worldwide is 150–180 m^3/ha in the small holding and estate sectors, respectively. The annual gross availability of rubber wood was 1.27 million m^3 during 1997–98, and the projected estimate for 2010 is 4.24 million m^3 (George and Joseph, 1994).

Rubber wood is a light to moderately heavy timber with medium density (515 kg/m^3), low shrinkage, straight grains, and attractive color, which makes it suitable for furniture manufacture (Reghu, 1998). The structural and anatomical features of rubber wood resemble those of other hardwood species with many physical and mechanical properties comparable to that of teak (George et al., 2000). However, tension wood formation is a natural defect causing a variety of drying, woodworking, and finishing problems. The distribution pattern of tension wood was reported by Reghu et al. (1989).

Since rubber wood is sensitive to biodeterioration, by various biological processes, processing has to be done very meticulously, and it is done in two stages: preservative impregnation and drying. Preservative treatments are aimed at either short-term and/or temporary protection or long-term and/or permanent protection. For temporary protection, the timber is dipped in wood preservatives, such as insecticides and fungicides. For long-term protection, the wood preservatives are allowed to penetrate deep into the timber by providing maximum dry salt retention and complete preservative penetration by the dip diffusion and pressure impregnation processes.

In dip diffusion, freshly sawed timber is immersed for an adequate period in a mixture of boric acid and borax in water (Gnanaharan and Mathew, 1982). In pressure impregnation, preservatives are impregnated into the wood by creating a vacuum under pressure and dried to optimum moisture levels in the timber to avoid dimensional variations in the end product. Development of latex timber clones (LTC) aimed at higher latex and timber yield to ensure economic viability of NR cultivation is gaining significance. Identification of genotypes with superior qualitative and quantitative timber characters is a priority, and work has already been initiated in this direction in different rubber-growing countries, including India (Reghu, 1998). In Malaysia, 20 fast-growing genotypes were identified as potential timber clones with a total wood volume ranging from 1.4 to 2.52 m^3/tree (RRIM, 1996).

The "Ecofriendliness" of Rubber

Rubber has turned out to be, after discovery and domestication dating back to more than a century, an ecofriendly and ecologically viable cash crop. Natural rubber, a deciduous forest tree, helps restoration of denuded and depleted forest lands. With a stand of 450 plants/ha, and a canopy that closes in less than 5 years, *Hevea* is a good candidate for afforestation of marginal and denuded soils. Rubber as an ecologically friendly crop contributes primarily to soil conservation, soil fertility, biomass generation, nutrient cycling, and as an alternate source to traditional timber. The last property is because rubber acts as a substitute and a viable option against deforestation, where many of the developing countries lose much timber from forest felling, leading to climate imbalance in addition to soil degradation. Western Ghats in the southern state of Kerala in India, and Amazon forests are classical examples of this environmental degradation, because of loss of much forest cover.

The cultivation of rubber involves measures to protect soil from erosion and preserve its fertility. The planting of rubber in hilly terrain in inwardly sloping contoured terraces and digging of silt pits in the terraces reduce soil erosion. Cultivation of leguminous crops helps to prevent soil erosion. Because rubber is a surface feeder, its root system helps to bind soil.

The recommended agromanagement practices in rubber are aimed at improving soil properties. Soil fertility is sustained through regular application of N, P, K, and Mg fertilizers, during mature and immature phases. The fertilizer inputs are optimized by discriminatory fertilizer recommendations to supply only the required quantity to the soil. N-fixing leguminous cover crops, such as *Crotalaria juncea*, when sown in between rubber plants improve soil fertility vastly. When turned in, the plants build up soil organic matter. The beneficial effects of these build up include enhancement of soil fertility, improvement in the soil physical properties, water holding capacity, and better water infiltration.

Mature rubber requires comparatively less fertilizer input to sustain high productivity, when compared to annual crops. The quantities of nutrients removed from the ecosystem by a latex crop is relatively small, when compared to other crops such as coconut or tea (Samarappuli, 1996).

Another feature of the rubber tree is that it is a potential solar energy harvester with an annual biomass production potential of around 35 tons of dry matter/ha/year, which is comparable to any fast-growing tropical forest tree species (Sethuraj, 1996). The biomass production potential of a plant species is related to its photosynthetic capacity/unit leaf area and the total leaf area produced by individual plant (Jacob, 2000). The high biomass production potential of *Hevea* makes it an excellent candidate for energy plantation and as a source of both timber and biofuel wood production. The Rubber Board of India, situated in Kerala State, has embarked on an ambitious research project on rubber as a potential climate control tree species.

The biomass growth, microflora, and understory vegetation in rubber plantations is comparable to that in teak plantations (Krishnakumar et al., 1991), indicating the ecological desirability of rubber in terms of habitat diversity, soil physical properties, and nutrient recycling. Considering the energy requirements, energy for the production of NR is derived mostly from sunlight through photosynthesis, whereas nonrenewable oil sources are used for the cost of production of synthetic rubber, which is about 11 times higher than that for NR (Mathew, 1996). Thus, NR has enormous potential in contributing to the cause of solving the energy crisis. Similarly, as an alternative timber source, wood from natural rubber has become the main nonforest timber resource, decreasing the logging pressure on natural forests and teak plantations in rubber-growing countries.

Useful Information on Research and Development

Organizations in Rubber Production and Use

Secretary General
Association of Natural Rubber-Producing Countries
7th Floor Bangunan Getah Asli (Menara)
148 Jalan Ampang, Kuala Lumpur 50450, Malaysia
Tel.: +60-3-2611-900; Fax: +60-3-2613-014

International Rubber Research and Development Board
PO Box 10150
Kuala Lumpur 50908, Malaysia

Executive Director
International Natural Rubber Organization
PO Box 10374, Kuala Lumpur 50712, Malaysia
Tel.: +60-3-248-6466; Fax: +60-3-248-6485
Email: inro@po.jaring.my

Director General
Malaysian Rubber Board
PO Box 10150, Kuala Lumpur 50808, Malaysia
Tel.: +60-3-456-7033; Fax: +60-3-457-4512
Email: igm.gov.my

Secretary General
International Rubber Study Group 8th Floor, York House, Empire Way
Wembley, Middlesex HA9 OPA, UK
Tel.: +44-181-903-7727; Fax: +44-181-903-2848
Email: irsg@compuserve.com

Director of Research
Tun Abdul Razak Research Centre
MRPRA, Brickendonbury
Hertford, SG 13 8 NL, UK
Tel.: +44-1992-584966; Fax: +44-1992-554837
Email: general@tarrc.tcxom.uk

Responsible du Programme hevea
Centre de Cooperation International en Recherché
GRONOMIQUE POUR LE DEVELOPMENT
Department des plantes perennes, programme hevea
2477, Avenue d'Agropolis, BP 5035
3402 Montpellier cedex I, France
Tel.: +33-4-67-61-71-55; Fax: +33-4-67-61-71-20
Email: banchi@cirad.fr

Professor President
Chinese Academy of Tropical Agricultural sciences
Baodao Xincun, Danzhou
Hainan 571 737, China
Tel.: +86-890-330-0157; Fax: +86-890-330-0157-0776
email: rcri@public.dzptt.hacn

Director
Shanghai Rubber Products Research Institute
381 Fanyau Road
Shanghai 200052, China
Tel.: +86-21-62523050; Fax: +86-21-62512948

Director
Rubber Research Institute, Department of Agriculture
Chatuchak, Bangkok 109900, Thailand
Tel.: +66-2-579-7557/8; Fax: +66-2-561-4744
Email: rk000051@netaccess.loxinfo.co.th

Director
Rubber Research Institute of Sri Lanka

Dartonfield, Agalawatta, Sri Lanka
Tel.: +94-34-47426/47383/23078; Fax: +94-34-47427
Email: director@rri.ac.lk

Rubber Research Institute of Vietnam
177, Hai Bai Trung Street, Ward 6, District 3
Ho Chi Minh City, Vietnam
Tel.: +848294139; Fax: +848298599

Executive Head
Institut de Recherches sur la Caoutchouc au Cambodge
Numero 48, Rue Neary Khlahan
Phnom Penh, Cambodia
Tel.: +855-23-27498; Fax: +855-23-27498

Executive Head
Centre de Cooperation Internationale en Recherché Agronomique pour le Development
Department des cultures perennes
2477, Avenue du Val de Montferrand, BP 5035
34032 Montpellier Cedex I, France
Tel.: +33-4-67-61-58-00; Fax: +33-4-67-61-56-59
Email: devernou@cirad.fr

Director
Indonesian Rubber Research Institute
Sungei Putih, PO Box 1415 Medan
Sumatera Utara 20001, Indonesia
Tel.: +62-61-958-045; Fax: +62-61-958-046
Email: rubbersp@indosat.net.id

Head of Station
Bogor Research Station for Rubber Technology
BRTK Bogor, Jalan Salak No. 1, Bogor
Jawa Barat 16151, Indonesia
Tel.: +62-251-319-817/324-048; Fax: +62-251-324-047
Email: bptkbgr@indonet.id

Executive Head
Indonesian Planter's Association for Research and Development
Jalan Tanjung, Karang No. 5
Jakarta 10230, Indonesia
Tel.: +62-21-324-473/325-478; Fax: +62-21-325-517

Director
Rubber Research Institute of India, Rubber Board
Kottayam-686009, Kerala State, India
Tel.: +91-481-353311; Fax: +91-481-353327
Email: rrii@vsnl.com

Chairman
Rubber Board
Kottayam-686002, Kerala State, India
Tel.: +91-481-571231; Fax: +91-481-571380
Email: rbktm@ker.nic.in

Executive Head
Rubber Technology Centre
Indian Institute of Technology
Kharagpur, West Bengal State, 721 302, India
Tel.: +91-3222-55221; Fax: +91-3222-55303

Director
Rubber Research Institute of Nigeria
PLMP 1049, Benin City, Bendel City
Bendel State, Nigeria
Tel.: +234-52-60-18-85; Fax: +234-52-60-18-85

Chief of Programme
National Rubber Research Programme
Institute of Agricultural Research for Development
PMP 25, Buea, Cameroon
Tel.: +237-32-2221; Fax: +237-32-2244

Amazon Agroforestry Research Centre
Km 28 da Rodovia AM-010
Manaus, Amazonas CP 31969011970, Brazil
Tel.: +55-926-222012; Fax: +55-926-221100

Director General
National Research Center for Genetic Resources and Biotechnology
SAIN-Parque Rural—CP 02372, Brasilia DF
CEP 70, 770-849, Brazil
Tel.: +55-61-340-3500/3501; Fax: +55-61-340-3502/3624
Email: valois@cenargen.embrapa.br

Director
Center of Research and Technological Development for the Rubber Industry
Av. General Paz entre Albarellos y Constituyentes
San Martin, Pcia. De Buenos Aires, C.P. 1650, Argentina
Tel.: +54-1-754-4141/45 or 5151/55; Fax: +54-1-753-5781
Email: Irehak@inti.gov.ar

Executive Head
Rubber Institute of Rosario
Argentina
Tel.: +54-41-587110; Fax: +54-41-264020

A Look into the Future of Rubber

The rubber plantation industry gained much through the research and development activities of the various rubber-growing countries in NR production. The industry became science-based and economically viable. Conventional breeding has brought about a tenfold increase in the productivity of rubber trees within a short period of time. Since fresh germplasm has been introduced to broaden the original narrow gene pool, incorporation of desired variability for yield and yield components is expected to bring in further genetic improvement. With pressure on expanding rubber cultivation to nontraditional rubber-growing areas in all rubber-producing countries, development of location-specific clones capable of withstanding various stress situations is a major priority. In this context, identification of molecular markers linked to specific plant traits, such as resistance to pests and diseases, drought, cold, and so on, are important. Marker-assisted selection will play a significant role in the selection of the desired variability in the future thereby enhancing the yield stability of high-yielding cultivars.

The tapping-panel dryness syndrome is an elusive physiological disorder, and hence, the search for its causes continues to be an important area of research. Basic physiobiochemical investigations on latex flow, environmental and stress physiology, rubber biosynthesis and biochemistry, and molecular biology of isoprene production are priority areas. Systems, such as controlled upward tapping, have been perfected for long-term exploitation of high panels. Low-frequency tapping systems are also in demand in view of increasing worldwide labor shortage, especially in developing countries like Indonesia and Malaysia, major producers of rubber in Asia. Optimization of exploitation by latex diagnosis is being perfected. Nutritional aspects and fertilizer recommendation systems based on soil nutrient analysis have been well established. Research on cropping systems, soil and water conservation, integrated nutrient and weed management, and so on are in progress. Effective control measures have been achieved for most of the serious diseases in the rubber-growing countries. However, identification of sources of durable resistance is the current priority in this area, and the wild germplasm will be a rich source of many valuable genes for disease resistance. Although SALB, the devastating disease of *Hevea brasiliensis*, is still confined to South America, sources of resistance must be located due to the potential threat of the disease in the Asian rubber-producing countries.

Though rubber production is the main focus, in the future, emphasis needs to be placed on NR latex, optimum exploitation of the ancillary products, like rubber timber products and more. Rubber wood is assuming considerable importance for various industrial applications. Considering the inherent natural defects of rubber wood, technology has to be perfected to increase the utilization of rubber timber in a more cost-effective manner. In this context, selection and development of timber latex clones is gaining importance as a breeding objective. Desirable quantitative and qualitative traits of timber should be identified from the available germplasm. The selections could be used directly for raising rubber forests or incorporated in breeding programs for developing latex timber clones. Being a deciduous forest tree, *Hevea brasiliensis* is basically an environment-friendly tree. However, more data need to be generated on the environmental impact of NR cultivation on the physical, chemical, and biological properties of rubber soils as well as water consumption by rubber plantations.

In fact, many milestones have been attained in rubber processing and rubber technology. Viscosity-stabilized rubber, thermoplastic natural rubber, chlorinated rubber, and liquid rubber are some of the modified forms of rubber that have already been produced commercially, the demand for which depends on the emerging market. In latex technology, the priorities include preservation of latex, prevulcanization of latex, radiation vulcanization, deproteinization, and so on. Since the global production of NR has almost reached a plateau due to a shortage of land and other reasons, it is quite likely that the global demand for NR in the next century will be higher than global production. Technologies to recycle rubber from used waste products (reclamation) and attempts to enhance the service life of conventional rubber products could lead to conservation of the polymer in the future.

On account of supply inelasticity, rubber market has widely fluctuated over the years, especially in the short run, like in the case of other perennial crops (pepper, for example). A major issue debated in the global NR market is the steep decrease in rubber prices that has occurred since 1996, following a record peak in 1995. As per projections of the World Bank, the NR price is expected to increase at a declining rate until 2010. The forecast is for a rise of 2.4 percent between 2006 and 2010, while earlier increases were 5.7 percent in 2002 followed by a 3.6 percent increase between 2003 and 2005. A recent forecast by the Economics Intelligence Unit of the International Rubber Study Group indicated NR consumption to outpace production and prices bound to shoot up substantially, which turned out to be true. The State of Kerala, where NR production has been quite significant, has benefited from this bounty, and many farmers are switching from other perennial crops like black pepper and coconut to rubber. The growth rate of NR consumption is forecast to be 4 percent, compared to 2.7 percent NR production for the year 2002. In absolute figures, the deficit increased from 80,000 to 355,000 tons in 2002. However, the IRSG forecast a significant and steady increase in NR production until 2005, with an increase in total world production from 6,750,000 tons in 1999 to 8,331,000 tons in 2005. In general, the projections indicate that the world elastomer industry can look forward to a relatively healthy growth rate in supply and demand for the rest of the first decade of twenty-first century.

Globalization provides an excellent opportunity for rubber producers to diversify the use of rubber into many end products, besides the principal rubber. However, the products must meet the customer quality requirements. And import substitution is a must. The R&D institutions and rubber-based industries across the world are gearing up to meet the challenges, which are nothing but opportunities. Whether they would meet them with success is something only time can tell.

9 Tea (*Camellia sinensis* L.)

Tea is mostly grown as a monocrop, as it has tremendous diversification, adapts to different conditions, and has varying customer requirements. Its various uses, through planting and processing techniques, make a wide array of teas and tea products. Through strategies for trading in a range of international markets to the varying preferences of the customers' palates, tea has become an enduring agricultural crop. From the earliest days of its cultivation, the requirements and convenience of trade and palate have driven the form and style of the tea crop and that of tea made from it. The old marketing cliche "the customer is king," was literally true only when princes and wealthy men and women could afford tea. Even now when tea is referred to as the "cheapest drink, next to water," the export trade in tea is still a buyers' market.

The Origin and History of Tea

Legend has it that tea originated by infusion to leaves from a tree falling by accident into drinking water intended for the Chinese Emperor, Shen Nung, in 2737 BC. This serendipitous discovery led to the monumental scale of consumption today, nearly 5000 years later, of an estimated 3.5 billion cups of tea consumed each day worldwide.

The plantation crop cultivated as tea (a name derived from the Chinese ideograph *ch'a*, first used about 725 AD), belongs to the genus *Camellia* (family *Theaceae*), which also contains nearly 90 species of wild and ornamental nontea forms. The genus appears to be native to the Southeast Asian landmass, and the "true" tea species, namely *Camellia sinensis* (L.) O. Kuntze, is native to Southwestern China from where it spread to central China and southern Japan 1000–2000 years ago.

A number of *herbal teas* are marketed around the world, but only the terminal shoots of *Camellia sinensis* are used to manufacture the ubiquitous and singular product, the "*tea sensu strictu*," which is preferred by a huge global market. In 1999, the world tea production was 2.8 billion kg. In the 1830s, tea plants for commercial cultivation were imported into Assam in northeast India and into Sri Lanka from China. Although indigenous, wild tea was already present in Assam, having been discovered in Upper Assam in 1823. Assamese tea seeds were introduced to the nurseries on the Sri Lankan islands by the British in the 1840s, and the first viable tea plantation was established there in 1876. In the 1820s, tea from Japan was planted in Java, which is now a part of Indonesia, by the Dutch.

The Agronomy and Economy of Important Tree Crops of the Developing World. DOI: 10.1016/B978-0-12-384677-8.00009-6
Copyright © 2010 by Elsevier Inc. All rights of reproduction in any form reserved.

After the discovery of Assam wild tea, bioprospecting in the adjacent areas of Assam and Myanmar led to the discovery of other germplasm, which gave rise to the tea stocks now used for commercial cultivation in the older and newer tea-producing countries of Asia, Africa, South America, and Australia. Now that a plateau has been reached in the yield of existing cultivars in major tea-producing countries, it seems time to try to resume exploration in other likely geographical locations, for example, in China and Myanmar, for new tea germplasm, such as natural polyploids, which may be more commercially productive and have better resistance to biotic and abiotic stresses. The official exchange of established, commercially grown germplasm between tea-growing countries is now difficult or impossible owing to governmental or industry restrictions.

The Different Practices of Tea Consumption

Ancient Chinese literature records the earliest tea cultivation and consumption dating back to 1100 BC. As it is still consumed today, tea was taken as an infusion after the leaves had been brewed in hot water. Tea consumption evolved into a part of Chinese religious symbolism and culture. In fact, the first monograph on tea, written about 780 AD, was called "*Ch'a Ching*," translated as the "Tea Scripture." This continues even today in the elegant and ornate "*tea ceremonies*" of far eastern countries. Green tea was brought to Japan in the thirteenth century by Zen Buddhist monks returning from China. *Chado*, meaning the "Way of Tea" (the discipline of tea preparation and consumption), arose in the fifteenth century. These rituals, still extant, have distinct reverential and aesthetic connotations.

The British Empire established formalized social and domestic tea rituals in the mother country and its outposts. Not the least of these is the *tea break* in the stately flow of the British game of cricket, now continued with aplomb in the former colonies. These far eastern and British rituals have survived and transmuted into the widespread, modern habit of drinking tea, either hot or cold, with or without milk and sugar, as a pleasant, relaxing, and beneficial beverage.

The Botany of Tea

The tea plant grows into a small- to medium-sized tree. The Chinese tree is small or shrublike with small, erect, dark-green leaves and single flowers. The Assam tree is taller with broad, light-green to yellowish leaves that droop at their ends and that have clustered flowers. The Cambodian type has oval, semierect leaves. The natural plants derived from seed (*seedling tea*) have a main stem and branches that grow from the leaf axils. Four to seven foliage leaves grow alternately, above two scale leaves, after which they become dormant. When the plant is under stress, resulting from low ambient temperature or lack of water, dormancy may ensue after the production of only two to three leaves. The small first scale leaf is shed; the second larger scale leaf either expands or becomes dormant, a so-called *banji* shoot. The crop of plucked shoots (the *flush*) consists of both actively growing and banji shoots. Desirable flush for the manufacture of tea is composed of the bud, the intervening portion of stem, and either two or three of the uppermost foliage leaves.

Seedling tea has deep anchoring roots and lateral roots, which produce a surface mat of feeder roots. The roots store starch as a food reserve. When cultivated, the plant is pruned to a low bush to obtain continuous vegetative growth, and its leaf and shoot morphology and growth habit, although not completely reliable for the purpose, has been used for categorization of bush types and varieties. The identification of tea bushes by their vegetative characteristics, which are plastic and show continuous variation, is therefore more or less subjective and not as reliable as biochemical or molecular characterization, although broad phenotypic characterizations useful for breeding programs have been achieved.

The reproductive characteristics in the flowers are less variable than the vegetative characteristics and are therefore taxonomically reliable. The flowers have a persistent calyx and usually have 5 sepals and petals, 100–300 stamens, and an ovary with 3 carpels. The styles and stigma are used for classification. Being almost self-sterile, tea flowers are usually crosspollinated by insects, and the flowers then develop into three-lobed capsules, each of which when mature produces one or two hard, spherical seeds. Their cotyledons are rich in oil.

Taxonomy and Genetics

Pragmatic and commercial imperatives to acquire the best cultivars for yield and quality have driven research in this sphere. Because the tender shoots are harvested for processing, only the biotypes and hybrids that possess the chemical compounds desirable in made tea are of economic interest. For this reason, hardly any extensive knowledge is available, based on modern techniques, on the evolutionary interrelationships of the taxa. Banding patterns that reveal isozyme polymorphisms and, more reliably, DNA isolated from leaves for random fragment length polymorphism (RFLP) and random amplified polymorphic DNA (RAPD) analysis have, however, been used in recent years for fixing interrelationships and constructing dendrograms.

Camellia sinensis is diploid ($2n = 30$ chromosomes), as most other *Camellia* species. This suggests a monophyletic evolutionary origin for the genus. It has also been suggested that the wild species are the result of widespread hybridization (with no real pure lines), in which case the genus could have a polyphyletic origin. Although mostly diploid, polyploidy may occur naturally in tea (up to $2n = 90$ chromosomes) or be induced by the alkaloid colchicines (colchiploidization) and by mutagens. Increased ploidy in plants, above the usual mono and diploid levels, impart vigor and hardiness and, in tea, would also give higher productivity. Rooting, leaf size, and dry weight are better in tri- and tetraploid tea plants.

For practical purposes, three types (or varieties) of cultivated teas are recognized: the China type (the original *Camellia sinensis*), the Assam type (considered a distinct species, *Camellia assamica*), and the Cambod type (considered a subspecies,

Camellia assamica sp. *lasiocalyx*). Tea has a high degree of self sterility and cross-pollination gives vigorous, better quality progeny than the parents. Biotypes cross freely with each other. In commercial seedling plantations, the crop consists of extremely mixed, hybridized populations of the China, Assam, and Cambod types, and perhaps even of *Camellia* species other than *Camellia sinensis*, particularly *Camellia irrawadiensis* from upper Myanmar. Even species such as *Camellia irrawadiensis*, which do not contain the compounds essential in made tea, may be used for hybridization.

Seedling teas therefore have high genetic variability and heterogenous anatomical, morphological, physiological, and biochemical characteristics, ranging in a wide spectrum from the typical Assam to the typical China, with merging and overlapping in between. As a result, many localized varieties of seedling tea have been recognized and given a plethora of ornate, vernacular names in different tea-growing countries.

Growing Conditions

Although primarily tropical and subtropical, the tea plant adapts to wide extremes: in latitude from 42° north (Georgia in the Commonwealth of Independent States) to 42° south (Tasmania, the most southern state of Australia); in elevation, from mean sea level (MSL) up to 2600 m; in different soil types (although well-drained acidic soils with pH 4.5–5.5 are optimal); in temperature, from −8°C to 35°C; and in annual rainfall, from <700 mm to >5000 mm. However, for optimum growth, data from different tea-growing regions indicate that 23–30°C and an annual rainfall of 2500–3000 mm are necessary. Uniform rainfall over the year (minimum 10–150 mm/month) gives better growth than seasonal rains, although seasonal stress conditions (bright sunshiny days, cold nights, and dry, desiccating winds) produce chemicals in the plant such as geraniol, linalool oxides, and methyl salicylate, that give a desirable flavor and aroma (called "seasonal quality") to manufactured tea. Tea is not an effective photosynthesizer, having only the Calvin-Benson or "C-3" cycle. C-3 plants are temperate in origin, and their photosynthetic rates increase when incident light is optimized by shading. Thus, shade trees are grown in tea plantations to attain maximum productivity.

Nutrient Requirements

The soils in which tea is cultivated are, in general, poor in nutrients, particularly in the tropics. Fertilization is therefore crucial and expensive (8–12 percent of the total cost of production).

Primary Nutrients

Of the major or primary plant nutrients (those required in relatively large amounts, namely N, K, and P), N and K leach quickly out of typical tea soils, the process being assisted by high rainfall. However, the low soil pH of tea soils serves to fix applied N and K and increase their uptake by the plant.

Relatively large quantities of N are lost in the harvested shoots and need to be replaced frequently by high rates of fertilizer application, such as urea or ammonium sulfate. The same high rate of replacement of K is not necessary for maintaining yield, although it is controversial whether rates of K applied to the soil are needed to withstand drought and for the development of sturdy bush frame. In contrast to N and K, applied inorganic P is held absorbed with the soil (clay) particles, mainly as Fe and Al phosphates, because of low soil pH. Though earlier unavailable to the tea plant, it may be readily taken up, owing to the presence of acid-producing organic salts (citrate and malate) released by the roots. Applied rock phosphate (calcium-bound P) is a good P source, because it is solubilized both by the acidic conditions near the roots and by application of N fertilizer in the acid—which promotes the ammonium form, rather than in the alkaline nitrate form. Notwithstanding the importance of inorganic NPK fertilizer, crop responses are optimized when NPK use is balanced with that of organic manure.

Secondary Nutrients

Calcium (Ca), magnesium (Mg), and sulfur (S) are secondary nutrients, required in smaller quantities compared to NPK. In tea soils, Ca and Mg are normally deficient because they are generally leached out of the soil by the impact of falling raindrops and running water, and also because of their displacement by applied K. Liming is therefore an essential operation in tea cultivation. Application of dolomitic limestone is particularly useful because it supplies both Ca and Mg in addition to raising soil pH. Sulfur content is also low in tea soils (even if the soils are high in organic carbon content), which means that there has to be supplemental addition.

Micronutrients

Micronutrients, which are needed only in very small amounts, such as zinc (Zn), iron (Fe), manganese (Mn), copper (Cu), boron (B), molybdenum (Mo), and chlorine (Cl), are adequate in most tea soils, with the exception of Zn. Translocation of Zn from the roots is poor, and is therefore applied to the foliage, rather than to soil. The source of Zn is mostly zinc sulfate, which is readily water-soluble and inexpensive.

Nursery Practices

Nursery tea plants are raised from both seeds and stem cuttings under shade. The seeds, after germination in sand beds or boxes, or the cuttings are transferred to the polyethylene bags filled with nematode-free soil (loamy soil in which "rehabilitation" grass has been grown for about 2 years, or forest soil) or soil substitutes. Growth in the nursery is faster under shade or in sealed polyethylene tunnels and when specially formulated nursery fertilizer mixtures are used. Following rooting (in 6–18 months), the bags are placed in planting holes in the field. Under rainfed conditions, the resulting bushes are ready for plucking in about 5 years. With irrigation, the period is considerably reduced. Although cuttings for vegetative propagation may be obtained from bushes at any phase of cultivation, mother bushes that are dedicated for the purpose and which have been spaced, pruned, and fertilized appropriately, are the best source of cuttings. Composite plants may be produced by grafting scions with desirable attributes, such as high-yielding potential, onto hardy (usually drought-resistant or nematode-tolerant) stocks.

Crop Management

Spacing

Cultivated tea is easily tailored to the convenience and requirements of growers. Tea bushes are grown at different planting densities: for the older seedling tea, from 3000 to 14,000 plants/ha; for VP tea, 12,500 to 18,000 plants/ha. Spacing of plants has to take into account such variables as soil, climate, habit, and growth pattern of the cultivars, the requirement for a continuous ground cover, access for different agricultural operations, and bush architecture suited to mechanical harvesting.

Pruning

Periodic pruning or removal of mature foliage is one of the major practices adopted to prevent the tea plant from becoming a tree, instead turning it into a low-spreading bush that is vegetatively vigorous and able to generate a continuous crop of fresh shoots at a convenient height for plucking. Pruning also ensures optimal utilization and productivity of the land available, as well as allowing enough space for other agricultural operations.

Pruning is done either manually or mechanically. The main pruning styles are low (clean or hard), high (cut across or light), and lung. In lung pruning, one or two leafy branches, or "lungs," are left for photosynthetic support of the bush during its post-pruning recovery. Bush debilitation, for example, from wood rot and cankers during dry weather may occur after a low prune, whereas during a high prune, dead or diseased branches or those affected by insect pests, live wood termites, or shoot hole borer cannot be satisfactorily removed. Lung pruning is a useful compromise. Apart from supplying carbohydrates, lung branches maintain the flow of nitrogenous beads and other material from the feeder roots to the developing shoots and function as a sink for toxic root metabolites.

Pruning just before the onset of the dry weather is avoided because wound healing and new growth are quicker during sap flow. A cessation of plucking (called "resting") for 2–3 months before pruning is necessary for increasing starch reserves. The frequency of pruning, or the length of the pruning cycle, is determined by the particular type of tea, the cropping pattern, and elevation. A cycle length of 2 years is suitable for tea near sea level, with the length increasing with elevation (6 years at

2000 m above MSL). In Sri Lanka, pruning cycles are 2–3 years in the low country, 4–5 years in the high country.

Harvesting and Plucking

The harvesting, or plucking, interval of the tender shoots or flush is adjusted according to environmental conditions and the need to maintain bush vigor, for obtaining the highest yield at the lowest cost and for the desired quality.

General climate, daily weather conditions, ambient temperature, day length, and bush vigor and nutrient status determine the rate of shoot growth, and this varies according to longitude and latitude. The average time period for a bud to develop into a shoot ready for plucking varies from 40 days in Southern Africa (high country) to 55–80 days in Sri Lanka (low country). In the tropics, although shoot growth fluctuates with annual variations in moisture conditions and temperature, plucking is done uniformly over the year at intervals of 4–10 days. In Sri Lanka, plucking rounds are 4 days for fast-growing, vegetatively propagated teas in the warm, low country, and 10 days for seedling teas under cold, high-country conditions. In Malawi (Southern Africa), rounds are 7–14 days. Shorter rounds give increased crop yield, but at higher plucking costs. In the tropics, dormant buds increase in number during unfavorable dry conditions, but when conditions become favorable, as with the onset of rains, "bud break" or a simultaneous growth of dormant buds occurs. This results in a peak or "rush" crop during the wet months of the year. In temperate regions (such as in Japan), a growing season takes place in the spring, when most of the shoots become harvestable, and cropping for the year is done over a continuous period of just 1–2 months.

In East Africa (exemplified by Kenya) with a high altitude and the annual rainfall evenly spaced, crop is taken every month although the yields fluctuate. By contrast, in Southern Africa, with lower altitudes, higher temperatures, and seasonal rain, 90 percent of the annual crop is taken in 4 months.

The top of the bushes can be trained by the plucking process to a level "plucking table" (for manual plucking) or to a dome shape (usually for mechanized plucking). The removal of the terminal bud with the plucked shoot causes the axillary bud just below it to produce a new shoot having at least four new leaves, after which leaf production on the shoot cases ceases temporarily. Flush consists of both actively growing and temporarily dormant (or *banji*) shoots.

The quality of the made tea is better under high-altitude (low temperature), slow-growing conditions. It is determined by the "leaf standard" (whether only 1, 2, or 3 of the top-most immature leaves and the bud are taken, or whether more mature leaves are included as well), by the care taken in handling and conveying the flush to the factory, and by the chemical constituents and fiber content of the shoots.

For manufacture of tea with good flavor or aroma (as in Sri Lankan "Ceylon" teas with Western High Grown or Uva quality), the leaf standard should not be less than 75–80 percent of shoots composed of the bud and two leaves. This finer, or lighter, plucking, however, reduces yield; quality and yield are therefore mutually exclusive. Coarser, or harder, plucking gives more yield and is appropriate wherever teas with more color and strength, or "cuppage" (as in Sri Lankan low-growns), are required by the trade. Crop management is therefore adapted to needs of the trade.

Propagation and Genetic Improvement of Tea Stocks

Seed and stem cuttings were the only source of new tea plants until the 1920s and 1930s, when the technique for vegetative propagation of tea from single-node cuttings was discovered and established. However, for accelerated progress in improving plants genetically, modern biotechnological methods have a much greater potential. Although conventional screening, selection, and breeding of tea clones will continue, biotechnology is now being used to incorporate a range of desirable attributes (including high yields; quality; tolerance to drought, pests, and diseases; and suitability for mechanized harvesting) into vegetatively propagated material.

Propagation from Seed

In plucking, only bushes with decreased vigor will flower and produce seeds; therefore, plants that give good yield and quality can only be produced from seeds developed in special "seed gardens." In these gardens, tea plants are allowed to develop out of the vegetative phase by growing into trees and becoming reproductive. The practice of the commercial seed gardens (called *bari* in Assamese native language) was innovated in Assam.

Selection of plants for seed gardens was based mainly on morphology and growth characteristics. Crosses between Assam and China-type "seed bearers" gave hybrids with improved yields and quality characteristics that were tolerant to poor soil fertility and suited to diverse climatic conditions. The hybrids came to be referred as Assam or China hybrids, depending on which parent they resembled more. A negative feature was that the large variation in morphology, yield potential, and so on of hybrid plants, resulting from an unpredictable admixture of parental attributes, was not conducive to crop management.

Vegetative Propagation

New genotypes were generated between the 1920s and 1930s by vegetative propagation in tea-growing countries, such as Japan, India, and Sri Lanka. These plants are raised by direct planting of vegetative material from carefully selected mother bushes. Since they have a genotype identical to that of the single-parent mother bush, these vegetatively propagated (VP) plants, or "clones," have attributes that are at least theoretically predictable and modified only by environmental variations and agricultural practices.

The development of an economical method of producing desirable planting material, by vegetative propagation from single-node leaf cuttings, was a giant step for the tea industry. VP tea has a much greater yield potential than seedling tea and greater tolerance against or resistance to drought, pests, and diseases. In addition,

VPs reduce erosion because of the spreading canopies of contiguous bushes, which gives a much better cover to the soil. Although considerable genetic improvements to tea stocks have been made by VP production (selection from good seedling teas and hybridization of the selected material), the process is empirical, slow, and laborious, with an improved VP taking about 15–20 years to produce good tea. VP production is frequently hampered by characteristics inherent in the tea plant, such as self-incompatibility, the high level of heterozygosity, and a long generative cycle. Grafted plants are also used in commercial cultivation. The stock is selected for vigor and drought resistance and the scion for quality and yield potential.

Characteristics of Seedling and VP Tea

Seedlings are heterozygous and have high genetic variability, and can therefore adapt to widely differing growing conditions and management and cropping systems much more efficiently than VP planting material. The productive life of VPs is 30–50 years, although senescence at 15–20 years or even earlier seems to occur. Bushes raised from seed last much longer—75–100 years, sometimes more.

VP cultivars comprise about 18 percent of the tea extent in China, although as much as 90 percent of the tea is VP in the Fujian Province. In Japan, about 80 percent of the tea is VP. In Sri Lanka, about 70 percent in smaller fields and about 45 percent in the tea estates are VPs.

Micropropagation

As a more rapid alternative to conventional vegetative propagation, tissue culture is used for the mass production of tea plantlets ("micropropagation") and plant regeneration. However, tissue culture is expensive (roughly fivefold costlier than producing plantlets by normal methods) and is therefore usually restricted to the multiplication of elite clones. Virtually any part of the plant (seed cotyledons, shoot meristems and tips, axillary buds, and nodal "explants") in a suitable medium containing growth hormones can give embryos *in vitro*, which develop into plantlets ("somatic embryogenesis"). To give an example of the order of the scale of production, one explant may give 500 genotypically identical plantlets in a relatively short period, about 8 months.

Following the formation of root hairs in another so-called rooting medium, followed by hardening or acclimatization, the plantlets are transferred to soil and placed in a nursery. After a year or so, the young plants can be transferred to the field in the normal manner, rooted, and grown, either for evaluating breeding lines or for developing bushes in bearing.

Biotechnological Screening for Desirable Traits

Cells or protoplasts in culture can be exposed to biotic and abiotic stresses, and those surviving because they are tolerant or resistant to these stresses may be regenerated into whole plants, hopefully with the same tolerance or resistance. However, although this technique has been successful in some plant species, this has not been the case with tea. Molecular marker technology involving DNA fingerprinting of clones using techniques of RFLP and RAPD can be used to make significant improvements to the existing genetic pool. Marker-assisted selection allows earlier evaluation of breeding lines and reduces the number of plants needed to be screened.

Embryo Rescue

In tea breeding, hybridizing species and genera usually fail because the hybrid embryos abort at an early stage of development. However, by excising the immature embryos and culturing them *in vitro*, it is possible to raise hybrid plants. This so-called embryo rescue technique is used for interspecific and intergeneric hybridization.

Anther and Microspore Culture

The production of homozygous diploids by inbreeding using conventional methods is, in practice, impossible in tea. Anther microspore culture has therefore been tried for the production of haploid plants, which could be used to develop material possessing hybrid vigor. There has only been limited success of anther culture in tea, and haploids have apparently not been used successfully in tea-breeding programs.

Somatic Hybridization Using Protoplast Fusion

Somatic hybridization between divergent tea species could give improvements in qualitative attributes of the resulting hybrids, such as growth rates and resistance to pests and environmental stresses. Protoplast fusion and plant regeneration have not been reported so far with tea. However, protoplasts have been isolated from leaf mesophylls and embryonic cell suspensions, and attempts are being made to achieve plant regeneration from those isolated protoplasts.

Cryopreservation

Although seedling and VP tea bushes with genotypes valuable for future hybridization are preserved in germplasm—or gene banks set up in the field conditions—the extended storage of germplasm (as shoot tips and axillary buds) under conditions of reduced growth *in vitro*, at temperatures below freezing, is employed as necessary.

The Evolution of Rational Pest and Disease Management

"Naturally" Grown Tea

In the earlier periods of tea cultivation, seedling tea was grown as a monocrop mostly in jungle clearings of southern Asia. The tea agroecosystem was therefore little more than an extension of the natural jungle ecosystem and subject to the same abiotic and biotic influences. Since 1920 in Sri Lanka, and for longer periods of time in the longer-established tea-growing countries (China and India), management practices based on organic manuring and mulching, with some minimal use of the inorganic compounds then available, must have kept the pests and diseases below economic injury levels, at least for the most part, by keeping their natural enemies and the crop in a state of ecological balance. Availability (yields of well under 500 kg/ha were the norm) and demand for the crop were small in comparison to modern times, inputs were financially modest, and therefore tea cultivation was a profitable enterprise.

The Downside of VP Tea Cultivation

With the development and introduction of VP planting material, yield potentials increased enormously, provided high levels of expensive, inorganic fertilizers were applied to meet the characteristically greater nutrient demands of the VPs. Typically, annual average yields vary from 1000 to 1500 kg/ha for seedling tea fields, 2500 per ha for rainfed VP fields at high elevations, considerably more at low elevations (as in Sri Lanka), and 9000 kg/ha for irrigated VP fields (as in Malawi). Although specific VPs could be bred for pest and disease resistance and tolerance (as well as for other desirable attributes), VPs in general, because of their lowered variability, were much more susceptible to pests and diseases than the hardier, more genetically variable seedling teas. The widespread planting of VPs has led, therefore, to an increase in the complex of pests and pathogens attacking tea and their absolute numbers.

Coincidentally, together with planting of VPs, a profusion of synthetic chemical pesticides became available after the Second World War for use in agriculture and for the control of human and animal disease vectors. Naturally, these began to be used for controlling tea pests as well. However, the synthetic pesticides also decimated beneficial organisms and natural enemies of the pests (parasitoids, predators, and pathogens), as a result of which pests that were of minor importance because they had been kept under check by natural enemies began to take on a new significance. On a worst case basis, losses in tea crop of 15–20 percent have been estimated from insect pests alone.

The Adaptation of Tea to Integrated Pest Management (IPM)

Just as the sustained use of inorganic, particularly nitrogen-based fertilizers has led to increasing of soil contamination and ground-water pollution, the use of synthetic, persistent pesticides and herbicides has led to the contamination of food, including tea, with the residues of pesticides. These unwelcome spinoffs from modern industrial type agriculture have assumed such alarming proportions that there are now national and international movements against the unregulated use of agrochemicals. The threat posed by agrochemicals to human and animal health and genetics, to soil and water, and to the whole natural order of food webs and interrelationships that sustain the planet has been realized.

The indiscriminate use of hazardous and persistent synthetic pesticides has now given way to integrated pest management (IPM) in many agricultural systems, including that of tea. In Sri Lanka, for instance, the IPM investigations led by the

Tea Research Institute has led tea growers to use a minimum of hazardous and non-persistent synthetic pesticides in combination with resistant VPs, as well as cultural methods which include agronomic operations timed to avoid or reduce pest damage. As a result, in February 1997, Sri Lankan tea was described by the Technical Sub Committee on Tea of the International Organization for Standardization (ISO as the "cleanest tea in the world," as far as persistence of pesticide residues were concerned in the manufactured tea. The description was repeated at the succeeding ISO conference in November 1999.

Though use of synthetic pesticides would still be needed to control pest outbreak in tea, there are indications that IPM and de-emphasis of synthetic chemicals will be a guiding principle in tea cultivation worldwide. The primary aim (apart from savings in foreign exchange—hard currency like the US dollar—arising out of curtailment in import costs, as the products have to be paid in foreign currency), is to move away from reliance on conventional, synthetic pesticides, which leave chemical residues in the manufactured tea, thereby lowering acceptability in the global market, especially as consumers in the west have become extremely health conscious. Everywhere, organic products are catching up, and tea is no exception. International tea drinkers, particularly in the developed countries, are becoming increasingly sensitive to additives and chemical residues, and demand is growing for tea as a natural health drink free of all extraneous chemicals.

The Variations in Tea Manufacture

There are three main types of tea manufactured from plucked shoots: black, green, and oolong tea. In each case, the processing methods employed are different. On delivery to the factory, shoots are either withered (for making black tea), steamed, or subjected to dry heat (for green tea).

Black Tea: Orthodox and CTC

To wither, batches of shoots are spread out on tats or netting or in troughs on mesh in layers less than 30 cm thick. Fans are used to force air through the shoots in order to prevent their being heated by internal respiration and to reduce their moisture content to an appropriate level, if necessary, by pre-heating the air stream depending on relative humidity. Withering for 6–20 hours releases flavor compounds, caffeine, and amino acids.

In black tea manufacture, the cells of the withered leaf (moisture content which is 55 percent) are either broken up in the machines called rollers (the "orthodox" process), or the leaf is cut into particles of less than 1 mm (the cut, tear, and curl [CTC] process). In Southern Africa, the Lawrie Tea Processor is used instead of the CTC machine. These processes cause the mixing of polyphenols and the polyphenol oxidase enzyme system from the leaf's palisade cells and epidermal cells, respectively. Polyphenol oxidase catalyzes the oxidation of the polyphenols by atmospheric oxygen, and more polyphenol units (or monomers) are added to the oxidized polyphenols to form dimers (two monomers) or polymers (more than two monomers). Although the chemical process is polymerization, it is incorrectly referred to in tea parlance as "fermentation."

Fermentation results in dimeric, orange-red theaflavins (TF) and predominantly polymeric, dark-brown thearubigins (TR). Black tea is 80–100 percent fermented, and its mixture of dimeric and polymeric pigments, TF and TR, may be referred to as oligomeric polyphenols. The tea maker adjusts fermentation time (45 minutes to more than 2 hours) and temperature, almost intuitively, in order to make black tea with the desired attributes. The proper control of temperature during fermentation can result in a TF:TR ratio of 1:10, which is known to give optimal color and brightness to the tea liquor.

The reaction of oxidized polyphenols with other leaf compounds, such as carotenes and amino acids, early in fermentation results in flavor compounds. The periods of fermentation are shortened during the "flavor seasons" in order to maximize flavor and minimize liquor strength. Firing, usually by using heated air in a machine called drier, deactivates the enzymes, stops fermentation, and removes most of the moisture.

Green Tea

In the manufacture of green tea, the initial heating or steaming of the shoots denatures the polyphenol oxidase at the outset, and no fermentation can occur. Green tea is thus rich in monomeric polyphenols (catechins).

Oolong Tea

Oolong tea is semi- or 50 percent fermented and contains polyphenol dimers. It is produced only in China (mainly in Fujian, Guangdong, and Taiwan).

Antioxidants in the Main Types of Tea

The percentage of catechins against total polyphenols in black tea is less than 20, in *oolong* tea (as well as in Sri Lankan *uvas*) it is 50–60, and in green tea it is above 65. Until recently, it was believed that the antioxidant activity, which makes tea beneficial to health, was due primarily to catechins. However, it is now known that the TFs, found exclusively in black teas, are equally effective antioxidants and even more beneficial to health as a counter to certain specific health conditions.

The Economics of Tea: Production and Global Trade

Early Tea Trade

The first international trade in tea began in China, more than a thousand years ago. The exports went overland through Asia as far as Western Europe. During the periods of the European sea voyages of discovery and commerce, from the 1500s onward, trading in tea and tea drinking were reintroduced by the Portuguese, the Dutch, and the British to Europe, by way of the sea route from China. China was the largest producer of tea, and supplied about 90 percent of global requirements in the latter part of the nineteenth century and for some years after, when fast sailing ships from Europe (the "tea clippers") would race with the precious commodity to their home markets.

Tea Production, Exports, and Consumption

World and Country Production

World tea production was stable during the first half of the 1990s at about 2.5 billion kg (1990–1995), but since 1996, there have been more or less regular increases in production. In 1999, production stood at 2.8 billion kg, an increase of 17 percent over the 1989 production level at 2.4 billion kg.

The major tea-producing countries during 1990s in volume terms were India, China, Sri Lanka, and Kenya. The fifth place was taken either by Turkey or Indonesia. Among the minor tea producers, Japan, Iran, Argentina, Bangladesh, and Vietnam were prominent with a distinct geographical group, the Eastern and Southern African countries, becoming also significant, along with Taiwan.

Production by Type

Of the black teas, more CTC is produced worldwide than orthodox tea. CTC is considered better for value addition, in particular, for making tea bags, because it consists of uniform particles. The main CTC producers are India and Kenya, with much smaller, though sizeable, volumes being produced in Bangladesh and Malawi. As with Malawi, the newer Eastern and Southern African tea-growing countries produce CTC tea only, to keep up with demand. The main orthodox producers are Sri Lanka, Turkey, and Indonesia. India (one-tenth of its total production), The Islamic Republic of Iran, Argentina, the other South American tea-growing countries, and China also produce orthodox tea. India and China are apparently looking for increasing their orthodox production, in order to have a larger share in the world market.

Sri Lanka produces 93 percent orthodox and 6 percent CTC tea, of which 93–95 percent is exported. Kenya produces about 99 percent CTC, more than 85 percent of which is exported.

Approximately one-fifth of world tea production is green tea, most of it produced internally and consumed, as for example in China, as in the earliest time in the country. This is also true in the case of other far eastern countries. Japan's production is 100 percent green tea, while China's is 71 percent and that of Vietnam is 75 percent. Indonesia, at 20 percent, produces far less comparatively. These countries together produce 98 percent of green tea in the world. Green tea from China is less expensive compared to that from Japan, and most of the green tea consumed in the United States is from China. In 2000, 62.3 percent of China's export was green tea and only 11.8 percent black tea. Most of Japan's green tea production is for internal consumption, but, it is also a net importer of orthodox tea.

Country Exports

The major tea exporting countries are Sri Lanka, Kenya, China, India, and Indonesia. The world's largest exporter of tea, Sri Lanka, keeps marginally ahead of Kenya. Sri Lankan export volumes have been climbing steadily, with records being set every year since 1993. In the year 2000, export volume reached an unprecedented 306 million kg.

China and India export less than the two top ranked, despite their larger production, because they meet a large internal demand with their huge populations. Approximately 80 and 65 percent of India's and China's production, respectively, are retained for internal consumption. It has been estimated that if every Indian drank an extra cup of tea every day, India would become a net importer of tea.

World and Country Consumption

The highest consumers are the CIS (the former USSR), the United Kingdom, Pakistan, the United States, and Egypt. The total imports for consumption in 1999 were 1.2 billion kg, up from 1.1 billion kg in 1989, an increase of 10 percent. The countries with the highest annual per capita consumption rates are Ireland, Turkey, Libya, the United Kingdom, Qatar, and Kuwait (above 2 kg). Many of the Arabic countries and Sri Lanka rank next. The highest consumers of tea in absolute volumes include India, China, the CIS, the United Kingdom, Turkey, and Japan.

Value Addition and Product Diversification in Tea

Variety is key, and diversification into innovative forms and presentations is an urgent requirement in tea marketing and promotion, among the affluent class in the developed countries and higher social strata of the developing countries. This is particularly true among the younger market, inasmuch as promotion of tea is concerned unlike in Asia, where tea is widely consumed by all social classes. Value addition has myriad faces, and is proving to be highly profitable. The tea-growing countries have traditionally exported their produce as a commodity, that is, as loose or bulk tea, with value addition done by commercial conglomerates in tea-producing countries. However, value addition is on the increase in the tea trade of producing countries as well, to different degrees and forms, by different companies and governmental agencies. Of the producing countries, Sri Lanka is ahead of India, China, Kenya, Indonesia, and Bangladesh in this area.

Packaging

Value addition to made tea involves packaging and imparting reliable information on quality and origin to international consumers. Packaging alone can increase the profit margin of made tea by as much as tenfold. For example, Sri Lankan tea was exported to the United States in 1997 at an average price of US$2.6 per kg and sold in supermarkets there, while the price shot up to US$23 per kg in 1998, through attractive and innovative packaging. The brand name on the package is a guarantee of quality, and the label can be used to provide information on origin, blending details, and natural health benefits. These are all essential for the discriminating consumer to make an informed choice between a bewildering shelf display of various teas and other beverages. The packaging may be in cardboard or foil packets or tins of loose tea.

Tea Bags

The introduction of the tea bag revolutionized tea-drinking habits in the United Kingdom, the United States, Japan, and in Western Europe, owing to their being hygienic, economical, and easy to handle (carrying and disposing of after use). Almost half the weight of tea sold in the developed countries is sold as tea bags. They are of various shapes and forms. A bag is simply dropped into a glass of boiled water, suspended by a thin string to steep.

Nearly 60 percent of CTC black tea production is suited to use in tea bags because of its uniform particle size, which makes infusion easy. However, orthodox and oolong teas are also adaptable to tea bags. Tea bagging uses expensive machinery and is a large industry in itself.

Instant Tea

An alternative to the tea bag and no less convenient is instant tea as powder or granules, manufactured from green leaf or fermented tea waste, which gives it a pronounced tea character. The tea solids are extracted with hot water, concentrated, and freeze-dried. It is available in a hot- or cold-water soluble form, plain or with milk and/or sugar incorporated into it, and is suitable for vending machines.

RTD Teas

Instant tea can also be a base for the ready-to-drink (RTD) tea, for sale in cans, bottles, or in tea packs. RTD does not have, in general, the tea characteristics that connoisseurs of tea look for. RTD serves only the temporary thirst of casual tea drinkers. However, a tea concentrate, appropriate for making RTD, developed by the Tea Research Institute of Sri Lanka in the late 1980s, with only 4–5 percent soluble tea solids, has now been improved to 10 percent of tea solids.

In the United States, 24–30 percent of RTD is consumed by 20–39 year olds, with those over 40 years of age comprising only 14–18 percent. Teenagers consume only 11 percent of RTD. On the other hand, a phenomenal growth in the production and consumption of RTD (canned and bottled, green, oolong, and orthodox) by all age groups has been observed in Japan, in the face of overwhelming competition from multinational soft drink and beverage giants.

Iced Tea Drinks

Soft drinks based on tea are consumed cold or iced, except in Japan, where they face strong competition from expensively promoted, branded, nontea soft drinks. At the present time, iced tea, instant or brewed, comprises about 80 percent of the tea drunk in the United States, most of it in the southern states. Presumably, it is being perceived as a component of a healthy, low-calorie diet and a pleasant, relaxing drink of convenience even for a more mature age group. In the United States, 24–27 percent iced tea is consumed by the 30 to over 50 age group, with 9–10 percent in the below-30 age group.

Scented and Flavored Tea

Teas containing essential oils to give flower scents are popular in Eastern countries, the Middle East, and in some Western countries. Teas of all kinds are easily treated with liquid flavors, either natural or artificial. Flavors in granular form are also incorporated into tea bags. The granules slowly release the flavor through the tea bag during infusion.

Alcoholic Tea Drinks

Tea wines and sherries are available in world markets. A tea-based alcoholic beverage with an alcohol content of 20 percent was developed by the Tea Research Institute of Sri Lanka.

Nonbeverage Tea Products

In Japan, a wide range of products based on green tea and its catechin components are available. Freshly picked leaves are used as a vegetable. As powder, green tea is incorporated into common food items such as noodles, rice, and bread, as well as into toothpastes. Tea extracts and extracted catechins are used to enrich drinks, bagged tea, chewing gum, laxatives, mouthwashes, deodorants, air-purifiers, soaps and cosmetics, sun-blocks, and molluscicidal sprays. Tea is reprocessed and used for making sweets, jams, jellies, and massage oils. New tea-based products emerge all the time. For instance, in Taiwan and elsewhere in Asia, and now in the United States, a new drink called bubble tea that contains glutinous tapioca balls (from cassava), shaken up with green or black tea, is becoming popular.

Nonfood Tea Products

Tea is used as a stuffing for pillows and bathing sponges, and its brew has been used even for making "tea-stain" paintings.

Research and Development Institutions Related to Tea

In most of the tea-producing countries, institutions are dedicated to tea research and development (R&D) or having a mandate for tea as well as for vegetable crops. Much of the research is concerned with studies on field production, but there is not much on the chemical and engineering aspects of tea processing.

Bangladesh

Bangladesh Tea Research Institute
Srimanga-3210, Moulavi Bazar, Bangladesh
Tel.: 0861-3328

Central Africa

Tea Research Foundation (Central Africa)
PO Box S 1, Mulanje, Malawi
Tel.: 462 261, 462 271

China

The International Institute of Tea Science and Culture
Tea Industry R&D Centre, Chinese Academy of Social Sciences
Tea Research Institute, Chinese Academy of Agricultural Sciences
Agricultural University of North-Western China
Agricultural University of North China
Agricultural University of Central China
Zhejiang Agricultural University
Hunan Agricultural University
Sichuan Agricultural University
Anhui Agricultural University
Fujian Agricultural University
Yunnan Agricultural University
Hangzhou Tea Research Institute
All China Supply and Marketing Cooperative
Tea Research Institute, Agricultural Scientific Academy, Hunan Province
Tea Research Institute, Agricultural Scientific Academy, Fujian Province
Tea and Silkworm Research Institute, Agricultural Scientific Academy, Jiangxi Province
Tea Research Institute, Guizhou Province
Tea Research Institute, Sichuan Province
Tea Research Institute, Guangdong Province
Fruit and Tea Research Institute, Hubei Province
Tea Research Institute, Agricultural Scientific Academy, Yunnan Province
Tea Research Institute, Agricultural Scientific Academy, Anhui Province
Tea Research Institute, Agricultural Scientific Academy, Wuxi, Jiangsu Province
Zhejiang Agricultural University
Hangzhou 310029, Zhejiang Province
Tel.: 42605
Department of Tea Science, Zhejiang Agricultural University

India

United Planters Association of South India (UPASI)
Tea Research Institute

Nirar Dam, Valparai 642127, Coimbatore District, Tamil Nadu
Tel.: 4253 71038
Fax: 4253 7102, 4253 71419
Tea Research Association (Tocklai)
Tocklai Experimental Station
785008, Assam
Tel.: 0376 320054
Fax: 0376 325589, 0376 327253
Institute of Himalayan Bioresource Technology
PO Box 6, Palampur, Himachal Pradesh 176 061
Tel.: 01894-3042-26, 3071
Fax: 91-1894-30433

Indonesia

Research Institute for Tea and Cinchona
Gambung, PO Box 148, Bandung 40010, Indonesia

Iran

State Tea Organisation, Ministry of Agriculture
202, Sornmaye Avenue, Tehran, Iran

Japan

National Research Institute of Vegetables, Ornamental Plants and Tea
2769 Kanaya, Shizuoka 428, Japan
Tel.: 05474-54101

Kenya

Tea Research Foundation of Kenya (TRFK)
PO Box 820, Kericho, Kenya
Tel.: 20598

Nepal

National Tea and Coffee Development Board
Singha Durbar, Kathmandu, Nepal
Tel.: 01-228638

Nigeria

Cocoa Research Institute of Nigeria
PMB 5244, Ibadan, Oyo State, Nigeria
Tel.: 022-412430

Pakistan

National Tea Research Institute
Shinkiari, Mansehra, NWFP, Pakistan
Tel.: 05992-2553, 2125

Russia

Russian Research Institute in Floriculture and Subtropical Crops
Fabritsius Str. 2/28, Sochi-354002, Russia

South Africa

Burgershall Research Station
Private Bag X501, Kiepersol 1241, 86 Hazyview, South Africa
Tel.: 0131 242

Sri Lanka

Tea Research Institute of Sri Lanka
St. Columbus, Talawakelle, Sri Lanka
Tel.: 94 51 22601
Fax: 94 52 58229

Taiwan

Taiwan Tea
Pushin, Yangmei, Taoyuan 32613, Taiwan
Tel.: 03-478 2059

Tanzania

Tanzania Agricultural Research Organization (TARO)
PO Box 9761, Dar es Salaam, Tanzania
Tel.: OS 1 74247
TARO Marikitanda Agricultural Research Institute
PO Box 8, Amani, Tanga, Tanzania
TARO Maruku Agricultural Research Institute
PO Box 127, Bukoba, Kafera Region, Tanzania

Vietnam

Tea Research Institute
Vinh Phu Province, Vietnam

A Look into Tea's Future

SWOT (Strengths, Weakness, Opportunity, Threat) can analyze the future of tea from the various perspectives, as a profitable crop in a variety of agricultural environments around the world and through the adaptability of the types and forms of tea to changing consumer requirements.

Strengths

The Traditions of Tea Drinking

Tea consumption is part of the traditional culture of Far Eastern Asian countries, such as China and Japan, and, of course, the diaspora of these two countries spread wide and far in the world. Apart from the ritualistic tea consumption enshrined in traditional ceremonies and banquets, the continuous sipping of green tea at ordinary meals and the sheer size of the populations involved in this age-old practice is bound to ensure a secure economic future for tea. It is unlikely that these Far Eastern Asian countries will ever give up this age-old custom, no matter what the onslaught of modernity is, as it is so much entwined in the people's culture. In other parts of Asia, notably in the Indian subcontinent, where black tea is the traditional, time-honored drink of millions, a huge market already exists. Large populations in west Asia and Russia are inveterate consumers of strong black tea, such as Indian and Sri Lankan tea. In the Arab region, small cups of black tea are consumed at all hours of the day and are the customary offering signifying hospitality and friendship.

The former colonies of the British Empire, and those countries influenced by the British mores, such as the United States, Canada, and Australia, have never abandoned the tea-drinking habit, which continues in their domestic and social settings. In addition, there has always been, in particular in the developed countries and in Japan, a coterie of connoisseurs of good teas, such as the Indian Darjeelings and Sri Lankan speciality teas (those well known ones, such as Nuwara Eliya, Dimbula, Uva, Uda Pussellawa, Kandy, and Ruhuna) grown under conditions unreproducible anywhere else. These conditions may be obtained in unique agroclimatic zones and subzones, or even in individual plantations, which only the ultimate connoisseur can distinguish. One of tea's main strengths is, therefore, its diversity and adaptability as a product for all manner and conditions of humankind (there are an estimated 3000 different types of tea), its long history as a traditional beverage, its relative inexpensiveness and availability, and an upmarket image among connoisseurs.

Tea Research Culture

There are a large number of research institutes in tea-growing countries as well as other institutions, including universities, both in the developed and developing worlds, where studies on all aspects of tea cultivation, processing, and marketing are going on apace. Some of these, as in India, Sri Lanka, and China, have been in existence for decades. This continuous examination of tea and the tea trade from all angles, and the findings continuously being disseminated, must be seen as another strength of the world tea industry.

The Favorable Environmental Impact of Tea

The adaptation of tea into a relatively environmentally benign, nonpolluting crop is being done within many tea-growing systems, based on the use of VPs, integrated nutrient management (IPM), organic farming, precision agriculture, microirrigation, and intercropping. As in certain areas of Sri Lanka, tea may be intercropped with rubber, coconut, coffee, spices, fruit crops, and legumes; this serves to increase the profitability to the farmer and also ensure biodiversity. Such widespread intercropping practices are uncommon in India, where tea is mostly grown as a monocrop in the hilly regions. A well-managed, modern tea plantation is a prime example of environmentally rational agriculture. It approximates to a natural forest in that plantations can be multicanopied—a thick tea canopy over which are the two canopies of the medium- and high-shade trees. Tea plantations, with their large biomass, cool the environment, encourage an animal biodiversity, and act as a buffer against environmental pollution by absorbing carbon dioxide.

Agricultural practices and factory processes connected with tea have evolved in such a manner that they now have a lesser impact on soil, water, and air, when compared to other intense production systems. Being completely worker-dependent, field operations conventionally use a minimum of machinery. Motorized harvesters and brush cutters are used but, with good management, the pollution arising from them is negligible.

Especially on uplands and on slopes where tea is cultivated, soil disturbance and erosion are permanent hazards, having both negative on- and offsite effects. However, researchers and tea growers can now adopt soil conservation measures that keep soil erosion well below the tolerance limit (9 tons/ha/year in Sri Lanka). These measures include bush spacing for maximum canopy cover, mulching and planting of cover crops during land preparation, and, in the first 2 years after planting, construction of leader and contour drains and contour terraces and the establishment of hedgerows in the sloping agricultural land technology system. A combination of manual, cultural, and chemical methods to manage weeds, rather than use of scrapers, also conserves the soil.

The use of soil fumigants against pests and diseases is kept to a minimum; a particularly damaging fumigant, methyl bromide, has been withdrawn internationally.

Tea manufacture makes use of fuel—wood, oil, or solar radiation—as energy sources. Trees are renewable, and if their removal for energy generation is controlled, this will not have a chronic environmental impact. The usual location of tea factories within large plantations and close to forests allows for the sequestration of greenhouse gases (GHG), mainly carbon dioxide. The small amounts of heat given out by scattered tea factories in these locations are not significant in adverse global-warming effects.

No chemicals are used in tea manufacture. Tea factory waste is merely reduced tea and wood ashes. Both can be composted and used as organic manure in tea plantations. Refuse tea can be used to make instant tea as well.

Weaknesses

The Oversupply Situation

Global tea consumption has been lagging behind production, with the result that oversupply is growing. The increase in overall production between 1989 and 1999 was 17 percent, although imports by consuming countries increased by only 10 percent.

Mismatch in Requirements

Currently, a significant weakness in tea industry is the mismatch between the types of beverages consumers seem to want and what is available in the form of tea and tea-based drinks. Even in traditional tea-drinking Western countries, such as the United Kingdom, a discernible reduction in tea imports has been observed. There is pessimism, in sections of tea trade in the United Kingdom, that tea has an outdated, old-fashioned image in the minds of the younger generation and that sales will inevitably decrease as a result. Tea is a perennial crop, with a production momentum that is difficult to slow down or redirect in order to cut back on supply adjust to product specifications. Therefore, short- to medium-term responses to market requirements are virtually impossible to accommodate. As a result, prices tend to move from boom to bust over relatively short periods of time, whatever the improvements in production and marketing efficiency.

Marginalization of Producers

Unless a seamless vertical integration occurs between production, marketing, and retail purchasing, with a "bush-to-cup" marketing strategy based on well-accepted national brand images, tea producers in both the corporate sector and small enterprise will continue to lose out on profit margins to middlemen. They will continue to be suppliers of a primary commodity to sophisticated, often multinational buying and export enterprises, with their own well-financed branding and distribution arrangements.

Agrochemical Residues in Made Tea

Consumers and international trade are becoming increasingly sensitive to additives and agricultural residues in made tea. The marketing of tea as a pleasant and beneficial health drink, free of contaminants, is jeopardized. In this connection, one has to specifically mention the use of Endosulfan, a banned insecticide used to control tea mites. This author has observed that its use is widespread in northeastern Indian tea plantations. The insecticide is banned in some parts of the country like Kerala State, but planters continue to use it.

At the technical Sub Committee Meeting on Tea of the International Organization for Standardization, held in Kolkatta in West Bengal State, India, in November 1999, the levels of pesticide residues in exported tea, as revealed by European surveillance during 1994–97 and 1997–98, were reported. Although most of the exporting countries did not exceed the maximum permissible residue limits set by the European

Union, it was indicated that certain Asian countries had incidence of residues in their teas that were at actual or potential violation of EU limits. The Technical Sub Committee advised that the use of all waste-soluble pesticides (whose residues find their way into the tea brew) should be stopped.

Opportunities

Optimizing on the Health Benefits of Tea

People in general are now aware that tea (all types, both black and green) is rich in polyphenols, which are highly effective antioxidants. Tea's health-promoting and beneficial effects have been scientifically and medically proven by independent research in different countries. Even a relatively slight increase in per capita consumption worldwide, under the impetus of renewed generic promotion efforts, based on the recent health findings, should serve to correct the present oversupply situation. For example, an increase in the current daily consumption of characteristically well-informed UK consumers from the present 3.2 cups per person to 4 cups daily would help in correcting the production–consumption imbalance.

The Emergence of New Markets

The opening up of new markets in countries where tea had not been commonly consumed hitherto is a new growth opportunity for the tea trade. Growing populations and higher incomes in developing countries entering the free market nexus have resulted in higher demands for inexpensive beverages. Thus, consumer preferences for tea in central and eastern European countries such as Poland, the Czech Republic, Slovakia, Hungary, and the CIS are growing.

With the emergence of new states of the CIS derived from the former USSR, and their opening to free trade, large Muslim populations in the CIS countries have become contiguous with the habitual tea drinkers of West Asia. The tea-drinking habit is therefore being reinforced in the CIS and has enormously expanded the outlook for tea.

Adapting to a New Generation of Consumers

There is guarded optimism in the global tea trade that, with value addition and product development moving away from traditional forms and styles of tea consumption, the present availability mismatch will reverse, and that tea will come to be accepted, particularly in Western countries in newer, more modern forms by young and old, for example, as a part of healthy diet.

The age-old adaptation of tea, this time to present-day consumerist trends of a new generation, will determine its future. There is a real opportunity for the world's tea industry to grow and remain viable, with value addition and new forms of the product adapting quickly to the convenience of consumers and their evolving tastes and lifestyles. The decreasing consumption by formerly dedicated tea drinkers, as in Western countries, could perhaps be reversed by creating new national branding images. Sophisticated and informed consumers would be willing to pay higher prices for a brand that gives assurance of quality and maintenance of product standards.

Cleaning Up the Product

There is now a welcome change in the mindset of both tea growers and exporters in Asia that to ensure delivery of good quality tea for the vigilant Western markets; chemical residues in tea have to be eliminated. This determination extends to calls for growing tea under stringent organic farming conditions as per internationally accepted norms.

Growing Tea Alongside Other Profitable Crops

The practice of growing tea with other tree crops, or intercropping, has emerged from field research efforts as being viable and profitable. Intercropping serves to maximize land productivity and minimize the economic and environmental risks involved in growing tea as a monocrop. Added advantages are favorable cost benefits and increased employment.

In Sri Lanka, for example, studies commencing in the early 1980s on intercropping tea with rubber proved successful at the lower elevations, where rubber is widely grown. Tea is also intercropped successfully with coconut in the low country. For many years prior to this, it was the practice for small holders in Sri Lankan mid-country to cultivate black pepper, coffee, and cloves mixed with tea. Intercropping tea with fruit crops and legumes has also proved profitable.

Threats

The external threat to tea trade comes from the global and high-pitch promotional campaigns of multinational beverage and soft drink marketers, which the tea promotional efforts cannot match. Hence, in most countries, the soft drink market outcompetes tea, with the only exception being Japan. Coffee is another competitor, and despite tea being a healthy drink, it cannot stand the competition from coffee in countries like the United States.

Flight of Plantation Workers

One of the major problems faced by tea planters in the developing countries is that with urbanization and expanding opportunities in education and employment, the labor force is pilling away from tea plantations where, more often than not, the working conditions are anything but satisfactory, either in the fields or in factories. This phenomenon is not restricted only to tea industry. It is equally applicable to others like the rubber, oil palm, coffee industries as well. The working conditions on the plantations are often hostile, with rough weather and limited physical and medical amenities. This is a phenomenon observed all over the developing world, be it India, Malaysia, Sri Lanka, Indonesia, Philippines, or Thailand. A recent survey in India indicated that given the option, more than 40 percent of the people engaged in agriculture would leave it. The only exception is China where a rigid state policy governs the exodus of workers.

SWOT: The Final Word

Although tea producers cannot afford to be sanguine at the present time, it is very clear that the inherent strengths of tea far outweigh the threats and weaknesses. The most enduring quality of tea is that it is not only a very affordable beverage, but it bestows enormous health benefits on the consumers. This will be the biggest positive point as far as tea industry globally is concerned and one can but hope the very best for tea in the decades to come, notwithstanding its current travails.

10 Wattle

Wattle belongs to *Acacia* (Family: Fabaceae), a large genus composed of more than 500 species found in the warmer and drier parts of the world, mainly Australia and Africa. Species with pinnately compound leaves are found throughout the tropics, and the phyllodineous ones are natives of Australia. In India, there are about 22 indigenous species, distributed throughout the plains. However, all species that yield wattle are exotic and are introduced from Australia.

The Different Kinds of Wattle

Acacia mearnsii *(Syn.* A. mollissima*)—Black Wattle*

It closely resembles *Acacia decurrens* in its habit and is sometimes considered a mere variety of the latter. In the Nilgiris, in Tamil Nadu, India, the area under wattle would be about 6,000 ha (Gupta, 1993). The principal difference from *Acacia decurrens* lies in the shape of the leaflets, which in *Acacia mearnsii* are short, obtuse, 5–7.5 cm long, and closely spaced. In *Acacia decurrens* they are narrow, 7.5–10 cm long, and widely spaced. The leaves of the former are dark green, whereas those of the latter are yellowish green. *Acacia mearnsii* is also a native of South Australia and Tasmania. It is the principal species grown in South Africa and is the chief source of the wattle bark of commerce. It was introduced into India around 1840, along with *Acacia dealbata*, mainly to provide fuel, of which a shortage existed at that time. It thrives at elevations of 1,665–2,330 m above MSL (mean sea level), where there is a well-distributed rainfall of about 150 cm per annum. It is now grown in the Nilgiris, the Pulneys, and in central Kerala State on a small scale. It also grows in the highlands of Sri Lanka, but does not appear to be suited to Myanmar. A few trees are also found in tea plantations as windbreaks and shade trees.

Acacia decurrens—*Green Wattle*

It is an evergreen tree, reaching to a height of 12 m. In general, the bark is olive green, but is dark gray in older trees. The pale-yellow flowers are less plentiful and less scented than those of *Acacia dealbata*. The bark is rich in tannin (36–41 at 11 percent moisture content).

The Agronomy and Economy of Important Tree Crops of the Developing World. DOI: 10.1016/B978-0-12-384677-8.00010-2
Copyright © 2010 by Elsevier Inc. All rights of reproduction in any form reserved.

Acacia dealbata—*Silver Wattle*

Acacia dealbata is an evergreen tree native to South Australia and Tasmania. It is now regarded as a variety of *Acacia decurrens*. It was introduced to the Nilgiris in 1840, where it became naturalized and is now a characteristic feature of the vegetation from 1,250 m upwards. It has also been planted in the Himalaya (Shimla, Nainital, and Almora Hills), chiefly between 1,800 and 2,500 m. It suffers considerably from snowbreak, but regenerates through its numerous root suckers.

On the Nilgiris, *Acacia dealbata* reaches a height of more than 12 m and a girth of more than 1.2 m. Its bark is thinner than that of *Acacia decurrens*, and is silvery gray in appearance. The young shoots and foliage are also of the same color. It blooms profusely, producing large quantities of yellow flowers. The pods are broader and less constricted between the seeds than those of *Acacia decurrens*.

The tree has extraordinary powers of reproduction through its spreading root suckers and is of great value for covering unstable hill slopes. The bark has only one-third the tannin content of that of *Acacia decurrens*. The tannin content of Nilgiri bark varies from 1 to 15 percent at a moisture content of 10 percent.

Acacia pycnantha—*Golden Wattle*

Acacia pycnantha is a native of South Australia where, along with *Acacia mearnsii*, it constitutes the chief source of tan bark. In size it is a smaller tree than the latter and its bark thinner. It is also said to be a less hardy species and has not found favor in South Africa. Experimental plantations are being raised on the Nilgiris. The species is very rich in tannin. Analyses showed up to 50 percent of tannin in the air-dried material. The best commercial bark has an average of more than 38 percent tannin in the Nilgiris.

The trees yield Australian gum, which is principally an arabogalactan. The extract of the bark is said to be equal to the best Indian catechu. In view of the keen demand for wattle and wattle extracts from the leather industry, large-scale plantations of black wattle were taken up in the Nilgiris and the Pulney hills in the 1950s. Nearly 18,000 ha have been brought under wattle plantation, and the annual production of bark at present is 8,000–10,000 tons (Sherry, 1971). Currently the area under wattle in Tamil Nadu is estimated to be 36,660 ha (Anonymous, 1999). A private factory in Mettupalayam, India, produces wattle extracts with an annual production capacity of 3,750 tons. Wattle bark contains tannins, nontannins, insolubles, and fibers. The tannins in wattle bark belong to the catechol group and consist of a complex mixture of polyphenols, of which catechin and its derivatives are the most important.

Growing Conditions

Climatic and Soil Requirements

Climate

Black wattle comes up well at elevations between 1,950 and 2,500 m above MSL. A well-distributed rainfall of 150 cm with 100 rainy days is essential for its growth.

In regions of high rainfall, the bark may be infested with lichen, resulting in deterioration of quality. Winds and frosts during the early part of the rainy season (June–July) adversely affect growth of the trees. In East Africa, it grows well where annual rainfall is 1,041–1,321 mm, which is about 75 percent between April and September. On the equator, where black wattle is grown in South America, the rainfall pattern is nearly the opposite: the mean annual temperature range is 17–23°C, with little seasonal variation, but considerable diurnal variation. At higher altitudes in South America, frost is a risk, and heavy snows may break tree limbs. Tannin content varies inversely with precipitation. Ranging from warm temperate dry through tropical thorn to tropical moist life zones, black wattle is reported to tolerate annual precipitation of 600–2,280 mm and annual mean temperatures of 14.7°C to 27.8°C. In South India, 8° to 11° north latitude, black wattle grows well at elevations above 1,500 m above MSL. At higher elevations, due to incidence of heavy frost and strong cold winds, the initial growth is very slow. Thereafter, growth (height and diameter) is rapid for 5–6 years and then falls off gradually. Except in the Nilgiris, wattle is planted in regions of heavy rainfall of around 2,500 mm and more (Rajagopalan, 1973). Black wattle plantations are confined to forest areas of the eastern highlands of Zimbabwe, where the mean annual rainfall exceeds 100 cm. Within this zone, the altitude varies from 700 m to more than 2,500 m, but, the higher collar regions are the most suitable. The mean temperature of the hottest month is 20–24°C in Zimbabwe and mountain mists are frequent.

Soil

Wattle grows on poor dry soils, but prefers deep, moist, and fertile soils. In Australia, black wattle occurs on soils derived from shells, mudstones, sandstones, conglomerates, and alluvial deposits. In South America, it grows well on red clay or sandy soils, which have suffered from severe erosion and depletion of fertility (ferruginous clay loams with little or no free silica). However, deep, well-drained soils are preferred for good growth of the tree. Sandy loams are deemed optimum. Clay pans and hard laterites are not suitable and should be avoided. In Zimbabwe, the soils of the main wattle-growing areas are derived from granite, quartzite, and dolerite with textures varying from sand to clay loam. Higher yields of wattle bark are obtained from deep, well-drained sites, irrespective of the parent rock. In the Nilgiri and Pulney hills of South India, growth of wattle in areas having adequate soil depth, protection from cold winds, and receiving an annual rainfall of around 150 cm is comparable with other wattle-growing countries. However, in areas receiving high rainfall (250 cm and more), wattle plantations are poorly grown and stocked thinly.

Pre-Treatment of Seeds

The common method of propagation of wattle is by seeds. Before sowing, the seeds are subjected to hot water treatment. After treatment, the seeds are washed four to five times with cold water to remove the gummy mucilage. They are then dried under shade and stored in lead-lined boxes. These seeds may retain viability for

6 months to 1 year. The pre-treatment involves boiling in a drum or a pot with watering about five times the volume of seeds. When the water begins to boil, the drum is removed from the fire and the seeds are put into the boiling water and allowed to cool for 24 h. Care is taken not to boil the water after putting the seed in. After cooling, the seeds are washed in cold water to remove the gummy mucilage and dried in the shade. Gupta and Thapliyal (1974) recommended that pre-germination treatments for *Acacia mearnsii* involve soaking the seeds in hot water for 12–14 h or stirring in concentrated sulfuric acid for 1 hour. This is sufficient to break dormancy.

Nursery Practices

Treated seeds are sown in raised standard beds of 10 m × 1 m; 300 g of seeds are sown per bed. The beds are covered with bracken ferns and watered. The object of providing bracken cover is to protect the beds from excessive evaporation and also to facilitate quick germination. Germination occurs in 10–15 days when the bracken cover is removed from beds.

Seeds are sown during October–November and seedlings are pricked out when they are 5 cm tall. Two-month-old seedlings should be pricked out from the bed and planted in small polythene bags (15 cm × 10 cm), filled with good soil, as done for any potting of nursery seedlings. The seedlings in the polyethylene containers should be watered twice daily for the next 2 months and then once daily on nonrainy days. In this condition, the seedlings are kept for 3–4 months. By this time, they reach a height of 25–30 cm and are ready for planting in the field. One square meter bed provides 400–500 seedlings. Nine- to 12-month-old seedlings are used for planting. About 450 g of seeds will give 3,500–4,000 seedlings, which can be planted in an area of 3–6 ha. The seedlings in the nursery have to be protected from frost during October to January and against rabbit attack by providing rabbit-proof fencing.

In Zimbabwe, root pruning is done during 3–4 months of the nursery period. It is essential to confine the roots either by drawing taut steel wires below the polyethylene bags or by lifting the plants individually and trimming the roots. Seedlings may also be shoot-pruned to the ideal planting size of 15–20 cm if planting is delayed.

Planting Practices

Planting of the container seedlings in the main field should be done at the onset of the monsoon season during May–June in pits of 30 cm^3 at a spacing of 2 m × 2 m. The plantations can accommodate 2,500 seedlings per ha. Wattle can also be propagated by stump cutting for which 7.5 cm of shoot and 22.5 cm of root have been found optimum. While planting polyethylene-container seedlings, care should be taken to ensure that the polyethylene bags are removed without disturbing the ball of earth at the bottom of the stems.

Preparation of Land

The pits may be filled with composted humus. Half kg of K fertilizer (normally muriate of potash) may be applied to each pit as an alternative practice. This helps to boost growth of the plants at the initial stages.

Planting and Aftercare

Nine- to 12-month-old seedlings are planted in the main field during June–July depending on monsoon. While planting, mossing of roots is done. A spacing of 3 m × 3 m or 4 m × 4 m is adopted. At the Wattle Research Institute, Pietermaritzberg, in The Republic of South Africa (Anonymous, 1960), it was found that the highest yield of tannin per acre was obtained from 11-year-old trees at a density of 600 trees per acre and from 9-year-old trees at a density of 800 trees per acre. Adequate windbreaks have to be provided for the plantations. Weeds have to be removed from around the young wattle seedlings. At higher elevations, seedlings may be provided with frost covers to guard against frost in the first year of planting.

Growth of seedlings raised by direct sowing is extremely slow during the first year of planting. Thereafter, growth increases both in diameter and height. In well-spaced plantations, the growth rate of wattle trees is higher in the Nilgiris than at Kodaikanal (Tamil Nadu, India) due to better soil and climatic conditions. In South America, fields are usually ploughed and harrowed in April or May. Seedlings are set out from May to November, but usually in winter, from June to August, after an initial shower. Plants are spaced 2 m × 2 m, at the rate of 2,500/ha. Propagation by cuttings is almost impossible without mist. Air layering is more promising. Two types of people grow acacia. The tanner or business owner plants 200 ha or so entirely to black wattle, usually one section at a time, so that he or she can plant and harvest within the same year and continue year after year. The farmer plants half or less of his or her land to black wattle and the rest to crops, such as corn, beans, cassava, sugarcane, vegetables, or pasture. The farmer plants 2–6 ha of acacias each year and thus evenly distributes work and production. Oxen may be useful for ploughing, but most work is done by hand. Usually only ploughs and hoes are used in cultivation. Intercrops may be grown the first year during which trees grow about 4–5 m in height and about 2.5 cm in girth (Duke, 1981).

Thinning

Wattle trees exhibit strong affinity to light and do not tolerate shade. Thinning is necessary in the third year of planting if plantations are closely spaced. However, in wider spacing, thinning is not necessary.

Manuring

Samraj and Chinnamani (1978) recommended that N, P, and K be applied to each plant of black wattle in the first year of planting to boost growth. Soil working should be done, leaving 15 cm around the plant, first in July–August and again in November–December. In the second year, only one working is carried out following the rains. They further suggested that 200 g NPK fertilizer (wattle mixture) be applied to each plant in the Nilgiris to boost initial growth. The NPK fertilizer should be made of a mixture of ammonium sulfate, superphosphate, and muriate of potash, which respectively supplies N, P, and K. Recent experiments conducted in the Nilgiris reveal that black wattle responds better to the application of 45 g ammonium sulfate, 90 g of super-phosphate, and 45 g of muriate of potash per tree, compared to other combinations of NPK fertilizers to boost initial growth. Hussain et al. (1980) found that application of NPK (45 g ammonium sulfate, 90 g superphosphate, and 45 g of muriate of potash per tree) gave the best results.

Aftercare

Wattle is susceptible to root competition from grasses; therefore, scrape weeding to a diameter of one meter around the plant after establishment (about 1 month) is essential. Subsequently, soil working is done leaving 15 cm around the plant, first in July–August and again in November–December. Casualties are replaced periodically in the first year along with weeding and soil working. In the second year, one soil working is given to the plants following the rains. No more care is necessary in the subsequent years, except protection against trampling by cattle, illicit cutting, and damage by fire and wind. For protection against damage by fire, clearing of fire lines 3 m wide around the plots during December–January in the first year of raising the plantation is the recommended practice. Rescraping of fire lines and maintenance by fire patrols should be done in subsequent years during December–January and every year thereafter until rotation age.

Management of Harvest

Throughout its growth period, black wattle sheds its twigs, leaves, flowers, and fruit. Being a leguminous tree, in addition to adding nutrients to the soil by atmospheric fixation of N, it also enriches soil fertility through soil organic matter level enhancement by litter accumulation. The total quantity of this leaf litter, recycled by trees, is estimated to be about 1.5 tons/ha/year under protected conditions. The average depth of litter accumulated was also found to be about 5 cm for a period of 10 years under black wattle cover. The annual dry weight of litter was 960 kg, made up of 22.1 kg N, 0.5 kg P, 3.4 kg K, 3.3 kg Ca, 0.9 kg Mg, and 662 kg organic matter (Venkataramanan et al., 1983). The maximum leaf fall is during June–July. Recycling of these nutrients keeps the land highly fertile. A comparison of data from *shola* forests in the Nilgiris showed that *shola* leaf litter contained more nutrients. The undergrowth consists mainly of wattle regenerating from the mother trees. Wherever the canopy is not too close, grasses and other herbs and shrubs also make their appearance. The leaf litter should be protected to safeguard the soil from erosion, whereas the twigs and lower branches can be pruned for use as fuel. This can provide periodic additional revenue.

The crop is ready for harvest in 10 years. The bark is first cut around the tree at the base and peeled off during the rainy season. The debarked tree is then cut 10 cm above ground level using a sharp axe or saw and then billeted into 1 m length pieces and stacked. The bark is dried and disposed of depending upon the quality of the bark and prevailing market price. The debarked wood is about fivefold more than the quantity of bark.

Natural Regeneration and Rotation

The second rotation crop is obtained mainly by profuse natural regeneration from seeds and also by root suckers. This renders black wattle particularly suitable for a silvicultural system of clear felling, followed by natural regeneration. Burning is an effective means of controlling grass and inducing germination of wattle seeds. After extraction of bark and wood, the remaining slash is spread evenly and burnt during February–March. Seeds are also sown in blanks before burning. By this method of controlled burning, profuse and adequate regeneration is achieved. On receipt of pre-monsoon showers, seeds germinate, and if any blanks are noticed, nursery raised container seedlings are planted. The resulting stand obtained by natural regeneration is usually very dense. Therefore, a mechanical thinning is done in the third year to reduce the stand to 2,500 trees/ha. Subsequently, a silvicultural thinning is done wherever necessary, taking into consideration growth and form of trees.

Recent observations indicate that the second rotation crop is more vigorous and healthier than the original planting. This may be due to accumulation of leaf litter, ash, and so on, resulting from burning, along with the absence of grass and other weeds.

Rotation

Rotation of wattle depends on the site and care given to plantations. A 12-year rotation is adopted in good quality sites in South Africa. In poor quality sites, an 8–10-year rotation is followed. Length of duration depends on conditions and growth of trees. In the Nilgiris, a 10-year rotation has been adopted. Samraj et al. (1977) have compared runoff and soil loss during 1958–1970 in the plots of (1) natural *shola* (wet mountain, temperate, evergreen) forest; (2) a blue gum (*Eucalyptus globulus*) plantation; (3) a black wattle (*Acacia mearnsii*) plantation; and (4) a mixed plantation of (2) and (3), on the Nilgiri plateau, Tamil Nadu. Little difference existed in runoff between the vegetation types: soil loss (slight) was observed only from (2). Annual litter accumulation was about 2.9, 2.3, and 1.3 tons per ha. for (1), (2), and (3), respectively. The yield of timber (at the rotation age of 10 years) from both *Eucalyptus globulus* and *Acacia mearnsii* was greater from mixed plantations and is therefore recommended for both timber production and soil and water conservation in the region.

Utility of Bark and Wood and Maintaining Soil Fertility

The bulk of the bark goes to industrial raw material for the manufacture of wattle extract. The average percentage of tannins from black wattle is 35. The debarked wood is used for the manufacture of rayon-grade pulp and paper pulp. It is a very good mine crop. It adds more than 1,600–2,000 kg/ha/year of leaf litter and enriches the soil with N-fixing bacteria through root nodule, and thus is a direct enhancer of soil fertility. The per hectare production of bark and wood in the Nilgiris is perhaps lower than it ought to be, though in good pockets the production can be as high as 40 tons of bark and 200 tons of wood. From this, it can be surmised that wattle cultivation is a good source of enhancing soil fertility because of the huge leaf litter it produces. Though it may not be possible to bring all the areas to this production level, it may be possible by 2010 or so to bring the bulk of the areas to produce at least 30 tons of bark and 150 tons of wood/ha on average. Intensive soil preparation, addition of optimum dosage of fertilizers, and strict protection against fire are measures to bring the poor quality wattle areas to a reasonably productive level. This is more important as the Nilgiris have a considerable area of land that is too steep for terracing, and so unfit for general agriculture (crop production), but good enough for wattle production. Both in terms of profit and the utilization of natural resources, the cultivation of black wattle is evidently an excellent investment in the Nilgiris.

Yield

Black wattle stands are felled at about 10 years of age. Sometimes it may be necessary to carry out a prefelling thinning on the clearing about 18 months before the end of rotation. The objective is to remove diseased and nonstrippable trees to facilitate the main felling operation. Felling is done by axe or bow saw during the time of the year when the bark is most easily stripped from the trees. Bark strippability is the best when the sap is rising, and this occurs during warm, wet summer months.

At clear felling, the trees will have the breast height diameter of 15–20 cm and height up to 18–22 m. The felled trees are debranched and the bark is stripped as far up the stem as possible, cut into lengths of 1.2 m, and tied in 50 kg bundles for transport to the tannin extracting factory. Thoroughly dried bark is arranged in bales of 75–80 kg when ready for transportation. Tanning power improves by about 10–15 percent in bark carefully stored for a season. Percent tannin does not differ between bark harvested in dry and wet seasons. The amount of bark on trees may be less on soils of poor fertility, than on soils of rich fertility. Tannin runs about 25–35 percent/kg of dried bark, on either poor or fertile soil. Prior to felling of the trees, horizontal cuts are given about a meter above the ground level. Then the bark is stripped down. This makes the complete removal of the bark on the stump. Then the trees are felled and the bark is stripped upward to the point at which immature bark is reached. The bark stripped from the tips of the trees is generally green in color, whereas that obtained from below is gray in color. The difference in quality between the two extracts is relatively slight (Grey and Jones, 1960).

The green bark is sent to the factory for processing, drying being the first step. During drying, the bark is spread with its outer surface exposed. The inner surface is not exposed to the sun because it may lead to discoloration. Drying is done until the bark becomes brittle and brown in color. Instead of sun drying, bark may be dried in hot air ovens. The dried bark is cut into small pieces of 15–20 cm length and pressed into bales and sent to tanneries. Bark should be as fresh as possible for delivery to the factory, and any delay may result in oxidation and excessively red-colored extract or the formation of fungal mold might result, which is an undesirable characteristic. The yield of bark varies from 8 to 30 tons/ha. In the Nilgiris, it varies from 8 to 17 tons/ha. With respect to wood for fuel, it ranges from 75 to 100 tons/ha.

An average plantation yields about 40 tons of bark/ha and 200 tons of wood/ha. Taking the 2001 market prices of bark and wood as the base line (bark at the rate of Indian rupees 500/ton and wood at rupees 125/ton – US$1 = approximately 46 Indian rupees as per prevailing exchange rate), and also taking into account the expenditure on conversion from wood to bark, supervision, transportation, and overhead charges (Indian rupees 100/ton of wood and rupees 100 for peeling of 1 ton of bark), a hectare of wattle plantation fetches about Rs. 21,000 on average (approximately US$460) for a rotation of 10 years. In the second rotation the returns will be much higher. As per Indian standards, this is not a small sum of money to be earned.

The percentage of tannin content is correlated directly with bark thickness. The diameter of the tree is correlated positively with bark thickness and tannin content. The larger the mean diameter of the tree, the thicker its bark will be (Rajagopalan, 1973).

At the Wattle Research Station, Ooty (Tamil Nadu), in India, it was observed that there was not much difference between the age group and tannin content, though the maximum percentage of tannin was observed in the ninth year (41.12 percent) and the minimum in the sixth year (37.19 percent) (Rajagopalan, 1973, see Table 10.1).

With age, although the increase in tannin content is negligible, the increase in bark weight is substantial. There is an increase of 136 percent in moisture-free bark weight when the age of the tree doubles from 4 to 8.

Experiments conducted in South Africa on quality and yield of black wattle bark with regard to different densities ranging from 1,000/ha up to 2,000/ha and managed on rotation ranging from 8, 10, and 12 years revealed that for the rotation lengths, the percentage of stripping decreased more or less regularly with decreasing stand density (Rajagopalan, 1973, see Table 10.2). When the density was doubled from 1,000/ha to 2,000/ha the tonnage of bark increased only by 18 percent (Table 10.2).

The mortality percentage, however, though following a similar trend on short rotation, behaved quite differently on rotations of 10 and 12 years. Mortality was invariably the lowest at a density 1,500 trees/ha and showed a more or less regular

Table 10.1 Tannin Content of the Bark from Trees of Different Ages

Details	**Age in Years**				
	4	**5**	**6**	**7**	**8**
Tannin (%)	36.10	37.70	38.90	39.30	39.40
Moisture-free bark (tons/ha)	4.09	5.93	6.50	7.76	9.64
Tannin (tons/ha)	1.48	2.23	2.51	3.04	3.76

Source: Rajagopalan (1973).

Table 10.2 Yield of Bark from Different Stand of Densities/ha

Number of Stems	Yield of Bark (t/ha)
2000	23.98
1750	23.98
1500	23.15
1250	22.32
1000	20.39

Source: Rajagopalan (1973).

size with successive density levels above and below this point. It was presumed to be due to the fact that mortality resulting from the suppression rises as stand density increases above 1,500 trees/ha, and mortality from diseases increases as densities fall below this level (Sherry, 1966). The average yield for the South African wattle industry is estimated at 18.83 t/ha. Experiments conducted at the Wattle Research Institute at Pietermaritzberg, to determine the optimum time of felling the wattle stand in tons of tannin content revealed that the tannin content of the bark increased significantly with age of the tree (Anonymous, 1964).

Except for some mangrove species, black wattle in pure stand produces more tannin/ha than most tanniferous plants. In South Africa, well-managed plantations produced the equivalent of 3 tons/ha tannin, about twice the average, when grown in rotations in excess of 12 years. Twelve trees produce 1 cubic meter of firewood. The wood of debarked trees is dried and used for pulpwood, fuel, and mine timbers. Moisture loss is rapid during the first 4 weeks after felling, then much slower. Wood weighs 708.7 kg/m^3. One tree produces up to 10 cwt of bark or about 5 cwt when striped. One ton of black wattle bark is sufficient to tan 2,530 hides, best adapted for sole leather and other heavy goods; the leather is as durable as that tanned with oak bark. One ton of bark yields 4 cwt of extract tar. Destructive distillation of the wood yields 33.2 percent charcoal, 9.5 percent lime acetate, and 0.81 percent methyl alcohol. As a source of vegetable tannin, black wattle shares a large portion of the world market for vegetable tannins with quebracho and chestnut. According to Sherry (1971), plantation-grown wattle in South Africa, Rhodesia, Tanzania, Kenya, and Brazil supplied about 38 percent of the global demand for tannin. *Eucalyptus grandis* produces more wood than wattle, but is inferior for fuel and charcoal. At one time in South Africa, 56 percent of the proceeds from wattle were from bark, the balance being from timber (Duke, 1981).

The Processing of Bark

Wattle bark as a tanning material appears to have entered the European market in 1908, and its exceptional merits were immediately recognized by the tanning industries in England and continental Europe. It contains astringent catechol tannin and lends itself particularly to sole-leather manufacture, but it can also be used very successfully for light leather. Wattle leather is firm and durable. Although classified as a rapid tanning material, the leather is much less red than that obtained from many other catechol tans. The solubility of wattle tannin compares very favorably with that of other commercial vegetable tannins, and the temperature and concentration of extraction are not such important factors as in quebracho. The tan liquors produce very little acid on fermentation, and as a consequence, do not plump well. Wattle, therefore, makes good blend with acid-producing tanning materials, such as myrobalans.

Gupta et al. (1981) reported that condensed tannin (used in adhesives) from *Acacia mearnsii* bark was extracted in boiling water and purified using hide powder. Crude yield was 39 percent and bark dry weight, and purified yield was 52 percent of this. MCL weight was 1.025 kg and the contents of total and phenolic hydroxyl groups were 26 and 14 percent, respectively.

The tannin content of dried wattle bark is 30–40 percent. Wattle bark extract is manufactured by a countercurrent extraction process (Williams, 1954) whereby shredded bark is leached with hot water in a series of vessels. The resulting liquor is then clarified and concentrated by evaporation in multistage evaporators and vacuum pans until a point is reached where it solidifies on cooling. It is then run to cooling racks where the product is left to harden. Any contact with iron must be carefully avoided from the process because of the discoloration that will occur when tannin comes in contact with materials containing iron due to oxidation of iron in aerobic conditions, which will impart a deep reddish-yellow tinge.

Pests and Diseases

The wattle bag worm (*Acathopsyche junodi* Hely) is a serious pest that can bring down yield due to damage of the tree. This pest causes complete defoliation of trees of all ages with consequent permanent stunting of the growth. The other principal insects attacking Brazilian wattle are *Molippa sabina*, *Achryson surinamum*, *Placosternus cycleme*, *Eburodacrys dubitata*, *Neoclytus pusillus*, *Oncideres impluviata*, *Oncideres saga*, and *Trachyderes thoracica*. Ants, termites, and borers also cause damage. The dauva ant, which attacks the leaves, is fought constantly with arsenicals and carbon disulfide. Nematodes reported on this species include *Meloidogyne arenaria*, *Meloidogyne incognita* sp. *acrita*, and *Meloidogyne javanica*.

The most serious disease is dieback, caused by *Phoma herbarum*. Other fungi attacking black wattle include *Chaetomium cochliodes*, *Daldinia* sp., and *Trichoderma viride*.

Ribeiro et al. (1988) found that several *Acacia decurrens* plants of the Capao Bonito region of Brazil were affected with symptoms of wilting, wood splitting, and gum exudation. Transversely cut wood showed ashy-colored pith, which liberated numerous perithecia when incubated in a humid chamber. The pathogen was identified as *Chaetomium fimbriata*. The 4-month old plants inoculated with isolate died after 14 days. In cross-inoculation tests, isolates from *Acacia* and mangoes were pathogenic to both hosts. Morris et al. (1988) recorded a rust fungus on *Acacia mearnsii* for the first time in South Africa. It is a uredial rust, and comparison with rusts on this host in Australia suggests that it is probably the uredial state of *Uromycladium alpinum*.

Bibliography

Abdul Khader, K.B., 1998. Effect of drip irrigation on yield attributes of arecanut. National Seminar on Drip Irrigation. Water Technology Centre, Tamil Nadu Agricultural University, Coimbatore, India.

Abdul Khader, K.B., Antony, K.J., 1968. Intercropping: a paying proposition for areca growers—What crops to grow? Indian Farming 18, 14–45.

Abdul Khader, K.B., Havanagi, G.V., 1991. Consumptive use of water in relation to cumulative pan evaporation (CPE) with and without mulching in arecanut. J. Plant. Crops 18 (Suppl.), 139–146.

Abdul Salam, M., Bhaskara Rao, E.V.V. (Eds.), 2001. Cashew Varietal Wealth of India. Directorate of Cashew and Cocoa Development, Government of India, Ministry of Agriculture, Cochin, India.

Abdullah, A., 1994. Technological Packages for Cashew Development. Upland Farmer Development Project. Directorate-General Estate, Ministry of Agriculture, Indonesia.

Abraham, K.J., 1974. Intercropping in arecanut helps to build up farmer's economy. Arecanut Spices Bull. 5, 73–75.

Abraham, S.T., 2000. Genetic parameters and divergence in certain wild genotypes of *Hevea brasiliensis* (Willd. ex Adr.de Juss.) Muell. Arg. Ph.D. Thesis submitted to Mahatma Gandhi University, Kottayam, Kerala State, India.

Abraham, A., Mathew, P.M., Ninan, C.A., 1961. Cytology of *Cocos nucifera* L. and *Areca catechu* L. Cytologia 26, 327–332.

Abraham, P.A., Wycherley, P.R., Pakianathan, S.W., 1968. Stimulation of latex flow in *Hevea brasiliensis* by 4-amino-3, 5, 6-trichloropicolinic acid and 2-chloroethane phosphoric acid. J. Rubber Res. Inst. Malaya 20, 291–305.

Abraham, S.T., Reghu, C.P., Madhavan, J., George, P.J., Potty, S.N., Panikkar, A.O.N., Saraswathy, P., 1992. Evaluation of *Hevea* germplasm I: variability in early growth phase. Indian J. Nat. Rubber Res. 5 (1 and 2), 195–198.

Abraham, S.T., Reghu, C.P., Panikkar, A.O.N., George, P.J., Potty, S.N., 1994. Juvenile characterization of wild *Hevea* germplasm. Indian J. Plant Genet. Resour. 7 (2), 157–164.

Abraham, S.T., Reghu, C.P., George, P.J., Nair, R.B., 1997. Studies on genetic divergence in *Hevea brasiliensis*. Paper presented at the Symposium on Tropical Crop Research and Development, Trissur, Kerala State, India, September 9–12.

Abrizah, O., 1995. Partial purification and characterization acyl-ACP thioesterase from oil palm mesocarp (*Elaeis guineensis*). M.Sc. Thesis. Universiti Kebangsaan Malaysia, Malaysia.

Abrizah, O., Lazarus, C.M., Stobart, A.K., 1999. Isolation of a cDNA clone encoding an acyl – acyl carrier protein thioesterase from the mesocarp of oil palm (Elaeis guineensis). J. Oil Palm Research Special Edition, 81–87.

Aerts, R.J., 1992. Detrimental effect of cinchona leaf alkaloids on larvae of the polyphagous insect. Spodoptera Chem. Ecol. 18 (11), 1955–1964.

The Agronomy and Economy of Important Tree Crops of the Developing World. DOI: 10.1016/B978-0-12-384677-8.00014-X
Copyright © 2010 by Elsevier Inc. All rights of reproduction in any form reserved.

Aerts, R.J., Snoeiger, W., Van der Moijden, E., Veerporte, R., 1991. Allelopathic inhibition of seed germination by cinchona alkaloids. Phytochemistry 30 (9), 2947–2951.

Afnor, 1991. Controle de la qualite des produits alimentaires analyse sensorielle, fourth ed. Association Francaise de normalisation (AFNOR-DGCCRF), Paris, France.

Aggarwal, J.S., 1954. The cashew nut and CSNL. Oleagineux 8 (8–9), 559–564.

Aggarwal, J.S., 1973. Resins, varnishes, and surface coatings from cashew nut shell liquid. Paint Manuf. 43 (8), 29–31.

Agnoloni, M., Giuliani, F., 1977. Cashew Cultivation. Library of Tropical Agriculture. Ministry of Foreign Affairs, Instituto Agronomico Per L'oltremare, Florence, Italy.

Agwanda, C.O., 1987. Amelioration du Cafeier *Coffea arabica* L. au Kenya: optimisation des strategies des selection en presence d'interaction genotype X. milieu. Ph.D. Thesis. University of Montepellier, France.

Agwanda, C.O., 1999. Flavour: an ideal selection criterion for the genetic improvement of liquor quality in *Arabica* coffee. In: Proc. 19th Int. Sci. Colloquium on Coffee, Helsinki.

Agwanda, C.O., Owuor, J.B.O., 1989. Toward a more sustainable and economic production of Arabica coffee (*Coffea arabica* L.) in Kenya through the exploitation of improved cultivars: a review. Kenya Coffee 54 (638), 735–743.

Agwanda, C.O., Lashermes, P., Trouslot, P., Combes, M.-C., Charrier, A., 1997. Identification of RAPD markers for resistance to coffee berry disease, *Colletotrichum kahawae* in *Arabica* coffee. Euphytica 97, 241–248.

Alamsyah, T.S., 1991. The effect of pod storage on the quality of dry cocoa beans. Buletin Perkebunam 22 (2), 137–145.

Allard, R.W., Bradshaw, A.D., 1964. Implications of genotype–environment interactions in applied plant breeding. Crop Sci. 4, 503–508.

Alvim, P.T., 1987. Cacao. In: Alvim, P.T., Kozlowski, T. (Eds.), Ecophysiology of Tropical Crops. Academic Press, New York.

Alvim, P.T., Machado, A.D., Vello, F., 1972. Physiological responses of cacao to environmental factors. In: Proc. IVth Int. Cocoa Research Conference. Cocoa Research Unit, Trinidad, pp. 210–215.

Ambika, B., Abraham, C.C., 1979. Bioecology of *Helopeltis antonii* Sign. (Miridae: Hemiptera) infesting cashew trees. Entomon 4, 335–342.

Amma, S.P.K., Mallika, V.K., Manoharan, S., Namboothiri, R., Nair, R.V., 2002. An insight into the pre selection method in cocoa. National seminar on technologies for enhancing productivity in cocoa (Ext. Sum.). Central Plantation Crops Research Institute, Regional Station, Vittal, Karnataka State, India.

Amritkumar, M.N., Sundaresan, K., Sampath, S.R., 1985. Effect of replacing cottonseed cake by rubber seed cake in concentrate of cows on yield and composition of milk. Indian J. Anim. Sci. 55 (12), 1064–1070.

Ananda, K.S., 1999a. Arecanut Varieties. Extension Publication, CPCRI, Kasaragod, India.

Ananda, K.S., 1999b. Genetic improvements in arecanut. In: Ratnambal, M.J., Kumaran, P.M., Muralidharan, K., Niral, V., Arunachalam, V. (Eds.), Improvement of Plantation Crops. Central Plantation Crops Research Institute (CPCRI), Kasaragod, India, pp. 52–57.

Ananda, K., 2000. Exploitation of a dwarf mutant in arecanut breeding. In: Muraleedharan, N., Raj Kumar, R. (Eds.), Recent Advances in Plantation Crops Research. Allied Publications Ltd, New Delhi, India, pp. 69–72.

Ananda, K.S., Thampan, C., 1999. Promising cultivars and improved varieties of arecanut (*Areca catechu* L.). Indian J. Arecanut Spices Med. Plants, 24–28.

Anderson, L., 1995a. Diversity and origin of Andean Rubiaceae. In: Churchill, S.P., Balselu, H., Farero, E., Luteyn, L.L. (Eds.), Biodiversity and Conservation of Neotropical Mountain Forests. Proceedings of the New York Botanical Garden, pp. 441–450.

Anderson, L., 1995b. Tribes and genera of the cinchoneae complex (Rubiacaeae): first international conference on the Rubiaceae. Ann. Missouri Bot. Garden 82 (3), 409–427.

Anderson, L., 1998. A revision of the genus *Cinchona* (Rubiaceae–Cinchonae). Memoirs NY Bot. Garden 80, 1–75.

Annamma, Y., Licy, J., John, A., Panikkar, A.O.N., 1989. An incision method for early selection of Hevea seedlings. Indian J. Natural Rubber Res. 2 (2), 112–117.

Annamma, Y., Marattukalam, J.G., Premkumari, D., Saraswathy Amma, C.K., Licy, J., Panikkar, A.O.N., 1990. Promising rubber planting materials with special reference to Indian clones. In: Proc. Planters Conference, Kottayam, Kerala State, India. Rubber Research Institute of India.

Anonymous, 1950. List of common names of Indian plant diseases. Indian J. Agric. Sci. 20 (1), 104–107.

Anonymous, 1960. Report of the Wattle Research Institute. University of Natal, Pietermaritzburg, South Africa.

Anonymous, 1964. Report of the Wattle Research Institute. University of Natal, Pietermaritzburg, South Africa.

Anonymous, 1985. Notes of the month: new CBD/leaf rust resistant variety launched. Kenya Coffee 50, 378–382.

Anonymous, 1989. An Atlas of Coffee Pests and Diseases. Coffee Research Foundation, Ruiru, Kenya.

Anonymous, 1993a. Indian cashewnut shell liquid—a versatile industrial raw material. Indian Cashew J. 21 (1), 8–9.

Anonymous, 1993b. The review of natural products. Quinine: facts and comparisons. Michael R. Riley Publishers, St. Louis, MO, USA, pp. 1–2.

Anonymous, 1994. Annual Report 1993–94. National Research Centre for Cashew, Puttur, India.

Anonymous , 1999. Forest Survey of India. Indian Council of Forest Research and Education, Dehradun, Madhya Pradesh, India.

Anonymous, 2001. Control of Coffee Berry Disease and Leaf Rust in 2001. Coffee Research Foundation. Technical Circular No. 804, Ruiru, Kenya.

Anuradha, S., 1999. Genetic resources and conservation strategies in arecanut (*Areca catechu*). In: Ratnambal, M.J., Kumaran, P.M., Muralidharan, K., Niral, V., Arunachalam, V. (Eds.), Improvement of Plantation Crops. Central Plantation Crops Research Institute (CPCRI), Kasaragod, India, pp. 48–51.

APCC (Asia Pacific Coconut Community), 1997. APCC codes and standard for aqueous coconut products. Asian Pacific Coconut Community, Jakarta, Indonesia.

Aradhya, M.K., Yee, L.K., Zee, F.T., Manshardt, R.M., 1998. Genetic variability in Macadamia. Genet. Resour. Crop Evol. 45, 19–32.

Aragundi, J., Suarez, C., Solorzano, G., 1987. Evidence of resistance of cocoa clone EET 233 to moniliasis and "witches broom". In: Proc. 10th Int. Cocoa Research Conference, Santo Domingo, Dominican Republic, Cocoa Producers Alliance, London, UK, May 17–23, pp. 479–483.

Archer, B.L. 1964. Site and mechanism of rubber biosynthesis from mevalonate. Proceedings of the National Rubber Producers Research Association Jubilee Conference, 1964, Cambridge, London pp. 101–112.

Archer, B.L., Barnard, D., Cockbain, E.G., Dickenson, P.B., McMullen, A.I., 1963. Structure composition and biochemistry of *Hevea* latex. In: Bateman, L. (Ed.), The Chemistry and Physics of Rubber-Like Substances. Maclaren & Sons, Ltd, London, UK, pp. 41–72.

Arif, M., Harwood, J.L., 2000. Purification and characterization of acyl-CoA: glycerol 3-phosphate acyltransferase from oil palm (*Elaeis*) tissue. Planta 210, 318–328.

Ariffin, D., 2000. Major diseases of oil palm. In: Advances in Oil Palm Research, vol. 1. Malaysian Oil Palm Board, Kuala Lumpur, Malaysia, pp. 596–622.

Ariffin, D., Idris, A.S., Marzuki, A., 1995. Development of a technique to screen oil palm seedlings for resistance to *Ganoderma*. In: Proc. 1995 PORIM National Oil Palm Conference Technology in Plantation—The Way Forward, pp. 132–141.

Arikiah, A., Tan, Y.P., Sharma, Y.P., Clapperton, J.F., 1994. Experiments to determine influence of primary processing parameters and planting material on the flavour of cocoa beans in Malaysia. Cocoa Growers' Bull. 48, 36–46.

Arokiaraj, P., Wan, A.R.W.Y., 1991. Agrobacterium-mediated transformation of *Hevea* cells derived from *in vitro* and *in vivo* seedling cultures. J. Nat. Rubber Res. 6 (1), 55–61.

Arokiaraj, P., Jaafar, H., Hamzah, S., Yeang, H.Y., Wan Abdul Rahaman, W.Y., 1996. Enhancement of *Hevea* crop potential by genetic transformation: HMGR activity in transformed tissue. In: Proc. Symposium on Physiological and Molecular Aspects of the Breeding of *Hevea brasiliensis*, Penang, Malaysia, November 6–7, 1995. International Rubber Research and Development Board, Hertford, England, pp. 74–82.

Arokiaraj, P., Yeang, H.Y., Cheong, K.F., Hamzah, S., Jones, H., Coomber, S., Charlwood, B.V., 1998. A gene transfer system for *Hevea brasiliensis* was established with *A. tumefaciens* GV 2260 (p35SGUSINT) and LBA4404 (pAL4404/pMON9793): CaMV 35S promoter directs beta-glucuronidase expression in the laticiferous system of transgenic *Hevea brasiliensis* (rubber tree). Plant Cell Rep. 17 (8), 621–625.

Ascenco, J.C., 1986. Potential of the cashew crop—1. Agric. Int. 38 (11), 324–327.

Ashburner, G.R., 1994. Characterization, collection, and conservation of *Cocos nucifera* L. in the South Pacific. Ph.D. Thesis. School of Agriculture and Forestry, University of Melbourne.

Ashburner, G.R., 1999. The application of molecular markers to coconut genetic improvement. In: Oropeza, C., Verdeil, J.L., Ashburner, G.R., Cardega, R., Santamarma, J.M. (Eds.), Current Advances in Coconut Biotechnology. Kluwer Academic Publishers, Boston, MA, USA.

Ashburner, G.R., Been, B.O., 1997. Characterization of resistance to lethal yellowing in *Cocos nucifera* and implications for genetic improvement of this species in the Caribbean region. In: Eden-Green, S.J., Ofori, F. (Eds.), Proc. Int. Workshop on Lethal Yellowing-Like Diseases of Coconut. NRI Chatham, Kent, UK, pp. 173–183.

Asokan, M.P., Sobhana, P., Sushmakumari, S., Sethuraj, M.R., 1988. Tissue culture propagation of rubber (*Hevea brasiliensis*) Wild. ex Adr. De Juss. Muell. Arg. Clone GT (Gondang Tapen) 1. Indian J. Nat. Rubber Res. 1 (2), 1–9.

Asokan, M.P., Kumari Jayasree, P., Sushmakumari, S., Sobhana, P., 1992. Plant regeneration from anther culture of rubber tree (*Hevea brasiliensis*). Abstract of International Natural Rubber Conference, Bangalore, February, Rubber Research Institute of India, Kottayam, Kerala State, India, p. 49.

Awoke, T.C., 1997. The culture of coffee in Ethiopia. Agroforestry Today 9 (1), 19–21.

Aziz, A., Rosnah, M.S., Mohamadiah, B., Wan Zailan, W.O., 1986. Relationship of the formation of storage oil with fatty acid composition in developing oil palm fruit. Proc. Mal. Biochem. Soc. Conf. 12, 147–151.

Bailey, L.H., 1958. The Standard Encyclopedia of Horticulture, seventeenth ed. vol. 1. MacMillan, New York, NY, USA.

Baker, C.J., 1970. Coffee bark disease in Kenya. Kenya Coffee July, 226–228.

Baker, C.J., 1972. *Fusarium solani* associated with a wilt of *Coffea arabica* in Kenya. E. Afr. Agric. For. J. 337, 137–140.

Baker, C.S.L., Gelling, I.R., Newell, R., 1985. Epoxidised natural rubber. Rubber Chem. Technol. 58 (1), 67–85.

Balasimha, D., 1986. Light penetration patterns through arecanut canopy and leaf physiological characteristics of intercrops. J. Plant. Crops 16 (Suppl.), 61–67.

Balasimha, D., Subramonian, N., 1984. Nitrate reductase activity and specific leaf weight of cocoa and light profile in areca–cocoa mixed cropping. In: Proc. Plantation Crops Symposium VI. Kottayam, Indian Society of Spice Crops, Kasaragod, India, pp. 83–88.

Balasimha, D., Abdul Khader, K.B., Bhat, R., 1996. Gas exchange characteristics of cocoa grown under arecanut in response to drip irrigation. All India Seminar on Modern Irrigation Techniques June 26–27, 96–101.

Balasubramanian, P.P., 2000. Cashew—the premier crop of Indian commerce. In: Mohanty, K.K., Lenka, P.C., Patnaik, A.K. (Eds.), Selected Articles on Cashew—State Level Seminar on Cashew. OSCDC, Bhubaneswar, India, pp. 3–11.

Basri, M.W., 1993. Life history, ecology, and economic impact of the bag worm, *Metisa plana* Walker (Lepidotera: Psychidae) on the oil palm *Elaeis guineensis* Jacquin (Palmae) in Malaysia. Ph.D. Thesis. University of Guelph, Ontario, Canada.

Basri, M.W., Norman, K., 2000. Insect pests, pollinators, and barn owls. In: Yusof, B., Jalani, B.S., Chan, K.W. (Eds.), Advances in Oil Palm Research. Malaysian Oil Palm Board, Kuala Lumpur, Malaysia, pp. 466–541.

Basri, M.W., Norman, K., Hamdan, A.B., 1995. Natural enemies of the bag worm *Metisa plana* (Lepidoptera: Psychidae) and their impact on host population regulation. Crop Prot. 14 (8), 637–645.

Basri, M.W., Simon, S., Ravigadevi, S., Othman, A., 1999. Beneficial plants for the natural enemies of the bag worm in oil palm plantations. In: Ariffin, D., Chan, K.W., Sharifah. (Eds.), Proc. 1999 PORIM Int. Palm Oil Congress—Emerging Technologies and Opportunities in the Next Millennium. Palm Oil Research Institute of Malaysia, Kuala Lumpur, Malaysia, pp. 441–455.

Baudouin, L., 1999. Genetic improvement of coconut palm. In: Oropeza, C., Verdeil, J.L., Ashburner, G.R., Cardega, R., Santanamarma, J.M. (Eds.), Current Advances in Coconut Biotechnology. Kluwer Academic Publishers, Boston, MA, USA.

Bauermeister, J., 1978. Grinding of cocoa mass. Industrie Alimentari 17 (5), 424–428.

Baulkwill, W.J., 1989. The history of natural rubber production. In: Webster, C.C., Baulkwill, W.J. (Eds.), Rubber. Longman Scientific & Technical, New York, NY, USA, pp. 1–56.

Baum, H., 1968. The coffee root mealy bug complex. Kenya Coffee 33, 175–178.

Bavappa, K.V.A., 1961. Some common intercrops in an arecanut garden. ICAC Monthly Bull. 2, 16–17.

Bavappa, K.V.A., 1963. Morphological and cytological studies in *Areca catechu* Linn. and *Areca triandra* Roxb. M.Sc. (Ag) Thesis. University of Madras, India.

Bavappa, K.V.A., 1966. Morphological and anatomical studies in *Areca catechu* Linn and *Areca triandra* Roxb. Phytomorphology 16, 436–443.

Bavappa, K.V.A., 1974. Studies in the genus *Areca* L. (cytological and genetic diversity of *A. catechu* L. and *A. triandra* Roxb.). Ph.D. Thesis. University of Mysore, India.

Bavappa, K.V.A., 1977. Mangala—a superior arecanut variety. Arecanut Spices Bull. 9, 55–56.

Bavappa, K.V.A., 1982. Future of arecanut. In: Nair, M.K., Prem Kumar, T. (Eds.), The Arecanut Palm. CPCRI, Kasaragod, India, pp. 319–325.

Bavappa, K.V.A., Nair, M.K., 1978. Cytogenetics of *Areca catechu* L., *A. triandra* Roxb. and their F1 hybrids (Palmae). Genetica 49, 1–8.

Bavappa, K.V.A., Nair, M.K., 1982. Cytogenetics and breeding. In: Bavappa, K.V.A., Nair, M. K., Prem Kumar, T. (Eds.), The Arecanut Palm. CPCRI, Kasaragod, India, pp. 51–96.

Bavappa, K.V.A., Pillai, S.S., 1976. Yield and yield component analysis in different exotic cultivars and species of Areca. In: Chadha, K.L. (Ed.), Improvement of Horticulture, Plantation and Medicinal Plants, vol. 1. Today and Tomorrow Printers and Publishers, New Delhi, India, pp. 243–246.

Bavappa, K.V.A., Ramachander, P.R., 1967. Improvement of arecanut palm, *Areca catechu* L. Indian J. Genet. 27, 93–100.

Bavappa, K.V.A., Raman, V.S., 1965. Cytological studies in *Areca catechu* Linn. and *Areca triandra* Roxb. J. Indian Bot. Sci. 44, 495–505.

Bavappa, K.V.A., Nair, M.K., Ratnambal, M.J., 1975. Karyotype studies in *Areca catechu* L., *A. triandra* Roxb. and *A. catechu* × *A. triandra* hybrids. Nucleus 18, 146–151.

Bavappa, K.V.A., Kailasam, C., Abdul Khader, K.B., Biddappa, C.C., Khan, H.H., Kasturi Bai, K.V., Ramadasan, A., Sundararaju, P., Bopaiah, B.M., Thomas, G.V., 1986. Coconut and arecanut based high density multispecies cropping systems. J. Plant. Crops 14, 74–87.

Beccari, O., 1919. The palms of the Philippine Islands. Philipp. J. Sci. 14, 295–362.

Beena, B., Abdul Salam, M., Wahid, P.A., 1995. Nutrient uptake in cashew (*Anacardium occidentale* L.). Cashew 9 (3), 9–16.

Beirnaert, A., 1953. Introduction a la biologie florale du palmier a huile *Elaeis guineensis* Jacq. Publis I.N.E.A.C. Ser. Sci. No. 5.

Beinaert, A., Vanderweyesn, R., 1941. Contribution a letude genetique et biometrique des varieties d'*Elaeis guineensis*. Jacquin. Publs. Inst. Natn. Etude Agron Congo Belge. Ser. Sci. 27.

Bell, G.D.H., Rogers, H.H., 1956. Cacao breeding at WACRI. In: Proc. Cacao Breeding Conference on West African Cocoa Research Institute, West African Cocoa Research Institute, Ghana, October 1–3, pp. 31–49.

Benjamin, M., Chee, K.H., 1995. Basal stem rot of oil palm—a serious problem on inland soils. MAPPS Newslett. 19, 3.

Berkley, M.J., 1869. Notes. Gardeners (cited in Kushalappa, 1989).

Bess, H.A., 1964. Populations of the leaf miner *Leucoptera meyricki* and its parasites in sprayed and unsprayed coffee in Kenya. Bull. Entomol. Res. 55, 59–82.

Bessa, A.M.S., Sardinha, R.M.A., 1994. *In vitro* multiplication of cashew (*Anacardium occidentale* L.) by callus culture. In: Proc. 8th Int. Congress on Plant Tissue Culture, Firenze, Italy, June 12–17, pp. 2–37.

Besse, P., Seguin, M., Lebrun, P., Chevallier, M.H., 1994. Genetic diversity among wild and cultivated populations of *Hevea brasiliensis* assessed by nuclear RFLP analysis. Theor. Appl. Genet. 88, 199.

Bhaskara Rao, E.V.V., 1994. Cashew Cultivation in Myanmar, Deputation Report—Project UNDP/MTA/86/018, ARCPC (Plantation Crops Division, Yangon, Myanmar).

Bhaskara Rao, E.V.V., 1996. Emerging trends in cashew improvement. In: Chopra, V.L., Singh, R.B., Varma, A. (Eds.), Crop Productivity and Sustainability—Shaping the Future, Second Int. Crop Science Congress. Oxford and IBH Publishing Co. Pvt. Ltd, New Delhi, India, pp. 675–682.

Bhaskara Rao, E.V.V., Bhat, M.G., 1996. Cashew breeding—achievements and priorities. In: Proc. Seminar on Crop Breeding in Kerala. Department of Botany, Kerala University, Trivandrum, India, pp. 47–53.

Bhaskara Rao, E.V.V., Nagaraja, K.V., 2000. Status report on cashew. In: Rethinam, P. (Ed.), Souvenir Int. Conf. on Plantation Crops, December 12–15. National Research Center for Oil Palm, Pedavegi, India, pp. 1–7.

Bhaskara Rao, E.V.V., Swamy, K.R.M., 1994. Genetic resources of cashew. In: Chadha, K.L., Rethinam, P. (Eds.), Advances in Horticulture, Volume 9: Plantation and Spice Crops, Part 1. Malhotra Publishing House, New Delhi, India, pp. 79–97.

Bhaskara Rao, E.V.V., Swamy, K.R.M., 2000. Cashew research scenario in India. In: Balasubramanian, P.P. (Ed.), The Cashew Industry in India. DCCD, Cochin, India. In press

Bhaskara Rao, E.V.V., Swamy, K.R.M, Bhat, M.G., 1998. Status of cashew breeding and future priorities. J. Plant. Crops 9 (4), 18–24.

Bhat, M.G., Kumaran, P.M., Bhaskara Rao, E.V.V., Mohan, K.V.J., Thimmappaiah, 1998. Pollination technique in cashew. Cashew 12 (4), 21–26. Cashew Export Promotion Council of India, 2003. Cashew Statistics, CEPCI, Cochin, India.

Bhat, M.G., Bhaskara Rao, E.V.V., Swamy, K.R.M., 1999. Genetic resources of cashew and their utilization in crop improvement. In: Ratnambal, M.J., Kumaran, P.M., Muralidharan, K., Niral, N., Arunachalam, V. (Eds.), Improvement of Plantation Crops. CPCRI, Kasaragod, India, pp. 91–98.

Bhat, P.S.I., Rao, K.S.N., 1962. On the antiquity of arecanut. Arecanut J. 13, 13–21.

Blaak, G., 1967. Oil palm prospection tour in the Bamenda highlands of West Cameroon. Intl. Report Unilever, London, UK.

Blaha, G., 1996. PNG's experience on diseases of cocoa: management for chemical treatments and for resistant plant material. In: Proc. 12th Cacao Research Conference, Salvador-Bahia, Brazil, November 17–23. Cocoa Research Center, Bahia, Brazil.

Blume, C.L., 1836. Rumhiasive Commntations Botanicae de Plantis Indiae Orientalis. Leyden.

Boatman, S.G., 1966. Preliminary physiological studies on the promotion of latex flow by plant growth regulators. J. Rubber Res. Inst. Malaya 19 (5), 243–258.

Bobilioff, W., 1919. Onderzoekingen over het onstan van latexvaten enlatex bij *Hevea brasiliensis*. Archive voor de Rubber Cult. 3 (43).

Bopaiah, B.M., 1991. Soil microflora and VA mycorrhizae in areca based high density multispecies cropping and areca monocropping systems. J. Plant. Crops 18 (Suppl.), 224–228.

Bopaiah, B.M., 1992. Deterioration of processed cocoa beans in storage and mycotoxin. Indian Cocoa Arecanut Spices J. 16 (1), 11–13.

Bouharmont, P., 1995. La selection du cafeier *Arabica* au Cameroon (1964–1991). Document de travail No. 1–95 (40 p.). CIRAD, France.

Bourdeix, R., Sangare, A., Le Saint, J.P., Meunier, J., Gascon, J.P., Rognon, F., Nuce de Lamothe, M., 1993. Coconut breeding at IRHO and its application in seed production. In: Nair, M.K., Khan, H.H., Gopalakrishnan, P., Bhaskara Rao, E.V.V. (Eds.), Advances in Coconut Research and Development. International Symposium on Coconut Research and Development. Oxford and IBH Publishing Co Pvt. Ltd, New Delhi, India.

Bourgoing, R., 1991. In: Benigno, D.R.A. (Ed.), 1991. Coconut—A Pictorial Technical Guide for Smallholders. IRHO/CIRAD, Paris, France, p. 301.

Bowler, W.W., 1953. Electrophoretic mobility of fresh *Hevea* latex. Indian Eng. Chem. 45, 1790.

Brahma, R.N., 1974. In North Bengal banana is a paying intercrop in areca gardens. Arecanut Spices Bull. 5, 80–81.

Broadbent, J.H., Turatti, J.M., Tocchini, R.P., Iaderos, M., 1997. Rural processing of cocoa beans in Brazil. Trop. Sci. 37 (3), 164–168.

Brockway, L., 1979. Science and Colonial expansion: the Role of British Botanical Garden. New York Academy Press, New York, NY, USA.

Brough, C.L., Coventry, J.M., Christie, W.W., Kroon, J.T.M., Brown, A.P., Barsby, T.L., Slabas, A.R., 1996. Toward the genetic engineering of triacylglycerols of defined fatty acid compositions: major changes in erucic acid content at the sn-2 position affected by the introduction of a 1-acyl-sn-glycerol-3-phosphate acyltransferase from *Limnanthes douglasii* into oilseed rape. Mol. Breed. 2, 133–142.

Browning, G., Fisher, N.M., 1976. High density coffee: yield results for the first cycle from systematic plant spacing designs. Kenya Coffee 41 (483), 209–217.

Bruncton, J., 1995. Pharmacology and Phytochemistry of Medicinal Plants. Intercept Ltd, Hampshire, England.

Burkill, 1935. Cinchona: In Dictionary of the Economic Plants of Malaya Peninsula, vol. 1, Oxford, UK, pp. 538–543.

Burotrop, 1992. Coconut and Oilpalms: World Production and Consumption. Map. BUROTROP, Montpellier, France.

Cannell, M.G.R., 1970. The contribution of carbohydrates from vegetative laterals to the growth of fruits on the bearing branches of *Coffea arabica* L. Turrialba 20, 15–19.

Cardena, R., Ashburner, G.R., Oropeza, C., 1999. Prospects for marker assisted breeding of lethal yellowing-resistant coconuts. In: Oropeza, C., Verdeil, J.L., Ashburner, G.R., Cardena, R., Santamarma, J.M. (Eds.), Current Advances in Coconut Biotechnology. Kluwer Academic Publishers, Boston, MA, USA.

Carron, M.P., Enjarlic, F., 1982. Studies on vegetative micropropagation of *Hevea brasiliensis* by somatic embryogenesis and *in vitro* microcutting. In: Proc. Fifth Int. Congress on Plant Tissue and Cell Culture, Tokyo, pp. 751–752.

Carron, M.P., Lardet, L., Dea, B.G., 1998. *Hevea* micropropagation by somatic embryogenesis. Plantation R es. Develop. 5, 187–192.

Carvalho, A., Monaco, I.C., 1969. The breeding of Arabica coffee. In: Ferwarda, F.P., Wit, F. (Eds.), Outlines of Perennial Crop Breeding in the Tropics. Landbouw, Wageningen, The Netherlands, pp. 198–216. Miscellaneous Paper 4.

Castillo, Z.J., Moreno, R.G., 1988. *La variedad Colombia* (The Colombia Variety). Cenicafe Publication, Bogota.

Central Plantation Crops Research Institute (CPCRI), 1983. Annual Report. Kasaragod, India.

Central Plantation Crops Research Institute (CPCRI), 1988a. Annual Report for 1986. Kasaragod, India.

Central Plantation Crops Research Institute (CPCRI), 1988b. Annual Report for 1987. Kasaragod, India.

Central Plantation Crops Research Institute (CPCRI), 1994. Annual Report. Kasaragod, India.

Chacko, E.K., 1993. Genetic Improvement of Cashew Through Hybridisation and Evaluation of Hybrid Progenies. A final report prepared for the Rural Industries R&D Corporation (mimeographed). CSIRO Division of Horticulture, Winnellie, Australia.

Chacko, E.K., Baker, I., Downton, J., 1990. Toward a sustainable cashew industry for Australia. J. Aust. Inst. Agric. Sci. 3 (5), 39–43.

Chan, K.W., Watson, I., Lim, K.C., 1981. Use of oil palm waste material for increased production. Planter 57, 14–37.

Chandramohanan, R., Kaveriappa, K.M., 1985. Epidemiological studies on inflorescence dieback of arecanut caused by *Colletotrichum gloeosporioides*. In: Proc. Silver Jubilee Symposium on Arecanut Research and Development, CPCRI, Vittal, India, 1982, pp. 104–106.

Chandy, B., Joseph, T., Mohanakumar, S., 1998. Commercial exploitation of rubber honey in India: report of a sample survey. Rubber Board Bull. 27 (3), 9–15.

Chapman, G.W., 1951. Plant hormones and yield in *Hevea brasilienssis*. J. Rubber Res. Inst. Malaya 13, 167–176.

Chappell, J., 1983. A revised sea level for the past 300,000 years from Papua New Guinea. Search 14, 99–101.

Charrier, A., 1982. Quelques reflexions sur les possibilities d'amelioration genetique de la qualite des cafes. In: 10th Int. Colloquium Coffee, Salvador, pp. 369–374. Association Scientifique International des Cafes, Paris. De couverture dans les cafeieres au Burundi. Cafe Cacao The xxxviii, 41–48.

Charrier, A., Eskes, A.B., 1997. Les Cafeiers. In: Charrier, A., Jacquot, M., Harmon, S., Nicolas, D. (Eds.), L'Amelioration des plantes tropicales. CIRAD and ORSTOM, Paris, pp. 171–196.

Chatterjee, S.K., 1982. Cultivation of cinchona in Darjeeling hills, West Bengal. In: Atal, C.K., Kapur, B.M. (Eds.), Cultivation and Utilization of Medicinal Plants. Regional Research Laboratory, Jammu, India, pp. 222–229.

Chatterjee, S.K., 1993. Domestication studies of some medicinally important exotic plants growing in India. Acta Hort. 331, 151–160.

Chatterjee, S.K., Bharati, P., Chhetri, K.B., Ramsong, A.F., 1988. Agrotechnology of cultivating cinchona and ipecac in Darjeeling hills. In: Proc. Fifth ISHA Symposium of Medicinal, Aromatic, and Spice Plants 188-A: Special Issue, pp. 25–27.

Cheah, S.C., Sambanthamurthi, R., Siti Nor Akmar, A., Abrizah, O., Mohamad Arif, A.M., Umi Salamah, R., Ahmad Parveez, G.K., 1995. Toward genetic engineering of oil palm (*Elaeis guineensis* Jacq.). In: Kader, J.C., Mazliak, P. (Eds.), Plant Lipid Metabolism. Kluwer Academic Publishers, Dordrecht, The Netherlands, pp. 570–572.

Chee, K.H., Holliday, P., 1986. South American leaf blight of *Hevea* rubber. Malay. Rubber Res. Develop. Board Monograph 13.

Cheesman, E.E., 1944. Notes on nomenclature, classification, and possible relationship of cacao populations. Trop. Agric. Tri. 21 (8), 144–159.

Chevallier, M.H., 1988. Genetic variability of *Hevea brasiliensis* germplasm using isozyme markers. J. Nat. Rubber Res. 3, 42–53.

Chinnappa, B., Hippargi, K., 2005. An economic impact assessment of drip irrigation technology in arecanut plantations: an empirical evidence from Karnataka. J. Plant. Crops 33 (2), 130–134.

Choo, Y.M., Yusof, B., 1996. *Elaeis oleifera* palm for the pharmaceutical industry. Palm Oil Research Institute of Malaysia, Kuala Lumpur, Malaysia. Technology Transfer No. 42.

Chowdappa, P., Balasimha, D., 1992. Non-stomatal inhibition of photosynthesis in arecanut palms affected with yellow leaf disease. Indian Phytopath. 45, 312–313.

Chowdappa, P., Balasimha, D., 1995. Chlorophyll fluorescence characteristics of arecanut palms affected with yellow leaf disease. Indian Phytopath. 47, 87–88.

Chowdappa, P., Balasimha, D., Daniel, E.V., 1993. Water relations and net photosynthesis of arecanut palms affected with yellow leaf disease. Indian J. Microbiol. Ecol. 3, 19–30.

Chowdappa, P., Balasimha, D., Rajagopal, V., Ravindran, P.S., 1995. Stomatal responses in arecanut palms affected with yellow leaf disease. J. Plant. Crops 23, 116–121.

Chowdappa, P., Saraswathy, N., Vinaya Gopal, K., Somala, M., 1999. Control of fruit rot of arecanut through polythene covering of bunches. In: Proc. Symposium on Plant Disease Management for Sustainable Agriculture. Indian Phytopathological Society. CPCRI, Kayamgulam, India, pp. 18–19.

Christou, P., Corley, R.H.V., Lee, C.H., Law, I.H., Wong, I.W., 1986. Abnormal flower development in oil palm clones. Planter 62, 141–240.

Chung, C.T.A., Staba, E.J., 1987. Effect of age and growth regulators on growth and alkaloid production in *Cinchona ledgeriana* leaf shoot organ cultures. Planta Med. 53 (2), 206–210.

Chung, G.F., 1990. Preliminary results on trunk injection of fungicides against *Ganoderma* basal stem rot in oil palm. In: Proc. *Ganoderma* Workshop. Palm Oil Research, Institute of Malaysia, Bangi, Kuala Lumpur, Malaysia, pp. 81–97.

Chung, G.F., 2000. Vertebrate pests of oil palm. In: Yusof, B., Jaslani, B.S., Chan, K.W. (Eds.), Advances in Oil Palm Research. Malaysian Oil Palm Board, Kuala Lumpur, Malaysia, pp. 542–595.

Claramma, N.M., Rajammal, G., Thomas, E.V., 1984. Graft co-polymerisation of acrylonitrile with natural rubber. Proc. Int. Rubber Conf. 2 (1), 13–18.

Clark, J.D., 1976. Prehistoric population and pressures favoring plant domestication. In: Harlan, J.R., de Wet, J., Stemler, A.B.L. (Eds.), Plant Population, Genetics, Breeding and Genetic Resources. Sinauer Associates, Sunderland, MA, USA, pp. 98–115.

Clement-Demange, A., Legnate, H., Chapuset, T., Pinard, F., Seguin, M., 1997. Characterization and use of the IRRDB germplasm in Ivory Coast and French Guyana: status in 1997. In: Proc. IRRDB Symposium on Natural Rubber, vol. 1. General, Soils and Fertilizations and Breeding and Selection Session, Ho Chih Minh City, October 14–15, pp. 71–88.

Clowes, M.*ST*.J., Logan, W.J.C., 1985. Advances in Coffee Management and Technology in Zimbabwe 1980–1985. Cannon press, Harare, Zimbabwe.

Clowes, T.J., 1949. The yellow-headed borer. Kenya Coffee October.

Coe, S.D., Coe, M.D., 1996. The True History of Chocolate. Thames and Hudson Ltd, London, UK.

Coleman, L.C., 1910. Diseases of the areca palm (*Areca catechu* L.) I. Koleroga or rot disease. Ann. Mycol. 8, 591–626.

Cook, O.F., 1928. Beginnings of rubber culture. J. Hered. 19, 204–215.

Cope, F.W., 1962. The mechanism of pollen incompatibility in *Theobroma cacao* L. J. Hered. 17, 157–182.

Coste, R., 1968. Les Cafeiers. Maisonneuve and Larose, Paris, France.

Council of Scientific and Industrial Research, 1992. Cinchona. Wealth of India: Raw Materials, revised edition, vol. 3, pp. 563–572.

D'Silva., D'Souza, L., 1992. *In vitro* bud proliferation of *Anacardium occidentale* L. Plant Cell Tissue Organ Cult. 29, 1–6.

Damodaran, V.K., 1977. F1 population variability in cashew. J. Plant. Crops 5 (2), 89–91.

Damodaran, V.K., Veera Raghavan, P.G., Vasavan, M.G., 1978. Cashew breeding. Agric. Res. J. Kerala 15 (1), 9–13.

Darlington, C.D., Janaki Ammal, E.K., 1945. Chromosome Atlas of Cultivated Plants. Allen & Unwin, London, UK.

Das, S., Jha, T.B., Jha, S., 1996. *In vitro* propagation of cashew nut. Plant Cell Rep. 15, 615–619.

Datta, M., Sah, K.D., Gupta, S., Banerjee, S.K., 1990. Characteristic of soils supporting cinchona plants in Darjeeling hills. Indian Agric. 34 (2), 73–77.

Dawson, R.F., 1991. Quinine and quinidine production in the Americas: a brief history. Hort. Technol. 1 (1), 17–21.

Dayrit, F.M., 1994. Determination of quinine content in the bark of cinchona trees grown in the Mt. Kilanglad. Bukidnon Philipp. J. Sci. 123 (3), 215–227.

De Castro, P., 1994. Summary of the study. In: Delogu, A.M., Haeuster, G. (Eds.), The World Cashew Economy. NOMISMA, L'Inchiostroblu, Bologna, Italy, pp. 11–12.

de Graaf, J., 1986. The economies of coffee. In: Economics of Crops in Developing Countries, No. 1, Pudoc, Wageningen, The Netherlands.

de Taffin, G., 1993. Le Cocotier (The Coconut). Maisonneuve et Larose, Paris, France, p. 166 (French language only).

De Vries, O., 1926. Superieur plantmateriaal (zaailingen en oculaties). De Bergcultures 1, 404.

Degani, C., Goldring, A., Gazit, S., 1989. Pollen parent effect on out crossing rate in "Hay" and "Fuerte" Avocado plots during fruit development. J. Am. Soc. Hort. Sci. 114, 106–111.

Devasahayam, S., Radhakrishnan Nair, C.P., 1986. The tea mosquito bug *Helopeltis antonii* Signoret on cashew in India. J. Plant. Crops 14, 1–10.

Devonshire, C.R., 1956. Explanation of the coffee report form. Coffee Board Kenya Monthly Bull 21, 186–187.

Dey, S.K., Thomas, M., Sathik, M.B.M., Vijaykumar, K.R., Jacob, J., Sethuraj, M.R., 1995. The physiological and biochemical regulation of yield in clones of *Hevea brasiliensis*. In: Proc. IRRDB Symposium on Physiological and Molecular Aspects of Breeding of *Hevea brasiliesis*, Malaysia, November, pp. 168–174.

Dharmaraju, E., Rao, P.A., Ayyanna, T., 1974. A new record of *Nephopteryx* sp. as an apple and nut borer on cashew in Andhra Pradesh. J. Res. 1 (4, 5, 198), Andhra Pradesh Agricultural University.

Dias, J.C., Avila, M.G.M., 1993. The effect of post harvest storage period of pods, turning process of the mass, and fermentation time on the acidity of cocoa and beans. Agrotropica 5 (2), 25–30.

Dickenson, P.B., 1965. The ultrastructure of the latex vessel of *Hevea brasiliensis*. In: Mullins, L. (Ed.), Proc. Natural Rubber Producers Research Association Jubilee Conference, 1964. MaClaren and Sons, Cambridge, London, pp. 52–66.

Dijkman, M.J., 1951. *Hevea*: Thirty Years of Research in the Far East. University of Miami Press, USA.

Directorate of Cinchona and Other Medicinal Plants, 1994. Res., Tech. Bull. 18, 3, Directorate of Cinchona and Other Medicinal Plants, Darjeeling (West Bengal), India.

Djekpor, E.J., Cilas, C., Paulin, D., 1994. Spotlight on cocoa breeding for resistance to swollen shoot. In: Proc. INGENIC Workshop on Cocoa Breeding Strategies, Kuala Lumpur, Malaysia, October 18–19, International Group for Genetic Improvement of Cocoa, University of Reading, UK, pp. 38–40.

Dollet, M., 1999. Conventional and molecular approaches for detection and diagnosis of plant diseases: application to coconut. In: Oropeza, C., Verdeil, J.L., Ashburner, G.R., Cardena, R., Santamarma, J.M. (Eds.), Current Advances in Coconut Biotechnology. Kluwer Academic Publishers, Boston, MA, USA.

Doraiswami, K., Venkatraman, K.P., 1982. Chemistry, production and marketing of cinchona alkaloids. In: Atal., C.K., Kapur, B.M. (Eds.), Cultivation and Utilization of Medicinal Plants. Regional Research Laboratory, Jammu, India, pp. 230–243.

Duke, J.A., 1981. Handbook of Legumes of World Economic Importance. Plenum Press, New York, NY, USA.

Echandi, E., 1958. La quema de las cafetas causada por *Phoma costarricensis* n.sp. Rev. Porol. Trop. 5, 81–108.

Echeverri, J.H., Fernandez, C.E., 1989. The PROMECAFE program for Central America. In: Kushalappa, A.C., Eskes, A.B. (Eds.), Coffee Rust: Epidemiology, Resistance, and Management. CRC Press, Inc., Boca Raton, FL, USA.

Edathil, T.T., 1986. South American leaf blight: a potential threat to the natural rubber industry in Asia and Africa. Trop. Pest Manag. 32 (4), 296–303.

Elliot, D.J., 1982. Natural rubber systems. In: Wheelan, A., Lee, K.S. (Eds.) Developments in Rubber Technology. Applied Science Publishers, London, pp. 221–223.

Engelmann, F., Lartaud, M., Chabrillange, N., Carron, M.P., Etienne, H., 1997. Cyropreservation on embryogenic calluses of two commercial clones of *Hevea brasiliensis*. Cyroletters 18, 107–116.

Enig, M., 1999. Coconut: in support of good health in the 21st century. In: Promoting Coconut Products in a Competitive Global Market: Proc. Thirty-Sixth COCOTECH Meeting of the Asia Pacific Coconut Community, Pohnpei, APCC Jakarta, Indonesia, p. 366.

Enig, M., 2000. Know Your Fats: The Complete Primer for Understanding the Nutrition of Fats, Oils, and Cholesterol. Bethesda Press, Silver Springs, MD, USA.

Enriquez, G.A., 1981. Early selection for vigor of hybrid seedlings. In: Proc. 8th Int. Cocoa Research Conference, pp. 535–539.

Eskes, A.B., 1983. Incomplete resistance to coffee leaf rust (*Hemileia vastatrix*). Ph.D. Thesis. Wageningen Agricultural University, The Netherlands.

Etherington, D.M., Mahendrarajah, S., 1998. Economic benefits of Direct Micro Expelling coconut oil in the South Pacific. In: Topper, C.P., Caligari, P.D.S., Kullaya, A.K., Shomari, S.H., Kasuga, L.J., Masache Khao, P.A.L., Mpumami, A.A. (Eds.), Trees of Life: The Key to Development: Proc. Int. Cashew and Coconut Conference, Dar es Salaam. Biohybrids International Ltd, Reading, UK, pp. 457–468.

Etherington, D.M., Zegelin, S., White, SI., 1999. Oil extraction from grated coconut: real-time moisture content measurement and its impact on oil production efficiency. Trop. Sci. 38, 10–19.

Figueiredo, P., Mariotto, P.R., Bonini, R., de Olieveira., Filho, N.L., Oliveira, D.A., 1981. Efeito do pyracarbolid e oxicarboxin applicados em misturas e in intercalados con fungicida cuprica no controle de ferrugem do cafeeiro (*Hemileia vastatrix* Berk and Br.). Biologico 47, 237–244.

Firman, I.D., 1965. Some investigations on a disease of *Coffea arabica* caused by *Ascochyta tarda*. Trans. Brit. Mycol. Soc. 48, 161–166.

Firman, I.D., and Waller, J.M., 1977. Coffee berry disease and other colletotrichum diseases of coffee. Phytopath. Pap.No. 20 C.M.I.

Flood, J., Brayford, D., 1997. Re-emergence of *Fusarium* wilt of coffee in Africa. In: Proc. 7th ASIC Conference, Nairobi, Kenya, July 20–25.

Foale, M.A., 1965. Nutrition du jeune cocotier dans les Iles Russell (Archipel des Salamon). Oleagineux 20, 1–4.

Foale, M.A., 1968a. The growth of coconut seedlings. Oleagineux 23, 651–654.

Foale, M.A., 1968b. The growth of the young coconut palm (*Cocos nucifera* L.). 1. The role of the seed and of photosynthesis in seedling growth up to 17 months of age. Aust. J. Agric. Res. 19, 781–789.

Foale, M.A., 1993a. Physiological basis for yield in coconut. In: Nair, M.K., Khan, H.H., Gopalasundaram, P., Bhaskara Rao, E.V.V. (Eds.), Advances in Coconut Research and Development. Oxford and IBH Publishing Co. Pvt. Ltd, New Delhi, India, pp. 181–189.

Foale, M.A., 1993b. The effect of exposing the germpore on germination of coconut. In: Nair, M.K., Kan, H.H., Gopalasundaram, P., Bhaskarao Rao, E.V.V. (Eds.), Advances in Coconut Research and Development. Oxford and IBH Publishing Co. Pvt. Ltd, New Delhi, India, pp. 247–252.

Foale, M.A., Ashburner, G.R., Friend, D., 1994. Canopy development and light interception of coconut. In: Foale, M.A., Lynch, P.W. (Eds.), Coconut Improvement in the South Pacific. Australian Centre for International Agricultural Research, Canberra, Australia, pp. 71–73.

Foltan, H., Ludders, P., 1995. Flowering, fruit set, and genotype compatibility in cashew. Angew. Bot. 69, 215–220.

Francies, R.M., 1998. Genetic analysis of certain clones, hybrids and inbreds in cocoa. Ph.D. Thesis. Kerala Agricultural University, Thrissur, Kerala, India.

Furtado, C.X., 1933. The limits of *Areca* Linn. and its sections. Repert. Sp. Nov. Regn. Veg. 33, 217–239.

Furtado, C.X., 1960. The philological origin of *Areca catechu*. Principes 4, 26–31.

Gallais, A., 1992. Bases genetiques et strategie de selection de L'adaptation generale. Le Selectionneur Francais 42, 59–78.

Ganesan, V., 1982. Development of small scale equipment for extraction of cocoa butter and production of cocoa butter. M.Sc. (Ag. Eng.) Thesis. Kerala Agricultural University, Thrissur, Kerala, India.

Garces, R., Mancha, M., Quinn, P.J., Harwood, J.L. (Eds.), 1990. Lipid, Biochemistry, Structure, and Utilization. Portland Press, Wye, London, pp. 387–389.

Geerlings, A., 1999. Alkaloid production by *Cinchona officinalis* Ledgeriana hairy root culture containing constructive expression constructs of tryptophan decarboxylase and strictosidine synthase DNA from *Catharanthus roseus*. Plant Cell Rep. 19, 191–196.

George, C.M., 1962. Mature rubber manuring. Effect of fertilizers on yield. Rubber Board Bull. 5, 202.

George, K.T., Joseph, T., 1994. Commercial exploitation of rubber honey in India: a preliminary assessment. Unpublished monograph, Kottayam, Rubber Research Institute of India, Kottayam, Kerala State.

George, P.J., 1989. Need for conservation of crop genetic resources with special emphasis on rubber. Rubber Board Bull. 24 (4), 13–17.

George, T.K., Reghu, C.P., Nehru, C.R., 2000. By products and ancillary sources of income. In: George, P.J., Jacob, C.K. (Eds.), Natural Rubber: Agromanagement and Crop Processing. Rubber Research Institute of India, Kottayam, Kerala State, India, pp. 507–520.

Ginting, G., Fatmawati, R.S., 1995. Propagation methodology of oil palm at Marihat. In: Rao, V., Henson, I.E., Rajanaidu, N. (Eds.), Proc. 1993 ISOPB Int. Symposium on Recent Developments in Oil Palm Tissue Culture and Biotechnology. Palm Oil Research Institute of Malaysia, Bangi, Kuala Lumpur, Malaysia.

Glendinning, D.R., 1966. The relationship between growth and yield in cacao varieties. Euphytica 9 (35), 1–35.

Gnanaharan, R., Mathew, G., 1982. Preservative treatment of rubber wood. Research Report. Kerala Forest Research Institute. 15, 1–16.

Goh, K.J., 2000. Climatic requirements of oil palm for high yields. Proceedings of the Seminar on Managing Oil Palm for High Yields: Agronomic Principles. Malaysian Society of Soil Science, Kuala Lumpur, Malaysia.

Goldblatt, P. (Ed.), 1984. Index to Plant Chromosome Numbers, 1981–1984. Missouri Botanical Garden.

Gomez, J.B., 1974. Ultrastructure of mature latex vessels in *Hevea brasilensis*. In: Proc. Eighth Int. Congress on Electron Microscopy. Australian Academy of Science, Canberra.

Gomez, J.B., 1983. Physiology of Latex (Rubber) Production. Malaysian Rubber Research and Development Board, Kuala Lumpur, Malaysia, pp. 71–98.

Gomez, J.B., Moir, G.F.J., 1979. The ultracytology of latex vessels in *Hevea brasiliensis*. Malaysian Rubber Research and Development Board Monograph, No. 4.

Gopalakrishnan, K.S., 1994. An insight into rubberised roads. Indian Road Congress, Seminar on Bituminous Roads: Design and Construction Aspects, New Delhi, India, August 25–26, pp. I.116–I.125.

Gramacho, K.P., Luz, E.D.M.N., Lopez, U.V., Paim, M.A., 1996. Inoculum density and inoculum suspension vehicle for evaluating for resistance of cocoa seedlings to *Crinipellis pernicosa*. In: Proc. 12th Int. Cocoa Research Conference, Salvador-Bahia, Brazil, November 17–23. International Group for Genetic Improvement of Cocoa, University of Reading, UK, pp. 91–101.

Greathead, D.J., 1965. Work in progress for the Coffee Research Foundation. The Commonwealth Institute for Biological Control, Kawada, Uganda, Kenya Coffee, January.

Greathead, D.J., 1995. Natural enemies in combination with pesticides for integrated pest management. In: Reuveni, R. (Ed.), Novel Approaches to Integrated Pest Management. Lewis Publishers, Boca Raton, FL, USA, pp. 183–197.

Grey, G.G., Jones, R.A., 1960. Report of the Wattle Research Institute. University of Natal, Pietermaritzburg, South Africa.

Griffiths, E., 1972. Negative effects of fungicides in coffee. Trop. Sci. 14, 79–89.

Guest, E., 1939. The international notation for tapping systems and its use in the tabulation of yield records. J. Rubber Res. Inst. Malaya 9 (239), 142–163.

Gunatilleke, D.I., Samaranayake, C., 1988. Shoot tip culture as a method of micropropagation of *Hevea*. J. Rubber Res. Inst. Sri Lanka 68, 33–44.

Gupta, B.N., Singh, S.P., Gupta, R.C., 1981. Studies on tannin from Indian wattle (*Acacia mearnsii*) bark. Indian J. Forestry 4 (1), 18–21.

Gupta, B.N., Thapliyal, R.C., 1974. Presowing treatment of black wattle (Acacia mearnsii De Wild) and Australian black wood (Acacia melanoxylon R.Br.) seed. Indian Forester 100 (12), 733–735.

Gupta, R., 1980. Cinchona. In: ICAR Aromatic and Medicinal Plants: Handbook of Agriculture. Indian Council of Agricultural Research, New Delhi, India, pp. 1190–1192.

Gupta, R.K., 1993. Multipurpose Trees for Agroforestry and Wasteland Utilisation. Oxford and IBH Publications, New Delhi, India.

Gurmit, S., 1991. Ganoderma—the scourge of oil palms in coastal areas. Planter 67, 421–444.

Guzman, N., 2000. The present status of clonal propagation of oil palm (*Elaeis guineensis* Jacq.) in Costa Rica by culture of immature inflorescences. *In vitro* Cell. Develop. Biol. 31, 79A.

Haggis, A.W., 1941. Fundamental errors in early history of cinchona. Bull. Hist. Med. Pt. 10, 568–592.

Hale, S.L., 1953. World production and consumption—1951–53. Report of Cocoa Conferences, London, UK, pp. 3–11.

Hardjosuwito, B., Hoesnan, A., 1976. Rubber seed oil analysis and its possible use. Merara Perkebunan 44, 225–229.

Hardon, J.J., Tan, B.K., 1969. Interspecific hybrids in the genus *Elaeis* 1. Crossability, cytogenetics, and fertility of F1 hybrids *Elaeis guineensis* × *Elaeis oleifera*. Euphytica 18, 312–379.

Haridasan, V., Gopalakrishnan, K.S., 1980. Economics of rubberized road. Rubber News May, 37–38.

Haridasan, V., Jayaratnam, K., Nehru, C.R., 1987. Honey from rubber plantation: a study of its potential. Rubber Board Bull. 23 (1), 18–21.

Harke, P.D.A., 1985. Influence of various media constituents on the growth of *Cinchona ledgeriana* tissue cultures and production of alkaloids and anthraquinones therein. Plant Cell Tissue Organ Cult. 4 (3), 199–214.

Harries, H.C., 1978. The evolution, dissemination, and classification of *Cocos nucifera* L. Bot. Rev. 44, 265–320.

Harries, H.C., 1990. Malesian origin for a domestic *Cocos nucifera*. In: Baas, P., Payme, J., Saridan Purwanisingh, A. (Eds.), The Plant Diversity of Malaysia. Kluwer, Dordrecht, The Netherlands, pp. 351–357.

Harries, H.C., 1999. Does clonal coconut material have a potential use in any agricultural system? In: Oropeza, C., Verdeil, J.L., Ashburner, G.R., Cardega, R., Santamarma, J.M. (Eds.), Current Advances in Coconut Biotechnology. Kluwer Academic Publishers, Boston, MA, USA, pp. 431–436.

Harrison, N., Cordova, I., Richardson, P., DiBonito, R., 1999. Detection and diagnosis of lethal yellowing. In: Oropeza, C., Verdeil, J.L., Ashburner, G.R., Cardena, R., Santamarma, J.M. (Eds.), Current Advances in Coconut Biotechnology. Kluwer Academic Publishers, Boston, MA, USA, pp. 184–196.

Hartley, C.W.S., 1988. The Oil Palm (*Elaeis guineennsis* Jacq.), third ed. Longman Group Limited, New York, NY, USA.

Hay, C.A., Anderson, C.A., Robert, M.F., Phillipson, J.D., 1986. *In vitro* culture of cinchona species: precursor feeding of *Cinchona ledgeriana* root organ suspension cultures with l-tryptophan. Plant Cell Rep. 1, 1–4.

Hayati, A., Wickneswari, R., Rajanaidu, N., Maizura, I., 2000. Genetic diversity of oil palm (*Elaeis guineensis* Jacq.) germplasm collections using izozyme markers. In: Proc. Int. Symposium on Oil Palm Genetic Resources and Utilization, Kuala Lumpur, Malaysia, June 8–10.

Hegde, M., Kulasekaran, M., Jayasankar, S., Shanmugavelu, K.G., 1993. *In vitro* embryogenesis in cashew (*Anacardium occidentale* L.). Indian Cashew J. 21 (4), 17–25.

Henson, I.E., 1991. Estimating potential productivity of oil palm. In: Proc. 1990 ISOPB Int. Workshop on Yield Potential in the Oil Palm. Palm Oil Research Institute, Bangi, Kuala Lumpur, Malaysia, pp. 98–108.

Hilditch, T.P., Williams, P.N., 1964. The Chemical Constitutions of Natural Fats. Wiley, New York, NY, USA. pp. 332–343.

Hocking, D., 1966. Brown blight (*Colletotrichum coffeanum* Noack) of *Arabica* coffee in East Africa. Ann. Appl. Biol. 58, 409–421.

Hodgson, R.A.J., Randles, J.W., 1999. Detection of cadang–cadang viroid-like sequences. In: Oropeza, C., Verdeil, J.L., Ashburner, G.R., Cardega, R., Santamarma, J.M. (Eds.), Current Advances in Coconut Biotechnology. Kluwer Academic Publishers, Boston, MA, USA, pp. 227–246.

Holmes, E.M., 1930. Three hundred years of cinchona. Chem. Drug 112, 827–832.

Hurov, V., 1960. Green bud strip budding of two to eight month old rubber plantation: a study of its potential. Rubber Board Bull. 23 (1), 18–21.

Hussain, A.M.M., Ponnuswamy, P.K., Viswanathan, M., 1980. Nutrition studies on black wattle 1: Effect of different NPK fertilizer combinations up to the 6th year. Indian Forester 106 (6), 397–402.

Idris, A.S., Ariffin, D., Swinburne, T.R., Watt, T.A., 2000. The identity of *Ganoderma* species responsible for BSR disease of oil palm of Malaysia—pathogenicity test. No. 77b. Malaysian Oil Palm Board, Kuala Lumpur, Malaysia.

International Rubber Research and Development Board (IRRDB), 1996. Quarterly report, International Rubber Research and Development Board, Hertford, UK.

International Trade Centre, 1982. Markets for Selected Medicinal Plants and Their Derivatives. International Trade Centre, Geneva, Switzerland, pp. 95–102.

Isaac, J.E., Robins, R.J., Rhodes, M.J.C., 1987. Cinchonine: NADPH oxido-reductase I and II novel enzymes in the biosynthesis of quinoline alkaloids in *Cinchona ledgeriana.* Phytochemistry 26 (2), 393–399.

Iwaro, A.D., Sreenivasan, T.N., Spence, J.A., 1997. Studies on black pod resistance in Trinidad. In: Proc. INGENIC Workshop on the Contribution of Disease Resistance to Cocoa Variety Improvement, Salvador-Bahia, Brazil, November 25–26, 1996. International Group for Genetic Improvement of Cocoa, University of Reading, UK, pp. 91–101.

Iyer, R.D., Damodaran, S., 1994. Improvements of coconut. In: Chadha, K.L., Rethinam, P. (Eds.), Advances in Horticulture, Volume IX: Plantation and Spice Crops, Part I. Malhotra Publishing House, New Delhi, India, pp. 217–242.

Jacob, C.K., 1996. *Phytophthora* diagnosis. IRDB Inf. Quart. 5, 14.

Jacob, C.K., 2000. Rubber tree, man, and environment. In: George., P.J., Jacob, C.K. (Eds.), Natural Rubber: Agromanagement and Crop Processing. Rubber Research Institute of India, Kottayam, Kerala State, India, pp. 599–610.

Jacob, J.L., Eschbach, J.M., Prevot, J.C., Roussel, D., Lacrotte, R., Chrestin, H., d'Auzac, J., 1986. Physiological basis for latex diagnosis of the functioning of the laticiferous system in rubber trees. Proc. Int. Rubber Conf. 3, 43–65.

Jacquemard, J.C., Meunier, J., Bonnot, F., 1981. Etude genetique dela production du croisement chex le plamier a huile *Elaeis guineensis*. Oleasgineux 36 (7), 543–552.

Jagoe, R.B., 1952. Deli oil palms and early introductions of *Elaeis guineensis* to Malaya. Malay. Agric. J. 35, 4–11.

Jalani, B.S., Cheah, S.C., Rajanaidu, N., Darus, A., 1997. Improvement of palm oil through breeding and biotechnology. Paper presented at the 88th American Oil Chemists Society (AOCS) Annual Meeting and Expo, May 11–14, Seattle, Washington, WA, USA.

Jessy, M.D., Mathew, M., Jacob, S., Punnoose, K.I., 1994. Comparative evaluation of basin and drip irrigation systems of irrigation in rubber. Indian J. Nat. Rubber Res. 7 (1), 51–56.

Jessy, M.D., Philip, V., Punnoose, K.I., Sethuraj, M.R., 1996. Multispecies cropping system with rubber: a preliminary report. IRRDB Symposium on Farming System Aspects of the Cultivation of Natural Rubber (*Hevea brasiliensis*), 1996. Beruwala, Sri Lanka. International Rubber Research and Development Board, Hertford, UK, pp. 81–89.

Johnson, D., 1973. The botany, origin, and spread of the cashew *Anacardium occidentale*. J. Plant. Crops 1, 1–7.

Johnson, D., 1982. Cashewnut processing in Brazil. Indian Cashew J. 12 (2), 11–13.

Jones, K.P., Allen, P.W., 1992. Historical development of the world rubber industry. In: Sethuraj, M.R., Mathew, N.M. (Eds.), Natural Rubber: Biology, Cultivation and Technology. Elsevier, Amsterdam, The Netherlands, pp. 1–25.

Jones, L.H., 1974. Propagation of clonal oil palm by tissue culture. Oil Palm News 17, 1–8.

Jose, N.R., Nair, R.V., Unnithan, V.K.G., 1998. Growth and yield analysis of top worked cocoa (*Theobroma cacao* L.). J. Plant. Crops 26 (1), 13–19.

Joshi, G.D., Prab, B.P., Relekar, P.P., Magdum, M.B., 1993. Studies on utilization of cashew apple waste. Abstract, Golden Jubilee Symposium on Horticultural Research—Changing Scenario, Bangalore, India, p. 372.

Kairu, G.M., 1991. Plant Pathology, The Coffee Research Foundation Annual Report 1991/92.

Kairu, G.M., Nyangena, C.M.S., Crosse, J.E., 1988. The effect of copper sprays on bacterial blight and coffee berry disease in Kenya. Plant Pathol. 34, 207–213.

Keeble, T.W., 1977. A cure for the ch: the contribution of Robert Talbor (1942–1981). J. Royal Soc. Med. 90 (5), 285–290.

Keene, A.T., Anderson, L.A., Philipson, J.D., 1983. Investigation of cinchona leaf alkaloids by high performance liquid chromatography. J. Chromatogr. 260 (11), 123–128.

Keshavachandran, R., 1979. Propagational studies in cocoa (*Theobroma cacao* L.). M.Sc. (Hort.). Thesis. Kerala Agricultural University, Thrissur, Kerala, India.

Keter, J.K.A., 1974. A study of soil phosphorus forms in Kikuyu red loam and their surface activity. University of Nairobi. M.Sc. Thesis.

Khairuddin, H., 1990. Results of four trials on *Ganoderma* basal stem rot of oil palm in Golden Hope estates. In: Proc. *Ganoderma* Workshop. Palm Oil Research Institute of Malaysia, Bangi, Kuala Lumpur, Malaysia, pp. 67–80.

Khaw, C.H., Ng, S.K., 1997. Performance of commercial scale clonal oil palm (*Elaeis guineensis* Jacq.) planting in Malaysia. Paper presented at the International Society of Horticulture Science Symposium, Brisbane, Australia, p. 8.

Khosla, P.K., Sareen, T.S., Mehra, P.N., 1973. Cytological studies on Himalayan Anacardiaceae. Nucleus 16, 205–209.

Kimemia, J.K., Cheya, J.A., Nyabundi, J.O., 2000. The effect of green manure application to coffee plants growth, yield and quality in Kenya. In: Proc. 8th Int. Colloquium on Coffee Science, Helsinki, Finland, pp. 426–429.

Kingori, P.N., Masaba, D.M., 1991. Distribution and persistence of benomyl (Benlate) resistance in populations of *Colletotrichum coffeanum* in Kenya. Kenya Coffee 56, 1071–1074.

Kingori, P.N., Masaba, D.M., 1994. Current status of coffee leaf rust in Kenya. Kenya Coffee 59, 1977–1987.

Knapp, A.W., 1926. Experiments in the fermentation of cacao. J. Soc. Chem. Ind. 45, 140–142.

Knutzon, D.S., Thompson, G.A., Radke, S.E., Johnson, W.B., 1992. Modification of *Brassica* seed oil by antisense expression of a stearoyl-acyl carrier protein desaturase gene. Proc. Natl. Acad. Sci. USA 89, 2624–2628.

Koblitz, Z., Koblitz, D., Schmauder, H.P., Groger, D., 1983. Studies on tissue culture of the genus *Cinchona* L. Plant Cell Rep. 2, 95–97.

Kridl, J.C., Davies, H.M., Lassner, M.W., Metz, J.M., 1993. New sources of fats, waxes, and oils: The application of biotechnology to the modification of temperate oilseeds. AgBiotech News and Information 5, 121N–126N.

Krishnakumar, A.K., Potty, S.N., 1992. Nutrition of *Hevea*. In: Sethuraj, M.R., Mathew, N.M. (Eds.), Natural Rubber: Biology, Cultivation and Technology. Elsevier, Amsterdam, The Netherlands, pp. 239–262.

Krishnakumar, A.K., Gupta, C., Sinha, R.R., Sethuraj, M.R., Potty, S.N., Eappen, T., Das, K., 1991. Ecological impact of rubber (*Hevea brasiliensis*) plantations in North east India: 2. Soil properties and biomass recycling. Indian J. Nat. Rubber Res. 4 (2), 134–141.

Kularatne, R.S., Shah, F.H., Rajanaidu, N., 2000. Investigation of genetic diversity in African natural oil palm populations and *Deli dura* using AFLP markers. In: Proc. Int. Symposium on Oil Palm Genetic Resources and Utilization, Kuala Lumpur, Malaysia, June 8–10.

Kumaran, K., Prasannakumari, S., Nair, P.C.S., 1981. Experiments on small scale fermentation of cocoa beans II. Effect of the different factors for aeration on the extent of fermentation and quality of cured cocoa beans. In: Proc. Int. Conference on Plantation Crops, Central Plantation Crops Research Institute, Kasaragod, Kerala, India. PLACROSYM III, Cochin, Kerala, India, pp. 124–137.

Kumararaj, K., Bhide, V.P., 1962. Damping off of cashewnut (*Anacardium occidentale* L.) seedlings caused by *Phytophthora palmivora* Butler in Maharashtra State. Curr. Sci. 31, 23.

Kumari Jayasree, P., Asokan, M.P., Sobha, S., Sankari Ammal, L., Rekha, K., Kala, R.G., Jayasree, R., Thulaseedharan, A., 1999. Somatic embryogenesis and plant regeneration from immature anthers of Hevea brasiliensis. Curr. Sci. 76 (9), 1242–1245.

Kuriakose, B., Thomas, K.T., 2000. Ribbed sheets. In: George, P.J., Jacob, C.K. (Eds.), Natural Rubber: Agromanagement and Crop Processing. Rubber Research Institute of India, Kottayam, Kerala State, India, pp. 386–398.

Kushairi, A., Rajanaidu, N., Jalani, B.S., Mohd Rafii, Y., Mohd Din, A., 1999. PORIM oil palm planting materials. PORIM Bull. 38, 1–13.

Laker, H.A., Sreenivasan, T.N., Kumar, D.R., 1987. The inheritance of some cocoa clones to *Crinipellis perncosa* in Trinidad. In: Proc. 10th Int. Cocoa Research Conference, Santo Domingo, Dominican Republic, May 17–23, pp. 637–641.

Lam, L.V., Ha, T.T.T., Ha, V.T.T., Hong, T., 1997. Studies of *Hevea* genetic resources in Vietnam: results of evaluation and utilization. In: Proc. IRRDB Symposium on Natural Rubber, vol. 1, General, Soils and Fertilization and Breeding and Selection Sessions, 3. Breeding and Selection, Ho Chi Minh City, October 14–15, pp. 89–100.

Lamin, K., Chong, T.C., Bong, C.L., Lee, M.T., Phua, P.K.O., 1996. Breeding for resistance to cocoa diseases in Malaysia with special reference to vascular streak dieback. In: Proc. INGENIC Workshop on the Contribution of Disease Resistance to Cocoa Variety Improvement, Salvador-Bahia, Brazil, November 25–26, International Group for Genetic Improvement in Cocoa, University of Reading, UK, pp. 189–193.

Lane, E.V., 1953. The life and work of Sir Henry Wickham. Indian Rubber J. 126, 25–27, 65–68, 95–98, 139–142, 177–180.

Lardet, L., Aguilar, M.E., Michaux-Ferriere, N., Berthouly, M., 1998. Effect of strictly plant-related factors in the response of *Hevea brasiliensis* and *Theobromo cacao* nodal explants cultured *in vitro*. *In Vitro* Cell. Develop. Biol. 34, 34–40.

Lashermes, P., Trouslot, P., Anthony, F., Combes, M.C., Charrier, A., 1966. Genetic diversity for RAPD markers between cultivated and wild accessions of *Coffea arabica*. Euphytica 87, 59–64.

Lashermes, P., Cros, J., Marmey, P., Chariier, A., 1993. Use of random amplified DNA markers to analyze genetic variability and relationships of *Coffea* species. Genet. Resour. Crop Evol. 40, 91–99.

Lashermes, P., Agwanda, C.O., Anthony, F., Combes, M.C., Trouslot, P., Charrier, A., 1997. Molecular marker-assisted selection: a powerful approach for coffee improvement. In: 17th Int. Colloquium on Coffee, Nairobi, Kenya, ASIC, Paris, France.

Lashermes, P., Combes, M.C., Robert, J., Trouslot, P., D'Hont, A., Anthony, F., Charrier, A., 1999. Molecular characterisation and origin of the *Coffea arabica* L. genome. Mol. Gen. Genet. 261, 259–266.

Lass, R.A., Wood, G.A.R., 1985. Cocoa production. Present constraints and priorities for research. World Bank Technical Report, 39, 95.

Le Pelley, R.H., 1932. On the contuly *Antestia Lineaticollis*. Star in coffee in Kenya colony. Bull. Entomol. Res. 23, 217–223.

Le Pelley, R.H., 1968. Pests of Coffee. Longmans, Green & Co, London, UK.

Leach, B.J., Foale, M.A., Ashburner, R.A. The wild-type coconut population of the Cocos (Keeling) Islands, Indian Ocean. Genetic Resources and Crop Evolution (in press).

Lebrun, P., Grivet, L., Baudouin, L., 1999. Use of RFLP markers to study the diversity of the coconut palm. In: Oropeza, C., Verdeil, J.L., Ashburner, G.R., Cardega, R., Santamarma, J.M. (Eds.), Current Advances in Coconut Biotechnology. Kluwer Academic Publishers, Boston, MA, USA, pp. 73–88.

Lee, C.H., 1996. Yield potential of Golden Hope D × P oil palm planting materials. Paper presented at the Seminar on Sourcing Oil Palm Planting Materials for Local and Overseas Joint Venture, July 22–23.

Lee, C.H., Toh, P.Y., 1991. Yield performance of Golden Hope D × P planting materials. Planter 47, 317–324.

Lespinasse, D., Rodier-Goud, M., Grivet, L., Leconte, A., Legnate, H., Seguin, M., 2000. A saturated genetic linkage map of rubber tree (*Hevea* spp) based on RFLP, AFLP, microsatellites, and isozyme markers. Theor. Appl. Mol. Genet. 100, 121–138.

Leung, A.V., 1980. Encyclopaedia of Common Natural Ingredients Used in Food, Drugs and Cosmetics. John Wiley & Sons, New York, NY, USA.

Licy, J., Panikkar, A.O.N., Premkumari, D., Sarawathy Amma, C.K., Nazeer, M., Sethuraj, M.R., 1998. Genetic parameters and heterosis in rubber (*Hevea brasiliensis*) Muell. Arg.: 4. Early versus mature performance of hybrid clones. In: Mathew, N.M., Jacob, C.K. (Eds.), Developments in Plantation Crops Research. Allied Publishers Limited, New Delhi, India, pp. 9–15.

Lievens, C., Pylyser, M., Boxus, P.H., 1989. First results about micropropagation of *Anacardium occidentale* by tissue culture. Fruits 44, 553–557.

Liu Kangde, Liang Shibang, Deng Suisheng, 1998. Integrated production practices of cashew in China. In: Papademetriou, M.K., Herath, E.M. (Eds.), Integrated Production Practices of Cashew in Asia. Food and Agricultural Organization of the United Nations, Regional Office for Asia and the Pacific, Bangkok, Thailand, pp. 6–14.

Liyanage, A., de S., Jayasinghe, C.K., Liyanage, N.I.S., 1991. Losses due to *Corynespora* leaf fall disease and its eradication. In: Proc. Rubber Research Institute of Malaysia Rubber Growers Conference, Malacca, pp. 401–410.

Liyanage, D.V., Abeywardena, V., 1958. Correlation between seednut, seedling, and adult palm characters in coconut. Bulletin No. 16. Coconut Research Institute, Ceylon.

Liyanage, D.V., Sakai, K.I., 1960. Heritabilities of certain yield characters of the coconut palm. J. Genet. 57, 245–252.

Longworth, J.F., Freeman, G.H., 1963. The use of trunk girth as a calibrating variate for field experiments on cocoa trees. J. Hort. Sci. 38, 61–67.

Louis, I.H., Rethinakumar, L., 1988. Genetic load in coconut palm. In: Silas, E.G., Aravindakshan, M., Jose, A.J. (Eds.), Coconut Breeding and Management: Proc. National Symposium on Coconut Breeding and Management. Director of Extension, Kerala Agricultural University, Trissur, Kerala, India.

Low, F.C., 1978. Distribution and concentration of major soluble carbohydrates in *Hevea* latex, the effects of ethephon stimulation, and the possible role of these carbohydrates in latex flow. J. Rubber Res. Inst. Malaya 26 (1), 21–32.

Lukman, R., 1983. Revised international tapping notation for exploitation systems. J. Rubber Res. I. Malay. 31 (2), 130–140.

Machado, O., 1944. Estudos novos sobre uma planta velha cajueiro (*Anacardium occidentale* L.). Rodrguesia 8, 19–48.

Madhavan, J., Abraham, S.T., Reghu, C.P., George, P.J., 1997. A preliminary report on two floral variants in the 1981 wild *Hevea* germplasm collection. Indian J. Nat. Rubber Res. 10 (1 and 2), 1–5.

Madon, M., Clyde, M.M., Cheah, S.C., 1999. Application of genomic *in situ* hybridization (GISH) on *Elaeis* hybrids. J. Oil Palm Res. (Special Issue), 74–80.

Maheran, A.B., Teng, A.W., Othman, A.Z., Weng, C.C., 1995. Vegetative propagation of oil palm (*Elaeis guineensis*) from laboratory to field—FELDA experience. In: Ariffin, B.S., Rajanaidu, N., Yaeb, M.D., Paranjothy, K., Basri, M.W., Henson, I.E., Chang, K.C. (Eds.), Proc. 1993 PORIM Int. Palm Oil Congress—Update and Vision. Palm Oil Research Institute of Malaysia, Bangi, Kuala Lumpur, Malaysia, pp. 99–113.

Mahesha, A., Abdul Khader, K.B., Ranganna, G., 1990. Consumptive use of irrigation requirement of areca palm (*Areca catechu* L.). Indian J. Agric. Sci. 60, 609–611.

Mahopatra, G., Bhujan, R., 1974. Land selection for cashew plantation—a survey report. Cashew Bull. 11 (8), 8–15.

Maizura, I., 1999. Genetic variability of oil palm (*Elaeis guineensis* Jacq.) germplasm collections using RFLP markers. Ph.D. Thesis. Universiti Kebangsaan Malaysia, Malaysia.

Maizura, I., Cheah, S.C., Raja Naidu, N., 1996. Genetic variation of oil palm (*Elaeis guineensis*) germplasm based on restriction fragmented length polymorphism. In: Proc. Second National Congress on Genetics, Genetics Society of Malaysia, Kuala Lumpur, Malaysia, pp. 338–344.

Majumder, S.K., 1964. Chromosome studies of some species of *Hevea*. J. Rubber Res. Inst. Malaya 18, 253–260.

Mallika, V.K, Amma, S.P.K., Abraham, K., Nair, R.V., Binimol, K.S., 2000. Evolution of cocoa varieties resistant to vascular streak die back through hybridization. In: Rethinam, P. (Ed.), Abs. Int. Conference on Plantation Crops. PLACROSYM X1V, December 12–15, 2000, Hyderabad, India. Central Plantation Crops Research Institute, Kasaragod, Kerala, India, pp. 6–7.

Mallika, V.K., Amma, S.P.K., Nair, R.V., 2002a. Crop improvement in cocoa (*Theobroma cacao* L.). In: Bhat, R., Balasimha, D., Jayasekhar, S. (Eds.), National Seminar on Technologies for Enhancing Productivity in Cocoa (Ext. Sum.). Central Plantation Crops Research Institute, Regional Station, Vittal, Karnataka, India, November 29–30. Central Plantation Crops Research Institute, Kasaragod, Kerala, India, pp. 19–27.

Mallika, V.K., Amma, S.P.K., Nair, R.V., 2002b. An overview of inbreeding in cocoa. In: Bhat, R., Balasimha, D., Jayasekhar, S. (Eds.), National Seminar on Technologies for Enhancing Productivity in Cocoa (Ext. Sum.). Central Plantation Crops Research Institute, Regional Station, Vittal, Karnataka, India, November 29–30. Central Plantation Crops Research Institute, Kasaragod, Kerala, India, pp. 50–51.

Mallika, V.K., Amma, S.P.K., Nair, R.V., Namboodiri, R., 2002c. Cross compatibility relationship within selected clones of cocoa. In: Bhat, R., Balasimha, D., Jayasekhar, S. (Eds.), National Seminar on Technologies for Enhancing Productivity in Cocoa (Ext. Sum.). Central Plantation Crops Research Institute, Regional Station, Vittal, Karnataka, India, November 29–30. Central Plantation Crops Research Institute, Kasaragod, Kerala, India, pp. 44–46.

ED and F. Man, 2002. Cocoa Market Report No. 370, April.

Manciot, R., Ollagnier, M., Ochs, R., 1979a. Nutrition minerale et fertilisation du cocotier dans le monde I (mineral nutrition and fertilization of the coconut around the world). Oleagineux 3, 499–515.

Manciot, R., Ollagnier, M., Ochs, R., 1979b. Nutrition minerale et fertilization du cocotier dans le monde II: Etude des differents elements (mineral nutrition and fertilization of the coconut around the world II: Study of different elements). Oleagineux 3, 563–580.

Manciot, R., Ollagnier, M., Ochs, R., 1980. Nutrition minerale et fertilisation du cocotier dans le monde II: Etude des differents elements—suite (mineral nutrition and fertilization of the coconut around the world II. Study of different elements—continued). Oleagineux 35, 13–22.

Mantell, S.H., Boggetti, B., Bessa, A.M.S., Lemos, E.P., Abdelhadi, A., Mneney, E.E., 1997. Micropropagation and micrografting methods suitable for international transfers of cashew. In: Proc. Int. Cashew and Coconut Conference, Dar es Salam, Tanzania, p. 24.

Marattukalam, J.G., Premakumari, D., 1981. Seasonal variation in budding success under Indian conditions in *Hevea brasiliensis* Muell. Arg. In: Proc. Fourth Annual Symposium on Plantation Crops, Mysore, India. Indian Society of Plantation Crops, Kasaragod, Kerala, pp. 81–86.

Marattukalam, J.G., Varghese, Y.A., 1998. Bench grafting with green buds in *Hevea brasiliensis*. Thirteenth Symposium on Plantation Crops, Coimbatore, India. United Planters Association of South India, Nalparai.

Marattukalam, J.G., Saraswathy Amma, C.K., George, P.J., 1980. Crop improvement through ortet selection in India. Abstract of Papers: International Rubber Conference. Rubber Research Institute of India, Kottayam, Kerala State, India. November.

Marechal, H., 1928. Observations and preliminary experiments on the coconut palm with a view to developing improved seednuts for Fiji. Fiji Agric. J. 1 (2), 16–45.

Markose, V.C., 1984. Biometric analysis of yield and certain yield attributes in the Para rubber tree: *Hevea brasiliensis* Muell. Arg. Ph.D. Thesis. Kerala Agricultural University, Thrissur, Kerala State, India.

Markose, V.C., Saraswathy Amma, C.K., Licy, J., George, P.J., 1981. Studies on the progenies of *Hevea* mutant. In: Proc. Placrosym IV, Mysore, India, pp. 58–61. Indian Coffee Research Institute, Chickmagalur, Karnataka State, India.

Marty, G., Le Guen, V., Fournial, T., 1986. Cyclone effects on plantations in Vanuatu. Oleagineux 41, 64–69.

Masaba, D.M., Kingori, P.N., Muthangya, P.M., 1990. Control of coffee berry disease (*Colletotrichum coffeanum*) using tank mixture of coppers and organic fungicides. Kenya Coffee 55, 949–954.

Massari, F., 1994. Introduction. In: Delogu, A.M., Haeuster, G. (Eds.), The World Cashew Economy. NOMISMA, L'Inchiostroblu, Bologna, Italy, pp. 3–4.

Mathew, N.M., Claramma, N.M., 2000. Latex preservation and concentration. In: George, P.J., Jacob, C.K. (Eds.), Natural Rubber: Agromanagement and Crop Processing. Rubber Research Institute of India, Kottayam, Kerala State, India, pp. 414–433.

Mathew, N.M., 1996. Natural rubber and the synthetics: Energy, environment, and performance considerations. In: Natural Rubber: An Ecofriendly Material. Rubber Board, Kottayam, pp. 77–99.

Mathews, P.M., Philips, K.O., 1979. Studies on South Indian Rubiaceae. Pt II. Cytology of *Cinchona* Linn. Nucleus (India) 22 (2), 125 128.

Maung Maung Lay, M., 1998. Integrated production practices of cashew in Myanmar. In: Papademetriou, M.K., Herath, E.M. (Eds.), Integrated Production Practices of Cashew in Asia. Food and Agricultural Organisation of the United Nations, Regional Office for Asia and the Pacific, Bangkok, Thailand, pp. 33–46.

Meheran, A.B., Chan, C.Y., 1993. Produktiviti klon lebih baik memberi pulangan yang lebih tinggi walaupun kos awalnya lebih mahal. Kemajuan Penyelidikan FELDA 21, 24–27.

Meunier, J., Renard, J.L., Quillec, G., 1979. Heredite de la resistance a la furiose chez la palmier Elaeis guineensis Jacq. Oleagineux 34, 555–559.

Meunier, J., 1989. Advances in Oil Palm Breeding: Progress and prospects. International Conference on Palms and Palms Product. National Institute for Oilpalm Research (NIFOR), Nigeria p. 15.

McDonald, J., 1926. A preliminary account of a disease of green coffee berries in Kenya colony. Trans. Brit. Mycol. Soc. 11, 145–154.

McKelvie, A.D., 1957. The polythene sheet method of rooting cacao cuttings. Trop. Agric. Tri. 34, 260–265.

Melchinger, A.E., 1990. Use of molecular markers in breeding for oligogenic disease resistance. Plant Breed. 104, 1–19.

Mercy, M.A., Abraham, S.T., George, P.J., Potty, S.N., 1995. Evaluation of *Hevea* germplasm. Observations on certain prominent traits in a conservatory. Indian J. Plant Genet. Resour. 8 (1), 35–39.

Meunier, J., 1969. Etude des populations naturelles d'*Elaeis guineensis* en Cote d'Ivoire. Oleagineux 24 (2), 195.

Milford, G.F.J., Paardekooper, E.C., Ho, C.Y., 1969. Latex vessel plugging, its importance to yield and clonal behavior. J. Rubber Res. Inst. Malaya 21, 274–282.

Minifie, B.W., 1968. Special cocoas—their manufacture and uses. Confec. Prod. 34 (2), 778–784.

Mitchell, H.W., 1970. The effect of manure and fertilizers on *Arabica* coffee in Kenya. I: Effect on yield and quality. Manure and fertilizer use seminar. National Agricultural Laboratories. Coffee Research Foundation, pp. 13–58.

Mitchell, H.W., 1976. Research on close spacing systems for intensive coffee production in Kenya. Annual Report 1974/75. Coffee Research Foundation, Kenya, pp. 13–57.

Mitchell, J.D., Mori, S.A., 1987. The cashew and its relatives (*Anacardium*: Anacardiaceae). Memoirs NY Bot. Garden 42, 1–76.

Mneney, E.E., Mantell, S.H., Tsoktouridis, G., Amin, S., Bessa, A.M.S., Thangavelu, M., 1997. RAPD—profiling of Tanzanian cashew (*Anacardium occidentale* L.). In: Proc. Int. Cashew and Coconut Conference, Dar es Salam, Tanzania, p. 24.

Mohamad, H., 2000. Utilization of oil palm biomass for various wood-based and other products. In: Yusof, B., Jalani, B.S., Chan, K.W. (Eds.), Advances in Oil Palm Research. Malaysian Palm Oil Board, Kuala Lumpur, Malaysia, pp. 1346–1412.

Mohan Rao, M., 1982. Introduction. In: Bavappa, K.V.A., Nair, M.K., Premkumar, T. (Eds.), The Arecanut Palm. CPCRI, Kasaragod, India, pp. 1–9.

Mohapatra, A.R., Bhat, N.T., 1982. Crop management. A. Soils and manures. In: Bavappa, K.V.A., Nair, M.K., Premkumar, T. (Eds.), The Arecanut Palm. CPCRI, Kasaragod, India, pp. 97–104.

Mohapatra, A.R., Vijaya Kumar, K., Bhat, N.T., 1973. A study on nutrient removal by the cashew tree. Indian Cashew J. 9 (2), 19–20.

Mohd Din, A., Rajanaidu, N., 2000. Evaluation of *Elaeis oleifera*, interspecific hybrids, and backcrosses. In: Proc. Int. Symposium on Oil Palm Genetic Resources and Utilization, Kuala Lumpur, Malaysia, June 8–10.

Mohd Hanif, H., 2000. Yield and yield components and their physiology. In: Yusof, B., Jalani, B.S., Chan, K.W. (Eds.), Advances in Oil Palm Research, vol. 1. Malaysian Palm Oil Board, Kuala Lumpur, Malaysia.

Mohd Hashim, T., Teoh, C.H., Kamaruzaman, A., Mohd Ali, A., 1993. Zero burning—an environmentally friendly replanting technique. In: Jalani, B.S., Ariffin, D., Rajanaidu, N., Mohd Tayeb, D., Paranjothy, K., Mohd Basri, W., Henson, I.E., Chang, K.C. (Eds.), Proc. 1993 PORIM Int. Palm Oil Congress. Palm Oil Research Institute of Malaysia, Kuala Lumpur, Malaysia, pp. 185–194.

Moir, C.F.J., 1959. Ultracentrifugation and staining of Hevea latex. Nature 161, 177.

Moreno, G., Moreno, E., Cadena, G., 1995. Bean characteristics and cup quality of the Colombian variety (*Coffea arabica*) as judged by international tasting panels. In: 16th Int. Colloquium on Coffee, ASIC, Paris, France, pp. 574–583.

Morton, J.F., 1977. Major Medicinal Plants. Charles C Thomas Publisher, Springfield, IL, USA.

Morris, M.J., Wingfield, M.J., Walker, J., 1988. First record of the rust on Acacia mearnsii in southern Africa. Transactions of British Mycological Society 90 (2), 324–327.

Mossu, G., 1992. The Tropical Agriculturist—Cocoa. Macmillan Press Ltd, London, UK.

Mugo, H.M., 1994. Coffee insect pests attacking flowers and berries in Kenya: a review. Kenya Coffee 59 (691), 1777–1783.

Mugo, H.M., 1995. A review of insect pests attacking coffee roots, stem, branches, and leaves in Kenya. Kenya Coffee 60 (708), 2095–2102.

Mukerjee, S., Chatterjee, S.K., 1960. A new method for propagation of *Cinchona*. Sci. Cult. 25 (11), 648–650.

Mukesh, S., Tan, Y.P. (1996). An overview of oil palm breeding program and performance of D × P planting materials at United Plantation Berhad. Paper presented at the Seminar on Sourcing Oil Palm Planting Materials for Local and Overseas Joint-Venture, July 22–23.

Murali Raghavendra Rao, M.K., 1999. Estimation of genetic diversity in cashew (*Anacardium occidentale* L.) cultivars by using molecular markers. M.Sc. Thesis. University of Agricultural Sciences, Bangalore, India.

Muralidharan, A., 1980. Biomass productivity, plant interactions, and economics of intercropping in arecanut. Ph.D. Thesis. University of Agricultural Sciences, Bangalore, India.

Muralidharan, A., Nayar, T.V.R., 1979. Intercropping in arecanut gardens. In: Nelliat, E.V., Shama Bhat, K. (Eds.), Multiple Cropping in Coconut and Arecanut Gardens. CPCRI, Kasaragod, India, pp. 24–27.

Murphy, D.J., 1994. Future perspectives for oil crops. In: Designer Oil Crops: Breeding, Processing, and Biotechnology. VCH, Weinheim, The Netherlands, pp. 297–310.

Murthy, K.N., Bavappa, K.V.A., 1960a. Floral biology of areca (*Areca catechu* Linn.). Arecanut J. 11, 51–55.

Murthy, K.N., Bavappa, K.V.A., 1960b. Morphology of arecanut palm—the shoot. Arecanut J. 11, 99–102.

Murthy, K.N., Bavappa, K.V.A., 1962. Species and ecotypes (cultivars) of arecanut. Arecanut J. 13, 59–78.

Murthy, K.N., Pillai, R.S.N., 1982. Botany. In: Bavappa, K.V.A., Nair, M.K., Premkumar, T. (Eds.), The Arecanut Palm. CPCRI, Kasaragod, India, pp. 11–49.

Muthappa, B.N., 1970. Studies on the role of *Colletotrichum coffeanum* in causing stalk rot of leaves and berries of *Arabica* coffee. Indian Coffee 34, 263–264.

Mwakha, E., 1980. Intercropping dry beans in high density Arabica coffee I. Preliminary observations on bean growth and yield. Kenya Coffee 45, 187–192.

Mydin, K.K., Nazeer, M.A., Licy, J., Annamma, Y., Panikkar, A.O.N., 1989. Studies on improving fruit set following hand pollination in *Hevea brasiliensis* (Willd.ex.Adr.de Juss) Muell. Arg. Indian J. Nat. Rubber Res. 2 (1), 61–67.

Mydin, K.K., Nair, V.G., Sethuraj, M.R., Saraswathy, P., Panikkar, A.O.N., 1992. Genetic divergence in *Hevea brasiliensis*. Indian J. Nat. Rubber Res. 5 (1&2), 120–124.

N'Goran, J.A.K., Kebe, I., Tahi, M., Paulin, D., Clement, D., Eskes, A.B., 1996. Pathology and breeding research on resistance to black pod in Cote d'Ivoire. In: Proc. INGENIC Workshop on the Contribution of Disease Resistance to Cocoa Variety Improvement, Salvador-Bahia, Brazil, November 25–26. International Group for Genetic Improvement of Cocoa, University of Reading, UK, pp. 135–139.

Nagabhushanam, S., Dasaradhi, T.B., Venkata Rao, P., 1977. Some promising cashew selections and hybrids in Andhra Pradesh. Cashew Bull. 14 (7), 1–4.

Nagaraja, K.V., 1987a. Lipids of high yielding varieties of cashew. Plant Foods Hum. Nutr. 37, 307–311.

Nagaraja, K.V., 1987b. Protein of high yielding varieties of cashew. Plant Foods Hum. Nutr. 37, 69–75.

Naidu, C.V.G., 1963. Seen a dwarf areca palm? Indian Farming 12, 16–17.

Naidu, G.V.B., 1962. Arecanut in Mysore. Department of Agriculture, Bangalore, India.

Nair, K.P.P., 1996. The buffering power of plant nutrients and effects on availability. Adv. Agron. 57, 237–287.

Nair, K.P.P., 1984. Towards a better approach to soil testing based on the buffer power concept. In: Pierre-Martin Prevel (Ed.), Proc. 6th Int. Colloquium for the Optimization of Plant Nutrition, vol. 4, Montpellier, France, September 2–8, pp. 1221–1228.

Nair, K.P.P., 1996. The buffering power of plant nutrients and effects on availability. Adv. Agon. 57, 237–287.

Nair, M.G., 1982. Intercropping with pepper. Indian Farming 32, 17–19.

Nair, C.P.R., Daniel, M., 1982. Pests. In: Bavappa, K.V.A., Nair, M.K., Premkumar, T. (Eds.), Arecanut Palm. CPCRI, Kasaragod, India, pp. 151–184.

Nair, K.P., Mengel, K., 1984. Importance of phosphate buffer power for phosphate uptake by rye. Soil Sci. Soc. Am. J. 48, 92–95.

Nair, K.P.P., 2002. Sustaining crop production in the developing world through the nutrient buffer power concept. In "proceedings of the 17th world soil science congress". Vol 2, 14–21. August, p. 652. Bangkok, Thailand.

Nair, K.P.P., 2006. The agronomy and economy of cardamom (elettaria cardamomum M.): the "queen of spices". Advances in agronomy 91, 179–483.

Nair, M.K., Nampoothiri, K.U.K., 1993. Breeding for high yield in coconut. In: Nair, M.K., Khan, H.H., Gopalasundaram, P., Bhaskara Rao, E.V.V. (Eds.), Advances in Coconut Research and Development. Oxford and IBH Publishing Co Pvt. Ltd, New Delhi, India, pp. 61–69.

Nair, M.K., Ratnambal, M.J., 1978. Cytology of *Areca macrocalyx* Becc. Curr. Sci 47, 172–173.

Nair, M.R.G.K., Das, N.M., 1962. On the biology of *Carvalhoia arecae* Miller and China, a pest of areca palms in Kerala. Indian J. Entomol. 24, 86–93.

Nair, V.K.B., George, P.J., 1969. The Indian clones: RRII 100 series. Rubber Board Bull. 10 (3 and 4), 115–140.

Nair, M.K., Bhaskara Rao, E.V.V., Nambiar, K.K.N., Nambiar, M.C., 1979. Cashew (*Anacardium occidentale* L.). Monograph on Plantation Crops—1. Central Plantation Crops Research Institute, Kasaragod, India. p. 169.

Nair, R.V., Mallika, V.K., Swapna, M., 1994a. A procedure for top working in cocoa. In: Int. Symposium on Plantation Crops (PLACROSYM XI). Abs. National Research Centre for Spices, Calicut, Central Plantation Crops Research Institute, Kasaragod, Kerala, India, p. 54.

Nair, R.V., Virakthmath, B.C., Mallika, V.K., 1994b. Management of cocoa. In: Chadha, K.L. (Ed.), Advances in Horticulture Plantation Crops and Spices. Part I, vol. 9. Malhothra Publishing House, New Delhi, India.

Nair, R.V., Mallika, V.K., Chandrika, C., 1995. Clonal propagation of cocoa. Indian Cocoa Arecanut Spices J. 19 (4), 120–122.

Nair, R.V., Mallika, V.K., Swapna, M., 1996. Response of cocoa to shade. J. Plant. Crops 24 (Suppl.), 99–101.

Nair, R.V., Mallika, V.K., Amma, S.P.K., 2002. Top worked trees: an appraisal of field performance. In: National Seminar on Technologies for Enhancing Productivity in Cocoa (Ext. Sum.). Central Plantation Crops Research Institute, Regional Station, Vittal, Karnataka, India, November 29–30. Central Plantation Crops Research Institute, Kasaragod, Kerala, India, pp. 74–75.

Nambiar, K.K., 1949. A survey of Arecanut Crop in Indian Union. Indian Central Arecanut Committee, Calicut, India.

Nambiar, K.K.N., Sarma, Y.R., Pillai, G.B., 1973. Inflorescence blight of cashew (*Anacardium occidentale* L.). J. Plant. Crops 1, 44–46.

Nambiar, M.C., 1974. Recent trends in cashew research. Indian Cashew J. 9 (3), 20–22.

Nambiar, M.C., 1977. Cashew. In: Alvin Paulo de T., Koslowski, T.T. (Eds.), Ecophysiology of Tropical Crops. Academic Press Inc, San Francisco, CA, USA, pp. 461–478.

Nambiar, M.C., Swaminathan, M.S., 1960. Chromosome morphology, microsporogenesis, and pollen fertility in some varieties of coconut. Indian J. Genet. 20, 200–211.

Nambuthiri, E.S., Shivshankar, S., 1987. Cocoa waste and its utilization. Indian Cocoa Arecanut Spices J. 8, 78–80.

Nandi, R.P., 1993. Cultivation of Cinchona and Production of Its Alkaloids in India. S. Nandi, Sevoke Road, Siliguri, India, pp. 1–67.

Nandi, R.P., Chatterjee, S.K., 1983. Effect of nitrogen, phosphorus and potassium on growth, development and alkaloid formulation in *Cinchona ledgeriana*. Indian J. Forestry 6 (3), 230–232.

Nandi, R.P., Chatterjee, S.K., 1991. Effect of shade on growth and alkaloid formation in *Cinchona ledgeriana* grown in Himalayas of Darjeeling. In: Trivedi, R.N., Saima, P.K.S., Singh, M.P. (Eds.), Recent Researches in Ecology, Environment, and Pollution, Series 5. Today and Tomorrow Printers and Publishers, New Delhi, India, pp. 181–184.

Nanjundaswamy, A.M., 1984. Economic utilization of cashew apple. Cashew Causerie 6 (2), 2–7.

National Research Center for Cashew (NRCC), 1995. Annual Report 1994–95. NRCC, Puttur, India.

National Research Center for Cashew (NRCC), 1998. Annual Report 1997–98. NRCC, Puttur, India.

Nawale, R.N., Salvi, M.J., 1990. The inheritance of certain characters in F-1 hybrid progenies of cashewnut. Cashew 4 (1), 11–14.

Nawale, R.N., Sawke, D.P., Deshmukh, M.T., Salvi, M.J., 1985. Effect of black polythene mulch and supplementary irrigation on fruit retention in cashewnut. Cashew Causerie July–September, 89.

Nayar, K.G., 1998. Cashew: a versatile nut for health. In: Topper, C.T., Caligari, P.D.S., Kulaya, A.K., Shomari, S.H., Kasuga, L.H., Masawe, P.A.L., Mpunami, A.A. (Eds.), Proc. Int. Cashew and Coconut Conference. Trees for Life—The Key to Development. BioHybrids International Ltd, Reading, UK, pp. 195–199.

Nayar, R., Selsikar, C.E., 1978. Mycoplasma like organisms associated with yellow leaf disease of *Areca catechu*. Eur. J. Forest Pathol. 8, 125–128.

Nayudamma, Y., Koteswara Rao, C., 1967. Cashew testa—its use in leather industry. Indian Cashew J. 4 (2), 12–13.

Nazeer, M.A., Saraswathy Amma, C.K., 1987. Spontaneous triploidy in *Hevea brasiliensis* (Willd. Ex. Agr de Juss.) Muell. J. Plant. Crops 15, 69–71.

Ngatchou, N.J., Lotode, R., 1971. Variability in precocity of the first hybrids obtained in the Cameroon Republic and search for a correlation between girth at a given age and precocity of the hybrids. Trop. Abstracts 26 (4), 232.

Ninan, C.A., Pillai, R.V., Joseph, J., 1960. Cytogenetic studies on the genus *Cocos*—I. Chromosome number in *C. australis and C. nucifera* L. vars *spicata* and *androgena*. Indian Coconut J. 13, 129–134.

Njoroge, J.M., 1985. Effect of nitrogen rates and frequency of application on coffee yield and quality in Kenya. Acta Hort. 158, 283–291.

Njoroge, J.M., 1986. New weeds in Kenya coffee—a short communication. Kenya Coffee 51, 353–355.

Njoroge, J.M., 1991. Management of Ruiru II—a review. Kenya Coffee 56, 1027–1035.

Njoroge, J.M., 1994. Weeds and weed control in coffee. Exptl. Agric. 30, 421–429.

Njoroge, J.M., Kimemia, J.K., 1989. A comparison of different weed control methods in Kenya coffee. Kenya Coffee 55, 863–870.

Njoroge, J.M., Kimemia, J.K., 1992. Use of foliage applied herbicides at low rates and volumes on annual weeds control in coffee. Kenya Coffee 57, 1273–1278.

Njoroge, J.M., Mwakha, E., 1983. Observations on the effects of weeding and cover crops on coffee yield and quality. Kenya Coffee 48, 219–224.

Njoroge, J.M., Waithaka, K., Chweya, J.A., 1992. The influence of tree training and plant density on growth, yield components, and yield of arabica coffee C.V. Ruiru 11. J. Hort. Sci. 67, 695–702.

Njoroge, J.M., Waithaka, K., Chweya, J.A., 1993. Effects of intercropping young plants of compact Arabica coffee hybrid cultivar Ruiru II with potatoes, tomatoes, beans, and maize on coffee yields. Exptl. Agric. 29, 373–377.

NOMISMA, 1994. In: Delogu, A.M., Haeuster, G. (Eds.), The World Cashew Economy. L'Inchiostroblu, Bologna, Italy.

Norman, H.K., Basri, M.W., 1997. Status of rhinoceros beetle, *Oryctes rhinoceros* (Coleoptera: Scarabaeidae) as a pest of young oil palm in Malaysia. Planter 73, 5–21.

Norman, H.K., Mohd Basri, W., Zaidi, M.I., Maimon, A. (1999). Population studies of *Oryctes rhinoceros* in an oil palm replant using pheromone traps. In: Proc. Int. Palm Oil Conference (PIPOC) Palm Oil Research Institute.

Northmore, J.M., 1965. Some factors affecting the quality of Kenya coffee. Turrialba 15, 184–193.

Nurhaimi, H., Darussamin, A., Dodd, W.A., 1993. Isolation of rubber tree (*Hevea brasiliensis* Muell. Arg.) protoplasts from callus and cell suspension. Menara Perkebunan 61 (2), 25–31.

Nyasse, S., Bidzanga, N.L., Blaha, G., Berruy, D., Cilas, C., Despreaux, D., Eskes, A.B., 1996. Update on the work on resistance of cocoa to *Phytophthora megakarya* in Cameroon. In: Proc. INGENIC Workshop on the Contribution of Disease Resistance to Cocoa Variety Improvement. Salvador-Bahia, Brazil, November 25–26, 1999. International Group for Genetic Improvement of Cocoa, University of Reading, UK, pp. 41–49.

Nyoro, J.K., Sprey, L.H., 1986. Introducing Ruiru II to estates and small holders. Kenya Coffee 51, 7–28.

Obasola, C., Menendez, T., Obisesan, J.O., 1978. Inflorescence and bunch abnormalities in the oil palm (*Elaeis guineensis* Jacq.). J. Nig. Inst. Oil Palm Res. 5 (20), 49–57.

Okuwuagwu, C.O., 1985. The genetic base of the NIFOR oil palm program. In: Proc. Int. Workshop on Oil Palm Germplasm, pp. 228–237.

Ong, S.H., Ghani, M.N.A., Tan, A.M., Tan, H., 1983. New *Hevea* germplasm: its introduction and potential. In: Proc. Rubber Research Institute of Malaysia, Planters Conference, 3, pp. 3–17.

Oo, K.C., Lee, K.B., Ong, A.S.H., 1986. Changes in fatty acid composition of the lipid classes in developing oil palm mesocarp. Phytochemistry 25, 405–407.

Oo, K.C., 1988. Biochemistry and biotechnology in research and development of oil palm. PORIM Bulletin 17, 22–30.

Oruko, B.A., 1977. Yield responses of Arabica coffee to fertilizers in Kenya: a review. Kenya Coffee 42, 227–239.

Pantzaris, T.P., Yusof, B., 1990. Techno-economic aspects of palm oil and palm kernel oil in oleochemicals. Palm Oil Dev. 14, 1–11.

Paramananthan, S., 2000. Soil requirements of oil palm for high yields. In: Proc. Seminar on Managing Oil Palm for High Yields: Agronomic Principles, Malaysian Society of Soil Science.

Parkin, J., 1910. The right use of acetic acid in the coagulation of *Hevea* latex. Indian Rubber J. 40, 752.

Patel, J.S., 1937. Coconut breeding. Proc. Assoc. Biol. 5, 1–16.

Paulin, D., Mossu, G., Lachenaud, P., Cilas, C., 1993. Cocoa breeding in Cote d'Ivoire. Performance analysis of sixty two hybrids in four localities. Cafe Cacao The 37 (1), 3–20.

Payne, J., Rhodes, M.J.C., Robins, R.J., 1987. Quinole alkaloids production by transformed cultures of *Cinchona*. Planta Med. 53 (4), 367–372.

Peixoto, A., 1960. Caju, Servico de Informacao Agricola. Ministerio de Agricltura, Rio de Janerio, Brazil.

Peries, R.R.A., 1993. Seedling selection in coconut based on juvenile and adult palm correlations. In: Nair, M.K., Khan, H.H., Gopalasundaram, P., Bhaskara Rao, E.V.V. (Eds.), Advances in Coconut Research and Development. Oxford and IBH Publishig Co Pvt Ltd., New Delhi, India, pp. 181–189.

Persley, G.J., 1992. Replanting the Tree of Life: Toward an International Agenda for Coconut Palm Research. CAB International, Wallingford, UK.

Petch, T., 1911. The Physiology and Diseases of *Hevea brasiliensis*. Dulau, London, UK.

Petch, T., 1914. Notes on the history of the plantation rubber industry in the east. Ann. Roy. Bot. Gardens, Peradeniya, Sri Lanka 5, 440–487.

Phadnis, A., Elijah, S.N., 1968. Development on cashewnut in Maharashtra state. Cashew News Teller 1 (8, 9 and 10) 25, 7–12.

Philip, V., 1997. Soil moisture and nutrient influence on growth and yield of *Hevea brasiliensis* in Tripura. Ph.D. Thesis. Indian Institute of Technology, Kaharagpur, India.

Phillips-Mora, W., 1996. Studies on resistance to black pod disease (*Phytophthora palmivora*) at CATIE. In: Proc. INGENIC Workshop on the Contribution of Disease Resistance to Cocoa Variety Improvement, Salvador-Bahia, Brazil, November 25–26, 1999. International Group for Genetic Improvement of Cocoa, University of Reading, UK, pp. 41–49.

Pieris, W.V.D., 1935. Studies on the coconut palm II: on the relation between the weight of husked-nuts and the weight of copra. Trop. Agric. 85, 208–220.

Piggot, C.J., 1990. Growing Oil Palm: An Illustrated Guide. Incorporated Society of Planters, Kuala Lumpur, Malaysia.

Pillai, G.B., 1975. Pests of cashewnut and how to combat them. Cashew News Teller, October–December, 31–33.

Pillai, G.B., 1979. Pests. In: Bhaskara Rao, E.V.V., Hameed Khan, H. (Eds.), Cashew Research and Development. Indian Society for Plantation Crops, CPCRI, Kasaragod, India, pp. 55–72.

Pillai, G.B., Dubey, O.P., Singh, V., 1976. Pests of cashew and their control in India: a review of current status. J. Plant. Crops 4, 37–50.

Pin, G., 1998. Quinine for cramps. Aust. Fam. Pharmacopea 27 (10), 922–923.

Pittier, H., 1993. Degeneration of cacao through natural hybridization. J. Hered. 36, 385–390.

Pomier, M., Bonneau, X., 1984. The development of the coconut root systems depending on environmental conditions in Cote d'Ivoire. Oleagineux 42 (11), 409–421.

Ponnamma, K.N., Solomon, J.J., Rajeev, K., Govindankutty, M.P., Karnavar, G.K., 1997. Evidence for transmission of yellow leaf disease of areca palm, *Areca catechu* L. by *Proutista moesta* (West Wood) (Homoptera: Derbidae). J. Plant. Crops 25, 197–200.

Popenoe, W., 1949. Cinchona cultivation in Guatemala. Econ. Bot. 3 (13), 150–157.

Posnette, A.F., 1951. Progeny trials with cacao in the Gold Coast. Emp. Agri. 19 (76), 242–251.

Pound, F.J., 1931. The genetic constitution of the cacao crop. Ann. Rep. Cacao. Res. Trinidad 1, 10–24.

Prem Kumar, T., Daniel, M., 1981. Studies on the control of soil grubs of arecanut palm. Pesticides 15, 29–30.

Premakumari, D., Panikkar, A.O.N., 1992. Anatomy and ultracytology of latex vessels. In: Sethuraj, M.R., Mathew, N.M. (Eds.), Natural Rubber: Biology, Cultivation and Technology. Elsevier, Amsterdam, The Netherlands, pp. 67–87.

Premakumari, D., George, P. J., Panikkar, A.O.N., Nair, V.K.B., Sulochanamma, S., 1984. Performance of RRII 300 series clones in the small scale trial. In: Proc. Fifth Annual Symposium on Plantation Crops, Kasaragod, Kerala State, India, 1982, pp. 148–157. Indian Society of Plantation Crops, Central Plantation Crops Research Institute, Kasaragod, Kerala State, India.

Premalatha, T., Mohanakumaran, N., 1989. Effect of storage before fermentation and after curing on quality attributes of cocoa. South Indian Hort. 37 (5), 277–281.

Proc. Silver Jubilee Symposium on Arecanut Research and Development. CPCRI, Vittal, 1982, pp. 33–37.

Punnoose, K.I., Kothandaraman, R., Philip, V., Jessy, M.D., 2000. Field upkeep and intercropping. In: George, P.J., Jacob, C.K. (Eds.), Natural Rubber: Agromanagement and Crop Processing. Rubber Research Institute of India, Kottayam, Kerala State, India, pp. 149–169.

Purseglove, J.W., 1968. *Theobroma cacao* L. In: Purseglove, J.W. (Ed.), Tropical Crops—Dicotyledons 2. John Wiley & Sons Inc, London, UK, pp. 571–599.

Pursglove, J.W., 1972. Tropical Crop Monocotyledons. John Wiley & Sons, New York, NY, USA.

Purseglove, J.W., 1988. Anacardiaceae. In: Tropical Crops—Dicotyledons. English Language Book Society, Longman, London, UK, pp. 18–32.

Pushparajah, E., Haridas, G., 1977. Developments in reduction in immaturity period of *Hevea* in peninsular Malaysia. J. Rubber Res. Inst. Sri Lanka 54, 93–105.

Quinn, P.J., Harwood, J.L. (Eds.), 1990. Lipid, Biochemistry, Structure, and Utilization. Portland Press, Wye, London, pp. 387–389.

Raghavan, V., 1957. On certain aspects of the biology of arecanut (*Areca catechu* Linn.) and utilization of its by-products in industry. D.Phil. Thesis. Gauhati University, India.

Raghavan, V., Baruah, H.K., 1958. Arecanut: India's popular masticatory—history, chemistry and utilization. Econ. Bot. 12, 315–345.

Rasid, O.A., Cheah, S.C., Mohd Arif, A.M., 1999. Isolation and sequencing of cDNA clones coding for oil palm (Elaeis guineensis) acyl carrier protein (ACP). J. Oil Palm Research 11, 88–95.

Rai, P.S., 1984. Hand Book on Cashew Pests. Research Publications, Delhi, India.

Rajagopalan, K.P., 1973. Seminar on Forest Institutes. Tamil Nadu Forest Department, Government of Tamil Nadu, Madras, India.

Rajagopal, V., Voleti, S.R., Kasturi Bai, K.V., Sivashankar, S., 1988. Physiological and bio-chemical criteria for breeding for drought tolerance in coconut. In: Silas, E.G., Aravindakshan, M., Jose, A.I. (Eds.), Proc. National Symposium on Coconut Breeding and Management. Director of Extension, Kerala Agricultural University, Thrissur, India, pp. 136–143.

Rajanaidu, N., 1985. The oil palm (*Elaeis guineensis*) collections in Africa. In: Proc. Int. Workshop on Oil Palm Germplasm and Utilization. Palm Oil Research Institute of Malaysia, Kuala Lumpur, Malaysia, pp. 59–83.

Rajanaidu, N., 1986. *Elaeis oleifera* collection in Central and South America. In: Proc. Int. Workshop on Oil Palm Germplasm Utilization. Palm Oil Research Institute of Malaysia, Kuala Lumpur, Malaysia, pp. 84–94.

Rajanaidu, N., 1987. Collection of oil palm (Elaeis guineensis) genetic material in Tanzania and Madagascar. PORIM Bulletin 15, 1–6.

Rajanaidu, N., Jalani, B.S., 1994a. Germplasm collection in Gambia. ISOPB Newsletter 10 (2), 1–16.

Rajanaidu, N., Jalani, B.S., 1994b. Oil palm germplasm in Senegal. ISOPB Newslett. 10 (1), 1–18.

Rajanaidu, N., Jalani, B.S., 1994c. Oil palm genetic resources—collection, evaluation, utilization, and conservation. A paper presented at the PORIM Colloquium on Oil Palm Genetic Resources. Kuala Lumpur, Malaysia, September 13.

Rajanaidu, N., Jalani, B.S., 1994d. Potential sources of lauric oil for the oleochemical industry. In: Applewhite, T.H. (Ed.), Proceedings of the World Conference and Exhibition on Lauric Oil: Sources, Processing, and Applications. AOCS, Manila, Philippines, pp. 47–50.

Rajanaidu, N., Rao, V., Tan, B.B., 1984. Analysis of fatty acid composition in *Ealeis guineensis*, *Elaeis oleifera*, and their hybrids and its implication in breeding. Palm Oil Research Institute of Malaysia, Malaysia.

Rajanaidu, N., Jalani, B.S., Kushairi, A., Rafii, Y., Mohd Din, A., 1996. Breeding for high kernel planting materials PORIM Series 3 (PS3). PORIM Information Series 59.

Rajanaidu, N., Jalani, S., Rao, V., Kushiari, A., 1991. New exotic palms for plantations. In: Proc. Int. Palm Oil Conference. Palm Oil Research Institute of Malaysia, Kuala Lumpur, Malaysia, pp. 19–27.

Rajanaidu, N., Jalani, B.S., Cheah, S.C., Ahmad Kushairi, D., 1993. Oil palm breeding: current issues and future developments. In: Proc. 1993 PORIM Int. Palm Oil Congress, pp. 9–23.

Rajanaidu, N., Kushairi, A., Jalani, B.S., Tang, S.C., 1994. Novel oil palm from exotic palms. Malay. Oil Sci. Technol. J. 3 (2), 22–28.

Rajanaidu, N., Rohani, O., Jalani, S., 1997. Oil palm clones: current status and prospects for commercial production. Planter 73 (853), 163–184.

Rajanaidu, N., Kushairi, A., Rafii, M., Mohd Din, A., Maizura, I., Jalani, B.S., 2000. Oil palm breeding and genetic resources. In: Yusof, B., Jalani, B.S., Chan, K.W. (Eds.), Advances in Oil Palm Research. Malaysian Palm Oil Board, Kuala Lumpur, Malaysia, pp. 171–237.

Ramakrishnan, T.S., 1955. Decline in cashewnut. Indian Phytopath. 8, 58–63.

Ramanayake, S.M.S.D., Kovoor, A., 1999. *In vitro* micrografting of cashew (*Anacardium occidentale* L.). J. Hort. Sci. Biotech. 74 (2), 265–268.

Ramle, M., Siti Ramlah, A.A., Basri, M.W., 1995. Histopathology of *Metisa plana* infected with *Beauveria bassiana*. Elaeis 8 (1), 10–19.

Randles, J.W., Wefels, E., Hanold, D., Miller, D.C., Morin, J.P., Rohde, W., 1999. Detection and diagnosis of coconut foliar decay disease. In: Oropeza, C., Verdeil, J.L., Ashburner, G.R., Cardega, R., Santamarma, J.M. (Eds.), Current Advances in Coconut Biotechnology. Kluwer Academic Publishers, Boston, MA, USA, pp. 247–258.

Rao, G.S.L.H.V.P., Giridharan, M.P., Naik, B.J., Gopakumar, C.S., 1999. Weather inflicted damage on cashew production—a remedy. Cashew 13 (4), 36–43.

Rao, M.M., 1974. A note on the Tamil Nadu Government Cinchona Department, pp. 1–21.

Rao, M.M., Veeraraghavan, R., 1954. Vegetative propagation of *Cinchona ledgeriana* by Gootees (Marcotte). South Indian Hort. 2, 71–84.

Rao, Y.P., Mohan, S.K., 1970. A new bacterial stripe disease of arecanut (*Areca catechu*) in Mysore State. Indian Phytopath. 23, 702–704.

Rau, M.K.T., 1915. The sweet arecanut, *Areca catechu* L. var.deliciosa. J. Bombay Nat. Hist. Soc. 23, 793.

Ravi, B., Sujatha, S., Balasimha, D., 2007. Impact of drip fertigation on productivity of arecanut (*Areca catechu* L.). Agricultural Water Management 90, 101–111 Science Direct Journal Homepage: www.elsevier.com/locate/agwat.

Rawther, T.S.S., Nair, R.R., Saraswathy, N., 1982. Diseases. In: Bavappa, K.V.A., Nair, M.K., Premkumar, T. (Eds.), The Arecanut Palm. CPCRI, Kasaragod, India, pp. 185–224.

Reghu, C.P., 1998. Indian rubber wood: a versatile timber for future. In: Kuriakose, K. (Ed.), Smrithimanjari. Rev. Fr. Paul Pazhampally Memorial Charitable Trust, Kottayam, Kerala State, India, pp. 259–264.

Reghu, C.P., Abraham, S.T., Madhavan, J., George, P.J., Potty, S.N., Leelamma, K.P., 1996. Evaluation of *Hevea* germplasm: variation in bark structure of wild Brazilian germplasm. Indian J. Nat. Rubber Res. 9 (1), 28–31.

Reghu, C.P., Premkumari, D., Panikkar, A.O.N., 1989. Wood anatomy of Hevea brasiliensis (Willd. ex Adr. de Juss.) Muell. Arg. 1. Distribution pattern of tension wood and dimensional variation in wood fibres. Indian J. Nat. Rubber Res. 2 (1), 27–37.

Ribeiro, I.J.A., Ito, M.F., Paradela Filho, O., De Castro, J.L., 1988. Gummosis of Acacia decurrens Wild. Caused by Ceratocystis fimbriata EII. and Halst. Bragantia 47 (1), 71–74.

Richards, A.J., 1986. Plant Breeding Systems. George Allen & Unwin, London, UK, pp. 189–232.

Rickson, F.R., Rickson, M.M., 1998. The cashewnut, *Anacardium occidentale* (Anacardiaceae), and its perennial association with ant: extrafloral nectary location and potential for ant defense. Am. J. Bot. 85 (6), 835–849.

Ridley, H.N., 1897. Rubber cultivation. Agric. Bull. Malay. Peninsula 7, 136–138.

Riedel, H.R., 1974. Effects of roasting on cocoa beans. Confec. Prod. 40 (5), 193–194.

Riedel, H.R., 1977. Roasting of cocoa beans. Confect. Prod. 43 (6), 240–241.

Rodrigues Jr., C.J., Bettencourt, A.J., Rijo, I., 1975. Races of the pathogen and resistance to coffee rust. Ann. Rev. Phytopath. 13, 49–70.

Rodriguez, M.J.B., 1999. Detection and diagnosis of coconut cadang–cadang. In: Oropeza, C., Verdeil, J.L., Ashburner, G.R., Cardega, R., Santamarma, J.M. (Eds.), Current Advances in Coconut Biotechnology. Kluwer Academic Publishers, Boston, MA, USA, pp. 221–226.

Rohani, O., Sharifah, S.A., Mohd Rafii, Y., Ong, M., Tarmizi, A.H., Zamzuri, I., 2000. Tissue culture of oil palm. In: Yusof, B., Jalani, S., Chan, K.W. (Eds.), Advances in Oil Palm Research. Malaysian Palm Oil Board, Kuala Lumpur, Malaysia, pp. 238–331.

Rozeboom, H.J., Budiani, A., Beintema, J.J., Dijkstra, B.W., 1990. Crystallization of hevamine, an enzyme with lysozyme/chitinase activity from *Hevea brasiliensis* latex. J. Mol. Biol. 212 (3), 441–443.

Rubber Research Institute of India (RRII), 1994. Annual Report 1992–93. RRII, Kottayam, Kerala State, India.

Rubber Research Institute of India (RRII), 1996. Annual Report 1995. RRIM, Kuala Lumpur, Malaysia.

Rubber Research Institute of India (RRII) 1998. Annual Report 1997–98, RRII, Kottayam, Kerala State, India.

Rubber Research Institute of Malaysia (RRIM). 1996. Annual Report 1995. RRIM, Kuala Lumpur, Malaysia.

Rudgard, S.A., Lass, R.A., 1985. Some economic information on cocoa production in South American countries affected by "witches broom" disease. Report of the Second Workshop of International "Witches Broom" Project, Wageningen, The Netherlands.

Saccas, A.M., Charpentier, J., 1969. L'antracnose des cafeiers robusta et excelsa due a *Colletotrichum coffeanum* Noack en republic centrafricaine. Café—Cacao—Thé 13, 221–230.

Sackey, S.T., 2000. Noel technologies for disease indexing and screening for CSSVD resistance. In: Proc. Int. Workshop on New Technologies and Cocoa Breeding, October 16–17, Kota Kinabalu, Sabah, Malaysia. International Group for Genetic Improvement of Cocoa, University of Reading, UK, pp. 149–156.

Sadanandan, A.K., 1974. Raise intercrops in arecanut plantation for higher returns. Arecanut Spices Bull. 9, 70–72.

Saha, J.C., 1972. A report on cinchona and quinine industry in India: present and future prospects. ICAR Scientific Panel on Medicinal Plants and Minor Crops, pp. 1–14.

Salvi, P.V., 1979. Cashew hybrids for increased production. Indian Farming 28 (12), 11–12.

Samah, O.A., Ibrahim, N., Alimon, H., Karim, M.I.A., 1993. Comparative studies on fermentation products of cocoa beans. World J. Microbiol. Biotech. 9 (3), 381–382.

Samarappuli, L., 1996. The contribution of rubber plantations toward a better environment. Bull. Rubber Res. Inst. Sri Lanka 33, 45–54.

Sambanthamurthi, R., Abrizah, O., Umi Salamah, R., 1999. Biochemical factors that control oil composition in the oil palm. Journal of Oil Palm Research 11, 24–33.

Sambanthamurthi, R.O., Abrizah, O., Umi Salamah, R., 1996. Toward understanding the biochemical factors that affect oil composition and quality in the oil palm. PIPOC ISOPB International Conference on Oil and Kernel Production in Oil Palm, September 27–28, Kuala Lumpur, Malaysia.

Sambanthamurthi, R., Oo, K.C., 1991. Thioesterase activity in the oil palm (*Elaeis guineensis*) mesocarp. In: Quinn., P.J., Harwood, J.L. (Eds.), Plant Lipid Biochemistry, Structure and Utilization: The Proceedings of the Ninth International Symposium on Plant Lipids. Portland Press Ltd, London, UK, pp. 166–168.

Sambanthamurthi, R., Abrizah, O., Umi Salamah, R., 1999. Biochemical factors that control oil composition in the oil palm. J. Oil Palm Res. 11, 24–33.

Sampath, S.R., Kumar, M.N.A., Sundareshan, K., 1990. Evaluation of cocoa bean by-products for ruminant feeding. Indian J. Anim. Nutr. 7 (1), 45–48.

Sampath Kumar, S.N., 1981. Bacterial leaf stripe disease of arecanut (*Areca catechu* L.) caused by *Xanthomonas campestris* pv. arecae. Ph.D. Thesis. Indian Institute of Science, Bangalore, India.

Sampath Kumar, S.N., Nambiar, K.K., 1990. Ganoderma disease of arecanut palm—isolation, pathogenicity, and control. J. Plant. Crops 18, 14–18.

Sampath Kumar, S.N., Saraswathy, M., 1994. Diseases of arecanut. In: Chadha, K.L., Rethinam, P. (Eds.), Advances in Horticulture, vol. 10. Malhotra Publishing House, New Delhi, India, pp. 930–967.

Samraj, P., Chinnamani, S., 1978. Black wattle cultivation in Nilgiris. Indian Farming 28 (1), 15–18.

Samraj, P., Chinnamani, S., Haldorai, B., 1977. Natural versus man-made forest in Nilgiris with special reference to runoff, soil loss, and productivity. Indian Forester 103 (7), 46–465.

Samsudin, A., Chew, P.S., Mohd, M.M., 1993. *Oryctes rhinoceros*: breeding and damage on oil palms in an oil palm replanting situation. Planter 69, 583–591.

Sandeep Nayak, J., 1996. Crop weather assessment for arecanut. All India Seminar on Modern Irrigation Techniques, Bangalore, India.

Sannamarappa, M., 1990. Spacing trial in arecanut. J. Plant. Crops 17 (Suppl.), 51–53.

Sannamarappa, M., Muralidharan, A., 1982. Multiple cropping. In: Bavappa, K.V.A., Nair, M.K., Premkumar, T. (Eds.), The Arecanut Palm. CPCRI, Kasaragod, India, pp. 133–149.

Santos, G., 1999. Potential use of clonal propagation in coconut improvement programs. In: Oropeza, C., Verdeil, J.L., Ashburner, G.R., Cardega, R., Santamarma, J.M. (Eds.), Current Advances in Coconut Biotechnology. Kluwer Academic Publishers, Boston, MA, USA, pp. 419–430.

Saraswathy, N., Koti Reddy, M., Nair, R.R., 1977. *Colletotrichum gloeosporioides*, causing inflorescence dieback, button shedding, and nut rot of betelnut palm. Plant Dis. Reptr. 61, 172–174.

Saraswathy Amma, C.K., George, P.J., Nair, V.K.B., Panikkar, A.O.N., 1980. RRII-200 series clones. International Rubber Conference. Rubber Board, Kottayam, Kerala State, India.

Saraswathy Amma, C.K., George, P.J., Panikkar, A.O.N., 1987. Performance of a few RRII clones in the estate trials. Rubber Board Bull. 23 (2), 5–9.

Saraswathy Amma, C.K., Markose, V.C., Licy, J., Panikkar, A.O.N., 1984. Cytomorphological studies in an induced polyploidy of *Hevea brasiliensis* Muell. Arg. Cytologia 49, 725–729.

Saraswathyamma, C.K., 2003. Advances in crop improvement in *Hevea* in the traditional rubber-growing tract of India. In: Jacob, C.K., Thomas, K.K. (Eds.), Global Competitiveness of Indian Rubber Industry—Rubber Planters' Conference. Rubber Research Institute of India, Kottayam, Kerala State, India, pp. 101–116.

Sarma, P., Chatterjee, S.K., 1987. Diseases of cinchona and other medicinal plants growing in Darjeeling hills of West Bengal. In: Raychudhari, S.P., Verma, J.P. (Eds.), Review of Tropical Plant Pathology, vol. 4. Today and Tomorrow Printers and Publishers, New Delhi, India, pp. 331–340.

Sarma, P.V.S., 1947. A short note on the rubber plantation industry in India. India Rubber Board, Kottayam, Kerala State, India.

Sastry, M.N.L., Hegde, R.K., 1985. Control of koleroga in arecanut. In: Proc. Silver Jubilee Symposium on Arecanut Research and Development. CPCRI, Vittal, India, pp. 86–91.

Schurer, H., 1957. Rubber, a magic substance of ancient America. Rubber J. 7 (April), 543–549.

Senanayake, M., Jansz, E.R., Buckle, K.A., 1997. Effect of different mixing intervals on the fermentation of cocoa beans. J. Sci. Food Agri. 74 (1), 42–48.

Seneviratne, P., 1995. Young budding: a better planting material. Bull. Rubber Res. Inst. Sri Lanka 32 (1), 35–38.

Seneviratne, P., Flegmann, A., 1996. The effect of thidiazuron on axillay shoot proliferation of *Hevea brasiliensis in vitro*. J. Rubber Res. Inst. Sri Lanka 77, 1–14.

Senguin, M., Besse, P., Lespinasse, D., Lebrun, P., Rodier-Goud, M., Nicholas, D., 1996. *Hevea* molecular genetics. Plantations, recherché, development 3 (2), 85–87.

Sereviratne, P., Wigesekare, G.A., 1997. Effect of episodic growth pattern on *in vitro* growth of Hevea brasiliensis. J. Plantation Crops 25, 52–56.

Sethuraj, M.R., 1981. Yield components in *Hevea brasiliensis*. Theoretical considerations. Plant Cell Environ. 4, 81.

Sethuraj, M.R., 1992. Yield components in *Hevea brasiliensis*. In: Sethuraj, M.R., Mathew, N.M. (Eds.), Natural Rubber: Biology, Cultivation and Technology. Elsevier, Amsterdam, The Netherlands, pp. 137–162.

Sethuraj, M.R., 1996. Impact of natural rubber plantations on environment. In: Natural Rubber: An Ecofriendly Material. Rubber Board, Kottayam, Kerala State, India, pp. 55–59.

Shama, B., et al., 1962a. Studies in the development of the fruit of arecanut. South Indian Hort. 10, 1–17.

Shama Bhat, K., 1974. Intensified inter/mixed cropping in areca garden—the need of the day. Arecanut Spices Bull. 5, 67–69.

Shama Bhat, K., 1978. Agronomic research in Arecanut—a review. J. Plant. Crops 6, 67–80.

Shama Bhat, K., 1983. Plant interactions in a mixed crop community of arecanut (*Areca catechu* L.) and cacao (*Theobroma cacao* L.). Ph.D. Thesis. University of Mysore, Mysore, India.

Shama Bhat, K., 1988. Growth and performance of cacao (*Theobroma cacao* L.) and arecanut (*Areca catechu* L.) under mixed cropping system. In: Proc. 10th Int. Cocoa Research Conference, Cocoa Producers' Alliance, Lagos, Nigeria, pp. 15–19.

Shama Bhat, K., Abdul Khader, K.B., 1970. Inter and mixed cropping in arecanut gardens. Indian Farming 20, 35.

Shama Bhat, K., Abdul Khader, K.B., 1982. Crop management—agronomy. In: Bavappa, K.V. A., Nair, M.K., Premkumar, T. (Eds.), The Arecanut Palm. CPCRI, Kasaragod, India, pp. 105–131.

Shama Bhat, K., Bavappa, K.V.A., 1972. Cacao under palms. In: Wastie, R.L., Earp, D.A. (Eds.), Cacao and Coconuts in Malaysia. Incorporated Soc. Planters, Kuala Lumpur, Malaysia, pp. 116–121.

Shama Bhat, K., Leela, M., 1968. Cacao and arecanut are good companions for more cash. Indian Farming 18, 19–20.

Shama Bhat, K., Leela, M., 1969. The effect of density of planting on the distribution of arecanut roots. Trop. Agric. 46, 55–61.

Shama Bhat, K., Krishna Murthi, S., Madhava Rao, V.N., 1962. Studies on certain aspects of floral biology of the arecanut. South Indian Hort. 10, 22–34.

Sharma, A.K., Sarkar, S.K., 1956. Cytology of different species of palms and its bearing on the solution of problems of phylogeny and speciation. Genetics 28, 361–488.

Shaw, D.E., 1977. Report on continuing eradication of coffee rust in two areas in Papua New Guinea. Papua New Guinea Agric. J. 28, 27–32.

Sherry, S.P., 1966. Report of the Wattle Research Institute for 1965–66. Wattle Research Institute, University of Natal, Pietermaritzburg, South Africa.

Sherry, S.P., 1971. The Black Wattle (Acacia mearnsii De Wild). University of Natal Press, Pietermaritzberg, The Republic of South Africa.

Siddiqi, M.A., Corbett, D.C.M., 1963. Coffee bark disease in Nyasaland: pathogenicity, description, and identity of the causal organism. Trans. Brit. Mycol. Soc. 46 (1), 91–101.

Siddiqi, M.A., Corbett, D.C.M., 1965. *Fusarium* bark disease of coffee. E. Afri. Agric. J. 31 (1), 11–15. Report on research carried out in Malawi.

Simmonds, F.J., 1970. A memorandum on the possibilities of biological control of cocoa-infesting mirids. Commonwealth Institute of Biological Control, Bangalore Report (unpublished).

Simmonds, N.W., 1954. Chromosome behaviour in some tropical plants. Heredity 8, 139.

Simmonds, N.W., 1976. Evolution of Crop Plants. Longman, London, UK.

Simmonds, N.W., 1989. Rubber breeding. In: Webster, C.C., Baulkwill, W.J. (Eds.), Rubber. Longman Scientific & Technical Education, England, pp. 86–124.

Singh, S., Sehgal, H.S., Pandey, P.C., Bakshi, B.K., 1967. Anthracnose disease of cashew (*Anacardium occidentale* L.), its cause, epidemiology and control. Indian Forester 93, 374–376.

Siti Nor Akmar, A., Cheah, S.C., Aminah, S., Leslie, C.L.O., Sambanthamurthi, R., Murthy, D.J., 1999. Characterization and regulation of the oil palm (Elaeis guineensis) stearoyl-ACP desaturase genes. J. Oil Palm Research 11, 1–17.

Siti Nor Akmar, A., Farida, H.S., Cheah, S.C., 1994. Detection of differentially expressed genes in the development of oil palm mesocarp. Asia Pac. J. Mol. Biol. Biotechnol. 2 (2), 113–118.

Siti Nor Akmar, A., Farida, H.S., Cheah, S.C., 1995. Construction of oil palm mesocarp cDNA library and the isolation of mesocarp-specific cDNA clones. Asia Pac. J. Biol. Biotechnol. 3 (2), 106–111.

Siti Ramlah, A.A., Basri, M.W., Ramle, M., 1996. Isolation and amplification of baculovirus as a biocontrol agent for bagworm and nettle caterpillar of oil palm. Elaeis 8 (1), 1–9.

Smith, N.J.H., Williams, J.T., Plucknett, D.L., Talbot, J.P., 1992. Rubber, oils, and resins. In: Tropical Forest and Other Crops. Cornell University Press, New York, NY, USA, pp. 207–263.

Sobhana, P., Sinha, R.R., Sethuraj, M.R., 1986. Micropropagation on *Hevea brasiliensis*: retrospect and prospect. In: International Congress of Plant Tissue and Cell Culture. MN, USA, p. 106.

Sogeke, A.K., 1998. Stages in the vegetative propagation of oil palm (*Elaeis guineensis* Jacq.) through tissue culture. J. Oil Palm Res. 10 (2), 1–9.

Soh, A.C., 1986. Expected yield increase with selected oil palm clones from current D × P seedling material. Oleagineux 41 (2), 51–56.

Southern, P.J., 1969. Sulphur deficiency in coconuts. Oleagineux 24 (4), 211–220.

Southorn, W.A., Yip, E., 1968. Latex flow studies: 3. Electrostatic considerations in the colloidal stability of fresh *Hevea* latex. J. Rubber Res. Inst. Malaya 20 (4), 201–215.

Spriggs, R., 1984. Early coconut remains from the South Pacific. Polynesian Soc. J. 93, 71–77.

Sridevi, R., 1999. Estimation of genetic parameters from specific crosses of cocoa (*Theobroma cacao* L.). M.Sc. (Hort.) Thesis. Kerala Agricultural University, Thrissur, Kerala, India.

Stebbins, G.L., 1958. Longevity, habitat, and release of genetic variability in the higher plants. Cold Spring Harb. Symp. Quant. Biol. 23, 365–378.

Stefansson, B.R., 1983. The development of improved rapeseed cultivars. In: Kramer, J.K.G., Sauer, F.D., Pigden, W.J. (Eds.), High and Low Erucic Acid Rapeseed Oils. Academic Press, New York, NY, USA, pp. 143–160.

Stobart, A.K., Stymne, S., Shewry, P.R., 1993. Manipulation of cereal protein and oilseed quality. In: Verma, D.P.S. (Ed.), Control of Plant Gene Expression, pp. 499–533.

Storey, H.H., 1932. A bark disease of coffee in East Africa. Ann. Appl. Biol. 19, 172–184.

Stumpf, P.K., Conn, E.E. (Eds.), 1987. Biochemistry of Plants: A Comprehensive Treatise. Academic Press, New York, pp. 121–136.

Subbaiah, C.C., Manikandan, P., Joshi, Y., 1986. Yellow leaf spot of cashew: a case of molybdenum deficiency. Plant Soil 94 (1), 35–42.

Sujatha, S., Ravi, Bhat, Reddy, V.M., Abdul Haris, A., 1999. Response of high yielding varieties of arecanut to fertilizer levels in coastal Karnataka. J. Plant. Crops 27, 187–192.

Sukarja, D., 1977. Correlation between some morphological characters and propagation capacity of cinchona trees. Menara Perkebunan 45 (6), 275–278.

Sukarja, D., Munir Supralo, A., 1977. Annual report on cinchona plantation. Menara Perkebunan 45 (5), 233–236.

Sumarno-Kertowardjono, S., Taha Surida, A., Ain-Tjasadihardja, N.M., 1976. Some influence of ethrel stimulation on properties of natural rubber and latex. Menara Perkebunan 44, 75–81.

Sundararaju, D., 1984. Studies on cashew pests and their natural enemies in Goa. J. Plant. Crops 12, 38–46.

Sundararaju, D., 1993. Compilation of recently recorded and some new pests of cashew in India. Cashew 8 (1), 15–19.

Sundararaju, D., 1999. Screening of cashew accessions to tea mosquito bug (*Helopeltis antonii* L. Sign.). Cashew 13 (4), 20–26.

Sundarraju, P., Koshy, P.K., 1988. Effect of intercrops on occurrence of *Radopholus similis* in arecanut palms. J. Plant. Crops 18 (Suppl.), 299–301.

Sushmakumari, S., Asokan, M.P., Sobha, S., Kumari Jayasree, P., Kala, R.G., Jayasree, R., Power, J.B., Davey, M.R., Cochking, E.C., 1998. Protoplast culture of *Hevea brasiliensis*: optimization of parameters affecting isolation and culture. International Tree Science Conference, New Delhi, India, p. 15.

Swamy, A.Y., 1953. Cinchona industry in Madras State with particular reference to Nilgiris. In: Krishnamurti, S. (Ed.), Horticulture and Economic Plants of Nilgiris, State Horticulture Department, Tamil Nadu, India, pp. 19–29.

Swamy, K.R.M., Bhaskara Rao, E.V.V., Bhat, M.G., 1997. Catalogue of Minimum Descriptors of Cashew (*Anacardium occidentale* L.) Germplasm Accessions—I. National Research Center for Cashew, Puttur, India.

Swamy, K.R.M., Bhaskara Rao, E.V.V., Bhat, M.G., 1998. Catalogue of Minimum Descriptors of Cashew (*Anacardium occidentale* L.) Germplasm Accessions—II. National Research Center for Cashew, Puttur, India.

Swamy, K.R.M., Bhaskara Rao, E.V.V., Bhat, M.G., 2000. Catalogue of Minimum Descriptors of Cashew (*Anacardium occidentale* L.) Germplasm Accessions—III. National Research Center for Cashew, Puttur, India.

Syed, R.A., 1980. Pollinating insects of oil palm. CIBC 1977–1980 Report. Commonwealth Institute of Biological Control, Bangalore, India.

Tan, G.Y., Tan, W.K., 1990. Additive inheritance to pod caused by *Phytophthora palmivora* in cocoa. Theor. Appl. Genet. 80, 258–264.

Tan, H., 1987. Strategies in rubber tree breeding. In: Abbot, A.J., Atkin, R.K. (Eds.), Improving Vegetatively Propagated Crops. Academic Press, London, UK, pp. 27–62.

Tan, H., Subramaniam, S., 1976. A five-parent diallel cross analysis for certain characters of young *Hevea* seedlings. In: Proc. Int. Rubber Conference, International Rubber Research Organization, Kuala Lumpur, Malaysia, 2, pp. 13–26.

Tarmizi, A.M., 2000. Nutritional requirements and efficiency of fertilizer use in Malaysian oil palm cultivation. In: Yusof, B., Jalani, B.S., Chan, K.W. (Eds.), Advances in Oil Palm Research, vol. 1. Malaysian Palm Oil Board, Kuala Lumpur, Malaysia.

Teteny, P., 1970. Intra-Specific Chemical Taxa of Medicinal Plants. Akademia Kiado, Budapest, Hungary.

Thakurta, A.G., Dutt, B.K., 1944. Vegetative propagation of *Cinchona ledgeriana* from Gootes (Marcotte) and cuttings by treatment with auxins. Sci. Cult. 9 (9), 401–402.

Thampan, P.K., 1982. Handbook on Coconut Palm. Second printing, Mohan Primlani, Oxford and IBH Publishing Co, New Delhi, India, p. 311.

Thankamma, L., 1974. Phytophthora nicotiane var. nicotiane on *Anacardium occidentale* in South India. Plant Dis. Reptr. 58, 767–768.

Thimmappaiah., 1997. *In vitro* studies in cashew (*Anacardium occidentale* L.). Ph.D. Thesis. Mangalore University, Mangalore, India, p. 239.

Thimmappaiah, Shirly, R.S., 1996. Micropropagation studies in cashew (*Anacardium occidentale* L.). Natl. Symp. Hort. Biotech. (Souvenir). Bangalore, p. 65.

Thimmappaiah, Shirly, Samuel, R., 1999. *In vitro* regeneration of cashew (*Anacardium occidentale* L.). Indian J. Exp. Biol. 37, 384–390.

Thomas, K.G., 1978. Crops diversification in arecanut gardens. Indian Arecanut Spices Cocoa J. 1, 97–99.

Thomas, K.T., Panikkar, A.O.N., 2000. Indian rubber industry: genesis and development. In: George, P.J., Kruvilla, J.C. (Eds.), Natural Rubber: Agromanagement and Crop Processing. India Rubber Research Institute of India, Kottayam, Kerala State, India, pp. 1–19.

Thomas, R.L., Watson, L., Hardon, J.J., 1969. Inheritance of some component of yield in the *Deli dura* variety of oil palm. Euphytica 18, 92–100.

Thorald, C.A., 1945. Elgon dieback disease of coffee. E. Afr. Agric. J. 10 (4), 198–206.

Thulaseedharan, A., Chattoo, B.B., Asokan, M.P., Sethuraj, M.R., 1994. Preliminary investigations on variations in RAPD profiles between TPD affected and normal plants from a seed propagated rubber plantation. In: Proceedings of the Workshop on Tapping Panel Dryness of Hevea. Chinese Academy of Tropical Agricultural Sciences, Hainan, China, pp. 26–28.

Toh, P.Y., Brown, T.P., 1978. Evaluation of carbofuran as a chemical prophylactic control measure for *Oryctes rhinoceros* in young oil palms. Planter 54, 3–11.

Tucker, R., 1983. The Palms of Subequatorial Queensland. Palm and Cycad Society of Australia, Milton, Queensland, Australia.

Tupy, J., Primot, L., 1976. Control of carbohydrate metabolism by ethylene in latex vessels of *Hevea brasiliensis* in relation to rubber production. Biologia Plantarum 18, 378–382.

Uhl, N.W., Dransfield, J., 1987. Genera Plantarum: A Classification of Palms Based on the Work of Harold E. Moore. Allen Press, Lawrence, KS, USA.

Valeriano, C., 1972. O cajueiro. Boletim do Instituto Biologico de Bahia (Brazil) 11 (1), 19–58.

van der Graaff, N.A., 1981. Selection of Arabica coffee types resistant to coffee berry disease in Ethiopia. Ph.D. Thesis. Wageningen Agricultural University, The Netherlands.

van der Vossen, H.A.M., 1985. Coffee selection and breeding. In: Clifford, M.N., Wilson, K.C. (Eds.), Coffee Botany, Biochemistry and Production of Beans and Beverage. Croom Helm, London, UK.

van der Vossen, H.A.M., Cook, R.T.A., 1975. Incidence and control of berry blotch caused by *Cercospora coffeicola* on Arabica coffee in Kenya. Kenya Coffee 40, 58–61.

van der Vossen, H.A.M., Walyaro, D.J.A., 1981. Breeding for resistance to coffee berry disease in *Coffea arabica* L. II. Inheritance of the resistance. Euphytica 29, 777–791.

van der Vossen, H.A.M., Walyaro, D.J.A., 1981. The coffee breeding program in Kenya: a review of progress made and plan of action for the coming years. Kenya Coffee 46 (541), 113–130.

van der Vossen, H.A.M., Cook, R.T., Murakaru, G.N.W., 1976. Breeding for resistance to coffee berry disease caused by *Colletotrichum coffeanum* Noack (Sensu Hindorf) in *Coffea arabica* L. I. Methods of preselection for resistance. Euphytica 25, 733–745.

Van Eijnatten, C.L.M., 1991. *Anacardium occidentale* L. In: Verheij, E.W.M., Coronel, R.E. (Eds.), Plant Resources of South-East Asia, No. 2. Edible Fruits and Nuts. Pudoc-DLO, Wageningen, The Netherlands, p. 446.

Vandermer, J., Andow, D.A., 1986. Prophylactic and responsive components of an integrated pest management program. J. Econ. Entom. 79, 299–302.

Vanderweyen, R., 1952. La prospection des palmeraies congolaises et ses premier results. Bull. Inf. INEAC 1, 357–382.

Vargas, E., Gonzales, L.C., 1972. La mancha mantecosa del cafe causada pas *Colletotrichum* spp. Turrialba 22, 129–135.

Varghese, Y.A., 1992. Germplasm resources and genetic improvement. In: Sethuraj, M.R., Mathew, N.M. (Eds.), Natural Rubber: Biology, Cultivation and Technology. Elsevier, Amsterdam, The Netherlands, pp. 88–115.

Varghese, Y.A., 1998. Random amplified polymorphic DNA (RAPD) technique and its applications. In: Varghese, J.P. (Ed.), Molecular Approaches to Crop Improvement. Church Mission Society (CMS) College, Kottayam, Kerala State, India, pp. 53–69.

Varghese, Y.A., Abraham, S.T., 1999. Germplasm conservation utilisation and improvement in rubber. In: Ratnambal, M.J., Kumaran, P.M., Muralidharan, K., Niral, V., Arunachalam, V. (Eds.), Improvement of Plantation Crops. Central Plantation Crops Research Institute, Kasaragod, Kerala State, India, pp. 124–133.

Varghese, Y.A., Mydin, K.K., 2000. Genetic improvement. In: George, P.J., Kuruvilla, J.C. (Eds.), Natural Rubber: Agromanagement and Crop Processing. Rubber Research Institute of India, Kottayam, Kerala State, India, pp. 36–46.

Varghese, Y.A., John, A., Premakumari, D., Panikkar, A.O.N., Sethuraj, M.R., 1993. Early evaluation in *Hevea*: Growth and yield at the juvenile phase. Indian J. Nat. Rubber Res. 6 (1 and 2), 19–23.

Varghese, Y.A., Knaak, C., Sethuraj, M.R., Ecke, W., 1997. Evaluation of random amplified polymorphic DNA (RAPD) markers in *Hevea brasiliensis*. Plant Breed. 116, 47–52.

Varghese, Y.A., Thulaseedharan, A., Kumari Jayasree, P., 2001. Rubber. In: Parthasarathy, V.A., Bose, T.K., Dekha, P.A., Das, P., Mitra, S.K., Mohandas, S. (Eds.), Biotechnology of Horticultural Crops, vol. 1. Naya Prakash, Calcutta, West Bengal State, India, pp. 630–635.

Vasil, V., Castillo, A.M., Fromm, M.E., Vasil, I.K., 1992. Herbicide resistant fertile transgenic wheat plants obtained from microprojectile bombardment of regenerable embryogenic callus. Bio. Technol. 10, 667–674.

Veeh, H.H., Veevers, J.J., 1970. Sea level at −175 m off the Great Barrier Reef 13,000 to 17,000 years ago. Nature (London) 226, 536–537.

Venkataramanan, C., Haldorai, B., Samraj, P., Nalatwadmath, S.K., Henry, C., 1983. Wattle cultivation. Indian Forester 109 (6), 370–378.

Venkatasubban, K.R., 1945. Cytological studies in Palmae. Part I. Chromosome number in a few species of palms in British India and Ceylon. Proc. Indian Acad. Sci. 22, 193–207.

Verghese, R., 1999. Standardization of selection criteria for cocoa hybrids. M.Sc. (Ag.) Thesis. Kerala Agricultural University, Thrissur, Kerala, India.

Vidyasagar, P.S.P.V., Shama Bhat, K., 1986. A pentatomid bug causes tendernut drop in arecanut. Curr. Sci. 55, 1096–1097.

Vijayagopalan, P.K., Gopalakrishnan, K.S., 1971. Epoxitation of rubber seed oil. Rubber Board Bull. 11, 52–64.

Vijayakumar, K.R., Dey, S.K., Chandrasekhar, T.R., Devakumar, A.S., Mohanakrishna, T., Rao, P.S., Sethuraj, M.R., 1998. Irrigation requirement of rubber in the sub humid tropics. Agric. Water Manag. 35, 245–259.

Vijayakumar, K.R., Thomas, K.U., Rajagopal, R., 2000. Tapping. In: George, P.J., Kuruvilla, J.C. (Eds.), Natural Rubber: Agromanagement and Crop Processing. Rubber Research Institute of India, Kottayam, Kerala State, India, pp. 215–238.

Voelker, T.A., Worrell, A.C., Anderson, L., Bleibaum, J., Fan, C., Hawkins, D.J., Radke, S.E., Davies, H.M., 1992. Fatty acid biosynthesis redirected to medium chains in transgenic oilseed plants. Science 257, 72–73.

Voigt, J., Biehl, B., Heinrich, H., Voigt, G., 1994. Formation of cocoa specific aroma precursors from reserve globulins of cocoa seeds. Symposium Tropische Nutzpflanzen, Hamburg, Germany, September 22–24, 1993. Angewandte Botanik Berichte No. 5, 1994, pp. 138–147.

von Martius, C.F.P., 1832–1850. Historia Naturalis Palmarum. Leipzig, Germany.

Von Uexkull, H.R., 1972. Response of coconuts to potassium chloride in the Philippines. Oleagineux 27 (1), 13–19.

Wahid, P.A., Kamalam, N.V., Ashokan, P.K., Nair, R.V., 1989. Root activity pattern of cocoa (*Theobroma cacao* L.). J. Nuclear Agric. Biol. 18, 153–156.

Wakker, E., 2000. Funding forest destruction. Available at www.forest-trends.org

Waller, J.M., 1972. Water borne dispersal in coffee berry disease and its relation to control. Ann. Appl. Biol. 69, 1–18.

Waller, J.M., 1987. Coffee diseases: current status and recent developments. Rev. Trop. Plant Pathol. 4, 1–33.

Walyaro, D.J.A., 1983. Considerations in breeding for improved yield and quality in arabica coffee (*Coffea arabica* L.). Wageningen Agricultural University, The Netherlands.

Walyaro, D.J.A., van der Vossen, H.A.M., 1977. Pollen longevity and artificial cross pollination in *Coffea arabica* L. Euphytica 26, 225–231.

Walyaro, D.J.A., van der Vossen, H.A.M., 1979. Early determination of yield potential in Arabica coffee by applying index selection. Euphytica 28, 465–472.

Wan, A.R., Ghandimathi, A., Rohani, O., Paran Jothy, K., 1982. Tissue culture of economically important plants. In: Rao, A.N. (Ed.), COSTED, Singapore, pp. 152–158.

Wang, Z., Zeng, X., Chen, C., Wu, H., Li, Q., Fan, G., Lu, W., 1980. Introduction of rubber plantlets from anther of *Hevea brasiliensis* Muell. Arg. *in vitro*. Chinese J. Trop. Crops 1, 25–26.

Wanjala, F.M., 1980. Effect of infesting green coffee berries in different population levels of *Antestiopsis lineaticollis* Stal (*Hemoptera: Pentatomidae*) in Kenya. Turrialba 30 (1), 109–110.

Wargadipura, R., Sutandjona, S., 1977. Nursery media for Cinchona: their influence on seedling growth. Menara Perkebunan 45 (4), 167–173.

Watson, G.A., 1989. Field maintenance. In: Webster, C.C., Baulkwill, W.J. (Eds.), Rubber. Longman Scientific and Technical, New York, NY, USA, pp. 245–290.

Watts, G., 1893. Dictionary of Economic Products of India, vol. 2.

Wheeler, B.E.J., Mepstead, R., 1988. Pathogenic variability amongst isolates of *Crinipellis pernicosa* from cocoa (*Theobroma cacao* L.). Plant Pathol. 37, 475–488.

Whitehead, R.A., 1963. The processing of coconut pollen. Euphytica 19, 267–275.

Whitehead, R.A., 1966. Sample survey and collection of coconut germplasm in the Pacific Islands. Ministry of Overseas Development (Overseas Research Publication No. 16), London, May 30–September 5, 1964, p. 78.

Wijnsma, R.Q., 1986. Anthraquinones in *Cinchona ledgeriana* bark infected with *Phytophthora cinnamomi*. Planta Med. 3, 211–212.

Wilkinson, H., 1929. Report Entomologist. Report Department of Agriculture of Kenya, Nairobi, pp. 172–186.

Williams, J.A., Akinwale, S.A., 1994. Breeding strategies adopted in Nigerian cocoa collections. In: Proc. INGENIC Workshop on Cocoa Breeding Strategies, Kuala Lumpur, Malaysia, October 18–19. International Group for Genetic Improvement of Cocoa, University of Reading, UK, pp. 33–37.

Williams, W., 1954. The natal wattle industry. World Crops 6, 417–419.

Witch, M., 1994. Herbal Drugs and Phytopharmaceuticals. CRS Press, Boca Raton, FL, USA.

Wong, G., Goh, T.H., Tan, L.C., Kerk, P.P., 1991. Recent developments in fractionation of palm oil. In: Proc. 1991 PORIM Int. Palm Oil Conference, Module 1, pp. 144–150.

Wood, B.J., 1968. Pest of oil palms in Malaysia and their control. Incorporated Society of Planters, p. 204.

Wood, B.J., 1976. Insect pests in South-East Asia. In: Corley, R.H.V., Hardon, J.J., Wood, B.J. (Eds.), Oil Palm Research. Elsevier, Amsterdam, The Netherlands, pp. 347–367.

Wood, B.J., 1984. A Brief Guide to Oil Palm Science. The Incorporated Society of Planters, Kuala Lumpur, Malaysia.

Wood, G.A.R., 1975. Cocoa, third ed. Longman Group Ltd, London, UK.

Wood, G.A.R., Lass, R.A., 1985. Cocoa, fourth ed. Longman Publications, New York, NY, USA.

Wooi, K.C., 1995. Oil palm tissue culture: current practice and constraints. In: Rao, V., Henson, I.E., Rajanaidu, N. (Eds.), Proc. 1993 ISOPB Int. Symposium on Recent Developments in Oil Palm Tissue Culture and Biotechnology. Palm Oil Research Institute of Malaysia, Bangi, Kuala Lumpur, Malaysia, pp. 21–32.

Wrigley, G., 1988. Coffee Tropical Agricultural Series. Longmans Scientific and Technical Series, London, UK.

Wunnachit, W., Pattison, S.J., Giles, L., Millington, A.J., Sedgley, M., 1992. Pollen tube growth and genotype compatibility in cashew in relation to yield. J. Hort. Sci. 67 (1), 67–75.

Wycherley, P.R., 1968. Introduction of *Hevea* to the Orient. The Planter 44, 127–137.

Wycherley, P.R., 1977. Motivation of *Hevea* germplasm collection and conservation. Workshop on inter-collaboration in *Hevea* breeding and the collection and establishment of materials from the neo-tropics. International Rubber Research and Development Board, Kuala Lumpur, Malaysia.

Wycherley, P.R., 1992. The genus *Hevea*: botanical aspects. In: Sethuraj, M.R., Mathew, N.M. (Eds.), Natural Rubber: Biology, Cultivation, and Technology. Elsevier, Amsterdam, The Netherlands, pp. 50–66.

Yadava, R.B.R., Mathai, C.K., 1972. Chlorophyll and organic acid contents of arecanut. Madras Agric. J. 59, 306–307.

Yadukumar, N., Mandal, R.C., 1994. Effect of supplementary irrigation on cashew nut yield. In: Satheesan, K.V. (Ed.), Water Management for Plantation Crops: Problems and Prospects. CWRDM, Calicut, India, pp. 79–84.

Yadukumar, N., Abdul Khader, K.B., Shama Bhat, K., 1985. Scheduling irrigation for arecanut with pan evaporation. In: Arecanut Research and Development.

Yogaratnam, N., Iqbal, S.M.M., Samarappuli, I.N., 1995. Intercropping of rubber with tea feasible. Rubber Asia 9 (5), 75–79.

Yamada, Y., 1991. Interaction: its analysis and interpretations. In: Proc. Int. Society for Oil Palm Breeding, Malaysia, International Workshop on Genotype–Environment Interaction Studies in Perennial Tree Crops, K.L. Hilton, Malaysia.

Young, A.M., 1994. The Chocolate Tree. A Natural History of Cacao. Smithsonian Institution Press, Washington, DC, USA.

Yong, Y.Y., Chan, K.W., 1996. Breeding oil palm for competitiveness and sustainability in the 21st century. Procedings of PORIM International Palm Oil Congress, pp. 19–31.

Yusoff, M.N.M., 1994. Panel products from rubber wood: Particle board, block board, and medium density fibre board. In: Hong, L.T., Sim, H.C. (Eds.) Rubber Wood: Processing and Utilisation. Forest Research Institute of Malaysia, Kuala Lumpur, Malaysia, pp. 185–200.

Zamzuri, I., 1998. Efficient rooting of oil palm *in vitro* plantlets using double-layer technique. PORIM Bull. 36, 23–36.

Zamzuri, I., Mohd Arif, S., Rajanaidu, N., Rohani, O., 1999. Commercial feasibility of clonal oil palm planting material production. PORIM Occas. Pap. 40, 1–52.

Zeven, A.C., 1964. On the origin of oil palm. Grana Palynol. 5, 50.

Zeven, A.C., 1967. The Semi-Wild Oil Palm and Its Industry in Africa. Pudoc, Wageningen, The Netherlands.

Zin Zawawi, Z., 2000. Agronomic utilization of wastes and environmental management. In: Yusof, B., Jalani, B.S., Chan, K.W. (Eds.), Advances in Oil Palm Research, Malaysian Palm Oil Board, Kuala Lumpur, Malaysia, pp. 1413–1438.

Zizumbo, D., Fernandez, N., Torres, N., Orpeza, C., 1999. Lethal yellowing resistance in coconut germplasm for Mexico. In: Oropeza, C., Verdeil, J.L., Ashburner, G.R., Cardega, R., Santamarma, J.M. (Eds.), Current Advances in Coconut Biotechnology. Kluwer Academic Publishers, Boston, MA, USA, pp. 131–144.

Zushum, M., 1986. An investigation on meteorological indices for coconut cultivation in China. Oleagineux 41, 119–128.

www.ingramcontent.com/pod-product-compliance
Ingram Content Group UK Ltd.
Pitfield, Milton Keynes, MK11 3LW, UK
UKHW020127290926
15471UKWH00003B/335

9 780323 165051